Alfred Bögli

Karst Hydrology and Physical Speleology

Translated by June C. Schmid

With 160 Figures and 12 Plates

Springer-Verlag
Berlin Heidelberg New York 1980

Prof. Dr. Alfred Bögli,
Geographisches Institut der Universität, CH-8033 Zürich

Translator:
June C. Schmid BA. MA., Im Schooren 13, CH-8713 Uerikon

ISBN 3-540-10098-9 Springer-Verlag Berlin Heidelberg New York
ISBN 0-387-10098-9 Springer-Verlag Berlin New York Heidelberg

Library of Congress Cataloging in Publication Data
Bögli, Alfred, 1912- Karst hydrology and physical speleology. Translation of
Karsthydrographie und physische Speläologie. Bibliography: p. Includes
index. 1. Hydrology, Karst. 2. Speleology. I. Title.
GB843.B6313 551.4'47 80-15361

This work is subject to copyright. All rights are reserved, whether the whole or part of the material is concerned, specifically those of translation, reprinting, re-use of illustrations, broadcasting, reproduction by photocopying machine or similar means, and storage in data banks. Under § 54 of the German Copyright law, where copies are made for other than private use, a fee is payable to the publisher, the amount of the fee to be determined by agreement with the publisher.

© by Springer-Verlag Berlin Heidelberg 1980.
Printed in Germany.

The use of registered names, trademarks, etc. in this publication does not imply, even in the absence of a specific statement, that such names are exempt from the relevant protective laws and regulations and therefore free for general use.

Offsetprinting: Julius Beltz, Hemsbach/Bergstr.
Bookbinding: Konrad Triltsch, Graphischer Betrieb, D-8700 Würzburg
2132/3130/543210

Preface to the German Edition

The present publication on karst hydrology and physical speleology combines two subjects which have up to now been treated separately. The two fields of knowledge have gone their separate ways, less as a result of differences in subject matter than of varying approaches. The focal point in karst hydrology lies in the description of subterranean water with its physical and chemical properties, whereas physical speleology describes subterranean cavities with their contents (air, water, and sediments), which generally have been created by water. Such cavities can be correctly interpreted only by means of a knowledge of karst hydrology, yet they in turn yield indications of the properties of karst water. Karst hydrology and physical speleology are thus two aspects of the subterranean karst phenomenon and should be viewed congruently.

This book addresses geologists, hydrologists, geomorphologists, geographers, and karstologists, above all speleologists, as well as all friends of caves, especially the cavers among them. Its contents must therefore appeal to two groups: on one hand to the academically trained, whether university faculty, graduates, or students, who as a rule have the necessary basic knowledge to be able to understand the theoretical comments; on the other hand to the laymen, who have first-hand experience from their own observations in caves, but who often do not dispose over the scientific foundation necessary for an understanding of the phenomena. Therefore occasionally more attention will be given to problems of a simpler nature and to questions of technical terminology.

During geomorphological studies in the Muota Valley in 1946 it became necessary to investigate the effects of limestone corrosion underground. I then penetrated the Hölloch Cave for the first time. (The Hölloch, Hell Hole, is located in the Muota Valley in Central Switzerland.) Very soon a contrast became apparent between the then predominant notion about subterranean water in karst areas as founded by O. Lehmann and the facts in the Hölloch. Especially the passages which also showed up on the bedding planes in the commercial cave section matched the theory very poorly, and this weighed even more heavily, as Lehmann's hypothesis was founded on accepted physical principles. Furthermore there were additional observations which put his hypothesis equally in question. These contrasts between theory and reality challenged me increasingly into the underground of the karst. There I had to recognize, even though after long resistance, that nature had gone ways which could by no means be explained by the then accepted hypothesis. The goal of solving this problem enticed me to intensify my work and soon the idea of a book arose. The book was to grow out of the natural facts of karst hydrography and to do justice to the evolution of subterranean water passages and thereby to speleology. The necessity of having also to learn about the karst underground in differently structured landscapes took me into many other caves, above all in table-land and arid zones. For both types the USA offered excellent and easily accessible examples, the exploration of which is in full swing, giving rise to valuable publications.

The Hölloch, which happens to lie in my personal geomorphological area of study, became an important element of this book. In 1946 4280 m of this cave system had been surveyed and were well known. Today, 1980, there are 140000 m. Between 1904 and 1946 all research on it was suspended and all known facts were based on a dissertation by Egli (1904). The new thrust into virgin areas occurred first in 1948/49 by a research group from the Swiss Society of Speleologists (SSS) and by a precursor of the Association for the Exploration of the Hölloch (AGH) which I joined. In August 1952 three collaborators and I were trapped in the cave by rising water. This brought me into especially intensive contact with karst hydrology and gave further impetus to my study of this phenomenon.

In 1974 *Karsthydrogeologie* by J. Zötl was published. It brought substantial progress to the understanding of karst water. For the first time it was shown with all desirability that the hypothesis of O. Lehmann was untenable. Thus the spell was broken which had blocked progress for a long time, but which, however, had also stirred up contradiction and thus encouraged new research.

It is in the nature of karst hydrology and of speleology that they comprise a broad spectrum of supporting sciences. Apart from the geosciences, among them mineralogy, geology, and geomorphology, particularly hydromechanics and the chemistry of the $CaCO_3$-CO_2-H_2O system are of central significance. The necessity of a unified approach to subterranean karst prompted me to risk the disadvantages against the advantages of a presentation of the whole scope of material by one single author. Prof. M. Frey, to whom I am indebted, assisted me with hydrodynamics.

Research in subterranean karst is always teamwork. The scientist depends on the assistance of the cavers, whether this is in a purely physical sense or in research by means of surveying, gathering material and observations, or of contributions to discussions. For this reason I owe gratitude to the AGH and its members, my colleagues and friends. The few who are mentioned here stand for the more than 100 collaborators who have helped in the course of three decades: H. Nünlist, the man of the first thrusts, technical director of the AGH during the first decade of exploration; B. Bärtschi, untiring helper and organiser since 1960; D. Krämer, who lost his young life in the exploration. I should like to thank the many friends all over the world, east and west, for reprints, for their permission to use sketches and findings from their field of work, for inspiring and fruitful discussions and for numerous guided trips through "their" caves. Of these I should expecially like to mention the members of the Cave Research Foundation, CRF, most particularly R. Brucker. They always received me in the USA with open arms and I am grateful to them for new knowledge and deep insight into North American karst hydrology. My thanks go also to the authorities of Mammoth Cave and Carlsbad Caverns National Parks for always granting me permission to enter the great caves and thereby actively supporting my undertaking. Assistants at the Geographical Institute of the University of Zürich helped in the last phase of the realization of this book by copying, sketching, and proof-reading, among them U. Groner and P. Frehner.

To conclude this preface I should like to express my thanks to the publishers, especially to Dr. K.F. Springer for the good layout and presentation of the book, and to Mrs. A. Seeliger for her agreeable co-operation.

November 1978   Alfred Bögli

Contents

1.	**Karstifiable Rocks**	1
1.1	Introduction	1
1.2	Evaporites	2
1.2.1	Anhydrite and Gypsum	2
1.2.2	Rock Salt Halite	3
1.3	Carbonate Rocks: Limestone and Dolomite	4
1.3.1	Limestone and Dolomite	4
1.3.2	Differences between Limestone and Dolomite	6
1.3.3	Limestones	7
1.3.4	Structure of Limestones	8
1.3.5	Permeability of Limestone Rock	9
1.3.6	Reefs	10
2.	**Processes of Dissolution of Karstifiable Rocks, Corrosion**	12
2.1	Dissolution of Gypsum and Rock Salt	13
2.2	Dissolution of Carbonate Rocks	15
2.2.1	CO_2 on Either Side of the Interface Air/Solution	18
2.2.2	The Kinetics of CO_2	24
2.2.3	Dissolution of $CaCO_3$	28
2.2.4	Influence of Other Ions	33
2.2.5	Mixing Corrosion	35
2.2.6	Cooling Corrosion and Thermal Mixing Corrosion	38
2.2.7	Pressure Dependence of Limestone Solutions	40
2.2.8	Comparison and Evaluation of the Types of Corrosion	40
2.2.9	Further Possibilities of Corrosion in the Phreatic Zone	45
2.3	Karst Denudation	46
3.	**A General View of Exokarst**	50
3.1	Karren, the Small Solution Feature	50
3.1.1	Introduction	50
3.1.2	The Genetic System of Karren Forms	53
3.1.3	Karren and Karst Hydrology	60

3.1.4	Pseudokarren	60
3.2	Small, Closed Hollows in Karst	60
3.2.1	Dolines	60
3.2.2	On the Morphology of Dolines	62
3.2.3	Cenotes	64
3.2.4	Karst Window and Karst Gulfs	64
3.2.5	Cockpits	65
3.2.6	Uvalas (Slovenic)	65
3.3	Corrosion Plains	66
3.4	Fluvial Karst Forms: Karst Valleys, Dry Valleys	66
3.5	Glacial Karst	68
3.6	Poljes	68

4. Endokarst and Karst Hydrology ... 73

4.1	Introduction	73
4.2	The Origin of the Water in Endokarst	74

5. Physical Behavior of Karst Water ... 77

5.1	Hydrological Perviousness – Karst Hydrological Activity – Velocity of Flow	77
5.2	Catchment Area – Local Base Level	79
5.3	Shallow and Deep Karst	81
5.4	Pressure Flow – Gravitational Flow; the Cave River	82
5.5	Piezometric Surface	83
5.5.1	Introduction	83
5.5.2	Static Karst Water-Body, Local Base Level	85
5.5.3	The Bernoulli Effect (Equation of Continuity)	85
5.5.4	Torricelli's Theory (Law of Outflow)	87
5.5.5	Loss of Pressure in Flowing Water	89
5.5.6	Losses Through Friction, Losses of Pressure	90
5.5.7	Analysis of an Inaccessible Water-Course	93
5.5.8	Cavitation	97
5.6	Poljes as Karst-Hydrological Regulating Factors	97

6. The Karst Hydrological Zones ... 99

6.1	Introduction	99
6.2	Vadose Zone	101
6.2.1	Inactive Vadose Zone	101
6.2.2	Feeders	101
6.2.3	High-Water-Zone	102
6.3	Phreatic Zone	102

Contents

7.	**Karst Water – Groundwater**	111
7.1	Introduction	111
7.2	Underground Water	112
7.3	"Karst Barré"	114
7.4	Blocked Karst	115
8.	**Underground Karst Levels**	116
8.1	Introduction	116
8.2	The Cave Level of the Piezometric-Surface Type – Evolution Level	117
8.3	Cave Levels According to the Type of River-Bed	118
9.	**Karst Springs**	120
9.1	Introduction	120
9.2	Classification of Karst Springs	121
9.3	Vauclusian Springs and Other Large Karst Springs	124
9.4	Periodic Springs – Ebb and Flow Springs (Intermittent Springs)	125
9.4.1	Periodic Springs	125
9.4.2	Intermittent Springs, Ebb and Flow Springs	126
9.4.3	Episodic Springs	128
9.5	Subaqueous Springs	129
9.5.1	Sublacustrine Springs	129
9.5.2	Submarine Springs – Vrulje	130
9.5.3	The Sea Mills of Argostoli	131
9.6	Physicochemical Properties of the Water of a Karst Spring	134
9.6.1	Discharge	134
9.6.2	Variations in Temperature	136
9.6.3	Chemistry of Spring Water	137
10.	**Tracers**	138
10.1	Tracers	138
10.2	The Tracer-Diagram	141
11.	**Incasion, Breakdown**	144
12.	**Speleomorphology, the World of Forms Created by the Subterranean Removal of Matter**	151
12.1	Introduction	151
12.2	Large Forms	151

12.2.1	Passages – Passages Cross-Sections 152
12.2.2	Ceiling Half-Tube Passage 155
12.2.3	Dome – Bell-Shaped Dome – Chamber 156
12.2.4	Shafts 156
12.3	Small Forms 158
12.3.1	Cave Karren 158
12.3.2	Potholes, Inverse Solution Pockets 160
12.3.3	Scallops and Ceiling Dents 161

13. Cave Sediments 165

13.1	Clastic Sediments 165
13.1.1	Coarse Clastic Sediments 166
13.1.2	Fine Clastic Sediments 166
13.1.3	Conditions of Sedimentation 168
13.1.4	Interpretation of a Sediment Profile 173
13.1.5	Clay Minerals – Heavy Minerals 174
13.1.6	Small Forms of Fine Clastic Sediments 175
13.2	Organic Sediments 177
13.2.1	Phytogenic Sediments 177
13.2.2	Coprogenic Sediments 178
13.2.3	Cave Phosphates 178
13.3	Chemical Sediments 180
13.3.1	Limestone Deposits 180
13.3.2	Cave Sulphates, Gypsum 195
13.3.3	Cave Minerals 197

14. Speleogenetics 200

14.1	The Role of Joints and Bedding Interstices in Speleogenetics 200
14.2	The Development from Interstice to Cave Passage Under Phreatic Conditions 201
14.3	The Development to a Cave Level 204
14.4	Primary and Secondary Vadose Cave Formation 207
14.5	Widening of Interstices 208
14.6	Phases in the Development of Cavities 211

15. Speleometerology – Speleoclimatology 214

15.1	Movement of Air in Caves 214
15.1.1	Exogenous Factors of Pressure Differences 215
15.1.2	Endogenous Causes of Pressure Differences 216
15.1.3	Cave Winds as a Result of Temperature Contrast Between Open Atmosphere and Underground Cavities 218

15.1.4	Air Movements Caused by Flowing Water	222
15.2	Cave Temperatures	223
15.3	Humidity of the Air	226

16. Ice Caves 227

17. Classification of Underground Cavities 231

17.1	Definition of Cave	231
17.2	Genetic Classification	231
17.2.1	Primary Caves	232
17.2.2	Secondary Caves	233
17.3	Geological-Petrographical Classification	234
17.4.	Classification According to Size	234
17.5.	Classification According to Prominent Characteristics	237

Appendix (A)
Conventional Cave Signs 238

References 244

Subject Index 261

Plates 271

1 Karstifiable Rocks

1.1 Introduction

The formation of karst landscapes and karst hydrography is related to the occurrence of specific rocks. These must be soluble and may leave but little residue, so that the interstices widened by the processes of solution remain open, which is a prerequisite for the characteristic underground drainage.

The following groups of rocks are karstifiable:

evaporites: gypsum, anhydrite, rock salt
carbonate rocks: limestone, dolomite
quartzite: only under conditions of extreme tropic humidity.

Silicate rocks are known to be nonkarstifiable even though in the humid tropics forms occur which resemble karst forms. In the humid, tropical climate they show a "solubility" which should rather be called a readiness to decompose. It is not the case, as it is with karst rocks, that individual mineral components are dissolved, but that the minerals themselves are disintegrated, destroyed — they are weathered. Large amounts of insoluble residues are thereby created, e.g., clays and perhaps quartz sand. For this reason joints within the rock do not widen; at most, existing open joints are made tight, and as a result there is no underground drainage of the karst type. The forms remain limited to the surface.

Quartzites are a special case. Up to now they have not been listed as karstifiable rocks — in certain cases this has been incorrect, as remains to be shown. According to Krauskopf (1956) and Siever (1962), amorphous silicic acid is soluble, 50-80 ppm at 0°C, 100-140 ppm at 25°C — only 6-14 ppm of quartz dissolve at 25°C. For the first time in literature White (1960) reported a "quartzite karst" in the Roraima Formation of the Guayana Shield in southeast Venezuela with karren, even karren fields, dolines, and enlarged joints. In a small cave he found silicic stalactites. Colvée (1973) describes the 395-m-long quartzite cave in the Sierra Autana in the same area, which can have been created only by corrosion. Urbani, the driving force behind karst research in Venezuela, and Szczerban (1974), as well as Szczerban and Urbani (1974), made a thorough study of quartzite karst, especially of subterranean quartzite karst. They report shafts, one of them with a depth of 370 m, and a 800-m-long cave river with ponor and resurgence (karst spring, re-exit). Precambrian quartzite (age 1700-1800 million years Gassner, 1974) consists, according to White (1960) and White et al. (1966), of rounded grains of quartz in a silicic acid matrix, sporadically mixed with up to 5% *feldspar*. The quartz is partially transformed into amorphous silicic acid (opal), which explains its high solubility. The processes of transformation and dissolution caused by precipitation are, however, extremely slow and a matter of geological epochs. Therefore Urbani et al. (1976) expressed the supposition that the quartzite matrix, e.g., opal, had been mobilized by the thermal water as a result of granite intrusions and that the rock had thereby been loosened. Afterwards the cavities were hollowed out by erosion along joints and bedding interstices. Seen from a distance, however, the question arises as to why the grains of quartz did not make tight the open joints analogous to the fine dolomitic sand in Yugoslavian dolomites. The result would be a lack of karstification. Conclusion: quartzites, and this means orthoquartzite as well as quartzitic sandstone, are only conditionally karstifiable under hydrothermal but also under extremely humid tropical conditions which remain the same for millions of years. Therefore quartzites will not be further discussed as too little is known about their capacity to karstify.

1.2 Evaporites

Under the conditions found in lagoons in warm, dry climates, the subtropics and marginal regions of the tropics, sea water evaporates. According to the composition and the solubility of the components evaporites are deposited: limestone, gypsum, anhydrite, rock and potash salt. According to Clarke (1924) sea water contains twenty times more Ca-sulfates (anhydrite ions) than limestone ions. Therefore as a rule limestone occurs only as a component of other evaporites. Potassium salt (KCl) does not come into question for a karst landscape because of its small, original concentration of K-ions, of the high solubility of all other potassium salts, and their high plasticity, as well as because of their rarity.

1.2.1 Anhydrite and Gypsum

As corresponds to their creation in lagoons on the surface of the earth, anhydrite, $CaSO_4$, and gypsum, $CaSO_4 \cdot 2H_2O$, occur far more rarely than limestones and dolomites, which are found around the world. They were deposited in large quantities mainly in the Permian, Triassic, and Tertiary Periods. The following are a few of the numerous regions where useful finds have been made: in Germany gypsum in the Muschelkalk (middle Triassic) of western Württemberg, in the Zechstein (upper Permian) of the southern Harz and around Segeberg, in the USSR in Podolia, in the USA in western Oklahoma and northern Texas, as well as in the Pecos Valley and in the White Sands (gypsum dunes) of New Mexico. The deposits in the USA and in the USSR surpass those of all other known regions in their size and quantity (see also Herak and Stringfield, 1972). In nature Ca-sulphate is deposited by freshwater only in the form of gypsum, since the lack of greater concentrations of other salts prevents anhydrite from forming at temperatures lower than 90°C. In concentrated sea water in the evaporating basins of lagoons anhydrite is formed primarily already at 25°C. If anhydrite comes into contact with fresh water, it joins with two molecules of *crystal water* and turns into gypsum. This process results in an increase in volume of 1.557 times (Biese, 1931). This causes the layers to be compressed and gypsum with a wavy structure is formed. On the earth's surface local gypsum *banks* are pressed up by the swelling of anhydrite and small cavities are formed beneath, e.g., the Zwerglöcher (Dwarf Holes) and the Waldschmiede (Forest Forge) in the southern Harz. The floor of the Waldschmiede, the largest such form, has a diameter of 7.5 m, while the height of its dome is 2 m. Biese (1931) calls such cavities swelling caves, but does not list them as karst formations as they were not created by karstifying processes.

Gypsum, the product of a transformation of anhydrite, is very jointed and tends for this reason and because of its high solubility to the formation of caves. The same qualities and in addition its softness, however, lead to its quick destruction. Smaller solution forms, especially gypsum karren, are easily destroyed again. However, wherever gypsum has been deposited directly and has been recrystallized, it is massive and only slightly jointed or not at all.

Underground solution effects a slow sinking of the capping rock (subsidence). On the surface subsidence zones form with solution subsidence dolines and/or atypical karstic troughs. Reuter (1973) measured rates of sinkage as high as 40 cm/y. Occasionally covering layers and loose formations on top break off and plunge into the depths. Such an Erdfall is to be classed as a collapse doline. Recently in the DDR land sinking has been observed in areas of larger settlements on top of gypsum and salt deposits. The cause was found to be water seeping in at various distinct points and dissolving the underground. The buildings were damaged.

Gypsum is frequently made impure to a high degree by clay. In underground cavities a coating of clay is the rule. This, however, does not hinder the dissolution of the rock.

Table 1.1. Physical properties of Ca-sulphates

	Gypsum	Anhydrite
Density (kg/dm^3)	2.5	2.9
Hardness (acc. to Mohs)	1.5-2 can be scratched with fingernail	3-3.5 cannot be scratched with fingernail
Pressure resistance (kp/cm^2)	?	420 (acc. to Jennings, 1971)

1.2.2 Rock Salt, Halite

Rock salt (halite), NaCl, is a coarsely crystalline rock of high solubility. Under pressure it becomes highly plastic, which leads to the formation of *diapirs* (salt dome). For this reason open cracks are lacking. Natural salt caves from inside salt domes are not known, either because underground cavities are not formed at all, or because they are quickly destroyed again after their formation. In the 2500-year-old Iron Age salt mine of Hallstatt (Hallstatt Culture), the old adits have disappeared and the only witnesses left are the well-preserved remains of mining tools. In contrast to this the adits in the salt mine of Wieliczka (Poland) dating from the 18th century have hardly been touched by this phenomenon.

As is the case with gypsum, subsidence is frequent, but results even more quickly. Land sinkage is frequent above salt deposits which are exploited with extraction by water. The exploitation of salt brine had to be halted under the suburbs of Rheinfelden (Switzerland) because the houses were in danger. In humid climates rock salt cannot remain on the surface. Precipitations dissolve it and the residue, a thick mixture of clay and gypsum, protects the deposit from further washing away. In completely arid zones, on the other hand, rock salt is found on the surface also, as salt karst, e.g., in Iran, Abyssinia, and in Death Valley (CA, USA). Salt karren are pointed and sharp and can scarcely be walked on, e.g., the Devil's Golf Course in Death Valley.

1.3 Carbonate Rocks: Limestone and Dolomite

1.3.1 Limestone and Dolomite

Limestone and dolomite are carbonate rocks which have very similar karst morphological properties.

Table 1.2 Comparison of limestone and dolomite

Rock	Mineral	Chemical formula
Limestone	Calcite	$CaCO_3$
Dolomite	Dolomite	$CaMg(CO_3)_2$ or $CaCO_3 \cdot MgCO_3$ double salt

When sea water is confined in warm, dry climates, first limestone is formed, but not dolomite, although the Mg-ions are three times more frequent than the Ca-ions (Ca^{2+}). Dolomite is formed afterwards when the Mg^{2+} substitutes the Ca^{2+} in the calcite crystal until an ion ratio of 1:1 is reached.

$$2\,CaCO_3 + Mg^{2+} \rightarrow CaMg(CO_3)_2 + Ca^{2+}.$$

Table 1.3 Dissolved substances (ions) in sea water (acc. to Keller, 1962, p. 179)

An ions	g/l	Kations	g/l
Cl^-	19.26	Na^+	10.71
Br^-	0.07	K^+	0.41
SO_4^{2-}	2.69	Mg^{2+}	1.29
HCO_3^-	0.15	Ca^{2+}	0.42
	Various 0.33		
Total	35.33 g/l sea water		

When there is a higher concentration of salt water in lagoons dolomite can be deposited spontaneously. Strachow et al. (1944, quoted by Fischer, 1961) observed in the Caspian

Sea, in Kara Bugas (bay on the east side of the Caspian Sea) that when the salt concentration was less than 7% only limestone was deposited, when 7%-8% dolomite was spontaneously deposited, and when the concentration was higher *magnesite*, $MgCO_3$, was deposited. Seidel (1958) assumes that the following reaction takes place:

$$2\,Ca(HCO_3)_2 + MgCl_2 \rightarrow CaCO_3 \cdot MgCO_3 + CaCl_2 + 2\,H_2CO_3$$

It would be more correct to write the reaction as an ion equation:

$$Mg^{2+} + Ca^{2+} + 2\,HCO_3^- + 2\,Cl^- \rightarrow CaMg(CO_3)_2 + 2\,Cl^- + 2\,H^+$$

Dolomites which are formed spontaneously belong to the evaporites.

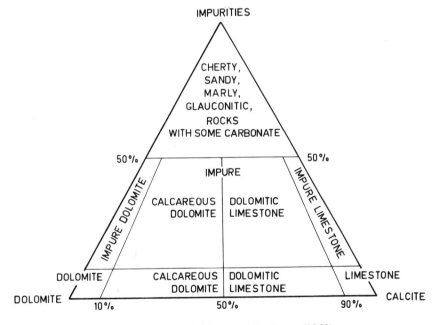

Fig. 1.1. Carbonate rocks according to Leighton and Pendexter (1962)

Table 1.4. Nomenclature for limestones and dolomites according to Pettijohn (1949)

Limestone: more than 90% $CaCO_3$	Calcareous dolomite: 50%-10% $CaCO_3$
Dolomitic limestone: 50%-90% $CaCO_3$	Dolomite: less than 10% $CaCO_3$

1.3.2 Differences Between Limestone and Dolomite

Limestone and dolomite are very similar, so similar that the nonexpert can scarcely distinguish one from the other. The surest method of distinguishing them without making a chemical analysis is to test the *density*. With a little practice one can make this test for a preliminary decision in the field. Dolomite can be distinguished from anhydrite, which is as dense and hard as dolomite, by its reaction in hot hydrochloric acid: it bubbles up.

Table 1.5. Comparison of limestone and dolomite

Properties	Pure limestone Calcite/$CaCO_3$	Pure dolomite Dolomite/$CaMg(CO_3)_2$
Dolomite content	See Table 1.4	See Table 1.4
Density (kg/dm³)	2.7, dolomitic limestone higher marly limestone 2.5 porous limestone under 2.3 chalk down to 1.6	2.9
Mohs hardness	3, with increasing clay content and porosity diminishing	3.5 otherwise like limestone
Pressure resistance (kp/cm²) acc. to Rinne (1928) acc. to Jennings (1971)	100-1900 340-3400	Up to 1200 620-3670
Solubility in cold hydrochloric acid	Good, even when in lumps bubbles up	Hardly at all when in lumps, noticeable when powder (with gas bubbles, only pure D)
Solubility in warm hydrochloric acid	Strong bubbling up	Slight bubbling up even in a lump
Solubility in acetic acid	Gas bubbles	No reaction

Special properties of dolomite which are not generally valid: certain dolomites are destroyed by weathering, breaking down into dolomite sand, e.g., in the Swabian-Franconian Jura; they are in general less resistant to frost than is limestone, thus large slopes of dolomite debris can be found (South Tyrolean Dolomites).

When limestone is mentioned hereafter dolomite will be included in that category. Wherever their properties differ special mention will be made of that difference. For instance dolomite in Yugoslavia is, in comparison to limestone, practically impermeable, because dolomite sand makes the interstices tight; in France underground water from dolomite is frequently heavily laden with dolomite sand.

1.3.3 Limestones

Most limestones are of marine origin, but there are also freshwater and terrestial limestones such as *calcarenites* from the sands of dunes of western Australia.

According to Clarke (1924) 500-1000 million tons of Ca-salts (sulfates and carbonates) are carried to the sea annually. These are deposited as limestone mainly on the *continental shelves*, but also as far down as in the bathyal zone (at a depth of 200-2000 m). In the abyssal plains it is found as globigerina ooze down to a depth of 5000 m (Murawski, 1963).

Limestone has been deposited partly detritically as gravel, sand and mud (fluvial, surf), partly by chemical deposition and partly organically (foraminifera, brachiopods, reefs etc.).

Chalk is a soft, friable, very porous limestone, consisting mainly of foraminifera. In a dry state its density varies between 1.6 and 2.0 kg/dm^3. Its great susceptibility to weathering and to any form of erosion causes karst forms to be destroyed as fast as they form. Therefore karren do not exist at all with the exception of solution pockets, which are related to them. There are isolated doline fields in chalk limestone such as in Picardy (Plateau of Baumetz). On the other hand there are frequent karst dry valleys which are of periglacial origin both in northern France and in southern England (Tricart, 1949; Pinchemel, 1954). Caves do occur rarely, e.g., near Rouen, Caumont caves (Avias, 1972). Underground drainage is developed to a medium degree and occasionally shows ponors and karst springs, also (Avre river in Normandy).

Karst landscapes occur on pure and compact limestones and dolomites. However, limestones are frequently encumbered with impurities, which reduce their ability to karstify or prevent them from karstification altogether. The two most common insoluble impurities are clay and silicic acid (opal, flint, quartz).

Table 1.6. Simplified list of mixtures of (a) limestone and clay, (b) limestone and silicic acid

a) Limestone/clay	Impurities in %	Ability to karstify
Limestone, very pure	0- 5	Excellent
Limestone	5- 10	Good
Marly limestone	10- 30	Karren formation is hindered, from 15% onwards only suggested. Dolines good. Small caves only as an exception up to 20%, more than 20% no karstification.
Marly/marly slate	30- 70	Impermeable, no karst forms
Marly clay	70- 85	Impermeable, plastic
Clay	85-100	Impermeable, plastic

b) Limestone/SiO$_2$	Impurities in %	Ability to karstify
Limestone	0- 10	Excellent to good
Cherty limestone	5- 30	Good
Limestone with fine quartz sand	10- 30	Hardly or not at all karstifiable
Siliceous limestone ("Kieselkalk")	30- 50	No karstification
Quartz sandstone with limestone matrix	50- 90	No karstification (matrix = basic mass, binding agent)
Quartz sandstone Quartzite	90-100	No karstification

If the silicic acid is finely distributed, e.g., in the Hauterivien siliceous limestone of the Helvetian nappes (Swiss Alps, Eastern Alps), karstification is completely prevented by it. If, however, it is concentrated in lumps and nodules of chert which are embedded in a pure limestone matrix, then karstification takes place according to the matrix. In one of the largest systems of caves in the USA, in Green Brier Caverns (WV, USA), there are many passages running through Paleozoic limestone with innumerable chert nodules and in Speak-Easy Cave near Springfield (MO, USA) lumps almost the size of a head stretch up out of the heavily corroded limestone of the Mississippian. Siliceous pebbles in Lower Tertiary limestone conglomerate on Mallorca do not at all prevent the development of rillenkarren (solution flutes), which are otherwise very sensitive to impurities.

Table. 1.7. Chemical analysis of some rocks according to Clarke (1924), Franz (1960), and Rinne (1928). CaO, MgO and CO_2 were recalculated according to the corresponding carbonates, (figures are percentages)

	1	2	3	4	5		6
$CaCO_3$	97.39	96.07	98.50	90.80	54.54		74.07
$MgCO_3$	1.23	1.17	0.42	1.09	45.52		0.25
Carbonate	98.62	97.24	98.92	91.89	100.06		74.32
CO_2 (remaining amount)	0	0	0	0.85	0.08	K_2O	0.12
						H_2O	1.56
Al_2O_3, Fe_2O_3	0.68	0.71	0	0.56	0.09		2.29
FeO	0	0	0	0.14	0.01		0
Insoluble residue	0.70	1.15	0.23	5.80	0.02		21.75
Total	100.00	99.10	99.15	99.24	100.26		100.04

1. Salem limestone, Mississippian, IN, USA — Clarke
2. Solnhofer limestone, upper Malm, Solnhofen, Germany — Clarke
3. Recent coral limestone, Bermuda — Clarke
4. Crinoid limestone, Jura, Hernstein, Austria — Franz
5. Dolomite, Trias, Pottenstein, Franconia, Germany — Franz
6. Pläner marl, Cenoman, Harz foreland, Germany — Rinne

1.3.4 Structure of Limestones

In petrography structure means the spatial arrangement and distribution of equal elements in rock. The English school especially attributes great importance to structure in the formation of limestone surfaces. Sweeting (1972, 1975), Oxford, is investigating this relationship.

Coarse crystalline limestones show less clear karst forms than do fine crystalline ones; thus karren are only poorly formed in the pure marble of the Apuanic Alps. Coarse crystalline parts in a fine crystalline matrix and calcite-filled tension cracks dissolve less easily and therefore project up above the limestone surface. The finer the grain, the larger the surface vulnerable to corrosion.

English literature has taken over the division according to Folk (1959) for petrolgeological investigations by karst researchers (Jennings, 1971; Sweeting, 1972). His theory is based on the fact that limestone mud is deposited in quiet, shallow seas. The resulting limestones he calls *micrites* since their basic mass is micro- to crypto-crystalline. The matrix appears gray in the microscope. If its deposits are disturbed by submarine slides or ploughing shellfish he speaks of *dismicrites*. The *allochems*, the parts brought in from the exterior, are embedded in this matrix. Detritic allochems are called *intraclasts*. They vary from the size of a fine grain of sand to that of pebbles. Ooliths, pellets (calcite particles of organic origin) and organisms (foraminifera, corals, brachiopods, molluscs, sponge needles, etc.) are further examples. The result is the following rock schema:

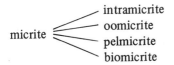

If the matrix consists of clearly distinguishable calcite crystals, its cleavability becomes obvious. This attribute is called spar. From spar Folk took the term sparite. The same terminology is valid here as with the micrites:

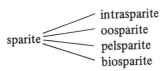

Sparites develop where cavities or large pores were already in existence primarily, or where limestone mud was washed away from between the allochems by water in motion, so that calcite crystals could develop in the spaces. There are, however, also autochthonous limestones, limestones which were already deposited on the site, e.g., reefs (coral, sponge and algae reefs). To these Folk applies the term bioliths, which Potonié (1905) initiated. As a consequence bioherm will be used in this work for reefs which tower up and biostrom for small flat reefs. Further differentiation will not be attempted.

1.3.5 Permeability of Limestone Rock

Three factors determine the permeability of carbonate rocks, their porosity, their jointing, and their bedding. Of these three the jointing especially is of great importance for karst morphology as well as for karst hydrology (see Chap. 14.1). The joints influence not only the appearance of the surface (grikes, bogaz, rows of dolines, directed karst), but also the underground water systems which, under vadose conditions, can come into existence only through joints which have opened. In the phreatic zone the water systems lie mainly in the bedding-planes. Joints are important also because they stand more or less vertically on bedding-planes, connecting them.

Bedding-planes occur in limestones in capillary to subcapillary widths. Their permeability is increased if they have served for tectonic movements; then they bear similarity to faults. Caves in folded mountains illustrate this. In Hölloch 90% of the 140 km of measured passages are bedding-plane controlled. However, this does not preclude that jointing also plays a part in it. Bedding planes have the advantage of being connected over long distances without interruption. Bedding planes are of special importance when they separate a capping limestone bed from the underlying, impermeable rock, e.g., from slate. The water courses form in the limestone and then, after they reach a certain size, they cut down, by means of erosion, into the softer, underlying rock. Frequently cave rivers follow those bedding planes. Compact carbonate rocks are practically impermeable to water. Since they are also brittle, even slight tectonic stress leads to the formation of joints, e.g., even in the tectonic domes in Tennessee and Ohio. On their flanks enormous caves have developed, among them the longest in the world, the Flint Mammoth Cave system. By virtue of the corrosive widening of jointings limestone becomes the most permeable rock; moreover a rock in which permeability increases with time.

Not all joints show the same properties. The main system of joints usually has closed pressure joints, standing perpendicular to the direction of pressure, and tension joints which are sometimes open. In such one can often find a permeable breccia, which is either still loose or consolidated by calcite. It can easily be attacked by water and cleared away. Porous limestones are permeable, even if only to a slight degree when compared to jointed limestone. Under a pressure head water flows slowly through the pores. This primary permeability is not so very dependent on the volume of the pores as on their width and the connections between them. Thus a sandstone with half the volume of pores that a marly clay has is nevertheless around 1000 times more permeable.

Table 1.8. The volume of pores in various rocks according to Martel (1921), p. 123 (quoted from Trombe, 1952) and in various loose materials according to Pettijohn (1949)

	Volume of pores (%)	
Granite	0.05- 0.86	
Slate	0.54- 0.70	pressed
Sandstone	3.32-39.8	
Compact sandstone	around 4	
Limestones	0.67- 2.55	Free of joints
Chalks	14.40-43.90	France; in G.B. max. 46% (Kendall et al., 1924)
Dolomites	1.50-22.15	Free of joints
Marbles	0.11- 0.59	Free of joints
Coarse sand	39 -41	
Fine sand	44 -49	
Loam	50 -54	

1.3.6 Reefs

Reef-forming organisms, corals, sponges, algae, build up unbedded limestone masses which can become dolomitized with time. They are called Massenkalke in Germany. Reef sands form slopes around such reefs. The sands become hardened to calcarenites with a primary bedding incline of up to 20° and 30°, e.g., the sponge reefs of the Swa-

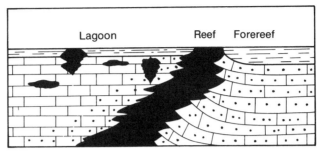

Fig. 1.2. Cross-section of a barrier reef

bian Jura. Such *bioherms* enclose only a delimited area and are completely local. In contrast to single reefs, atolls embrace a small, barrier reefs a great extent. The Australian barrier reef is 1800 km long. Such reefs, originating in the Paleozoic Era, have been geologically and karst-morphologically studied in detail; e.g., the Permian Capitan-Reef in the Guadalupe Ridge near Carlsbad (NM, USA; Newell et al., 1953; Motts, 1972) and the Devonian reef of the Napier Range in West Kimberley, Australia (Jennings and Sweeting, 1963; Playford and Lowry, 1966). The forereef facies correspond to those surrounding single reefs: calcarenites. The reef limestones themselves are dolomitic and unbedded. Interpolated calcarenitic nests give evidence of former reef caverns which filled up with reef sand. In the backreef zone where the surf becomes calm again calcarenites form which gradually turn into well-banked micrites with interpolated bioherms and biostroms. If it is a barrier reef, the content of impurities increases as it approaches the mainland.

The two examples of barrier reefs mentioned have essential differences in their karstification: the Napier Range shows surface karst with few small caves, the Guadalupe Ridge shows slight surface karstification but has, on the other hand, spacious caves such as the Carlsbad Caverns.

2 Processes of Dissolution of Karstifiable Rocks, Corrosion

Corrosion, used in the geomorphological sense, is the dissolution of rocks. The term chemical erosion, which used to be applied to this phenomenon, is too narrow and causes confusion because it inadequately expresses the basic processes. Today it is rarely used any more.

There are three types of corrosion:

a) The corrosion of carbonate rocks is a reversible chemical reaction (French: corrosion, German: Korrosion), which gives rise to the formation of carbonate karst in relatively pure limestones and dolomites.

b) The corrosion of gypsum and rock salt is a reversible physical process (French: dissolution, German: Korrosion) which results in gypsum or salt karst.

c) In warm, humid climates irreversible chemical processes take place in silicate rocks, that is a decomposition of rock-forming minerals. The results are partly soluble products, e.g., salts of Na^+, K^+, Ca^{2+}, Mg^{2+}, and partly insoluble ones such as SiO_2-gel (silicic acid), clays, and quartz sand as an unchanged residue. These are purely processes of weathering. The forms which are thereby created can be similar to those of karst (pseudokarst), e.g., pseudorinnenkarren, or they can take on their own shapes like the tafonis (Klaer, 1956; Wilhelmy, 1958). The actual processes involved are insufficiently known in detail, but seem to depend on the fact that silicates are unstable in the presence of water with a higher CO_2 content and organic acids in such temperatures as prevail in the humid tropics (25°-35°C and higher). Even if individual minerals can be regenerated in the resulting soils, it is still impossible to speak of reversibility in the sense of the creation of a rock with at least the same chemical properties as the original had.

The rate of dissolution v_L is important for all three types. It depends on numerous factors, e.g., on the diffusion coefficient D, the momentary concentration C, the concentration of saturation C_s, and on the surface A, as well as on exterior conditions, such as the temperature, which determines the rate of reaction, the removal of the dissolved material from the corroded surface, or on the rate of diffusion, the speed of flow, the turbulence of the water, and the roughness of the rock's surface. The exterior factors are defined in the constant of proportionality k, which is determined empirically:

$$v_L = \frac{dC}{dt} = k \cdot A \cdot D (C - C_s) \quad \text{(Feitknecht, 1949, p. 163)}$$

In chemical dissolution chemical reactions enter the picture with the diffusion or convection of active substances to the rock's surface, or to a rock component which has already been physically dissolved, e.g., in corrosion of lime where the rate of diffusion of atmospheric CO_2 into water is important because it becomes a limiting factor owing to its smallness.

The dissolution of rock is the decisive process in the formation of karst. Because of the difficulty in determining k, which shows a different value in every new case, v_L has till now not been used in geomorphological calculations. k is dependent on temperature and therefore v_L must be taken into consideration qualitatively as a factor in climamorphology.

2.1 Dissolution of Gypsum and Rock Salt

Anhydrite, $CaSO_4$, gypsum, $CaSO_4 \cdot 2H_2O$, and rock salt (halite), NaCl, form ionic lattices which the ions leave during dissolution without any chemical change: i.e., physical dissolution. When anhydrite, gypsum, or rock salt crystallize, the same crystal lattice forms again.

Anhydrite and gypsum: $\quad CaSO_4 \rightleftharpoons Ca^{2+} + SO_4^{2-}$
Rock salt: $\quad NaCl \rightleftharpoons Na^+ + Cl^-$

Dissolution occurs in a closed system and approaches continuously in an exponential curve the point of saturation.

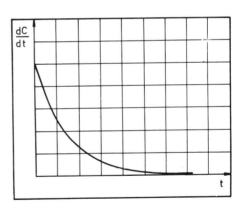

Fig. 2.1. Curve of Dissolution

The curve of saturation for gypsum reveals a relatively slight dependence on temperature. At 0°C there are 1.76 g of $CaSO_4$ in 1 kg of solution, at 35°C there are 2.105 g and at 45°C there are 2.100 g. Thus the curve runs through a maximum.

Gypsum deposits resulting from the cooling of saturated calcium-sulphate water are not common. The deposits, which are several meters in size, found in the Big Room of Carlsbad Caverns (NM, USA) have been explained in this way (Bretz, 1949; Good,

1957). As a rule, however, gypsum deposits are the result of the evaporation of water (evaporite).

The solubility of gypsum is 10 to 30 times greater than that of limestone (Gmelin, 1961). In addition, the rate of dissolution is higher because it is only a matter of the ions' detaching from the crystal lattice.

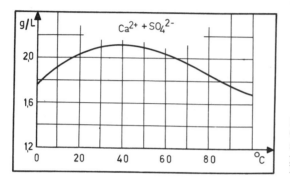

Fig. 2.2. Curve of saturation for Ca^{2+} and SO_4^{2-} with gypsum as sediment (acc. to Niggli, 1926, p. 110, Trombe, 1952, p. 139 and Freien, 1978, p. 15)

Gypsum is therefore more exposed to corrosion by precipitation than is limestone. Karren form quickly and without the participation of other substances (Priesnitz, 1969). They are sharp and pointed but because of the softness of the rock they are soon destroyed again. Caves are in general short, rarely longer than 4 km (Heimkehle, DDR). This is a result of physical dissolution which effects that the water's capacity to dissolve is exhausted after a certain distance of underground flow. There are no reactions which could interrupt the dissolution of gypsum for a definite time — nor can the concentration be decreased except by the introduction of fresh water capable of dissolving. Among gypsum caves those of Podolia are sole exceptions: Optimističeskaja Peschtschera with 119 km and Ozernaja Peschtschera with 103 km. There is still little known about the conditions of their formation. They are labyrinths.

Notable among the corrosion forms in the caves are the horizontal roofs created by dissolution and the facets created during their formation. These present an image of the layering of the water according to its density (Gripp, 1912). Brandt et al. (1976) pursued this problem further and won new insight into this process. Yet Völker (1973) did not exclude the possibility of their creation by erosive denudative processes.

Fig. 2.3. Roof created by dissolution (L) and facets (F) in Numburg Cave (Kyffhäuser, DDR) according to Biese (1931), in outline

Dissolution of Carbonate Rocks

The dissolution of rock salt proceeds according to the same laws. Yet its solubility is 180 times greater and reaches 360 g/l. Its dependence on the temperatures which occur in nature is very slight and practically negligible between 0° and 40°C.

2.2 Dissolution of Carbonate Rocks

The dissolution of limestone and dolomite is the central problem of superficial and underground karstification. It takes place in a system of substances of the type CO_2 – H_2O – $MeCO_3$ (Me: Ca or else Mg). It comprises numerous physical and chemical processes, in which all three aggregate states – gas, liquid, and solid – participate. This causes a mass transfer through the interfaces air/water and water/rock (Fig. 2.4).

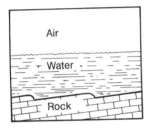

Fig. 2.4.

The processes at the interfaces are physical (mass transfer, diffusion), within the solution they are chemical.

In the system of substances CO_2 – H_2O – $CaCO_3$ essentially the following steps take place:

1. During precipitation atmospheric CO_2 diffuses into the water:

$$CO_{2 \text{ air}} \rightleftharpoons CO_{2 \text{ physically dissolved}}$$

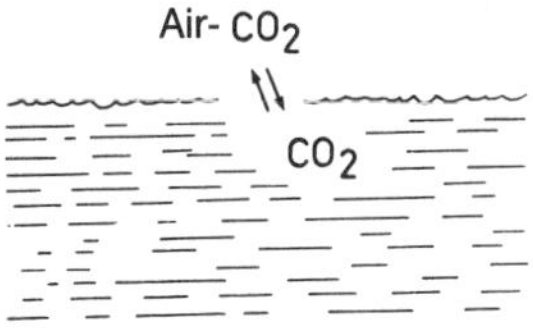

Fig. 2.5. See text at Eq. (1)

2. The physically dissolved CO_2 is 0.75% hydrated at 4°C (carbonic acid):

$$CO_{2 \text{ phys. dissolv.}} + H_2O \rightleftharpoons H_2CO_3, CO_2 \text{ chem. dissolv.}$$

Processes of Dissolution of Karstifiable Rocks, Corrosion

Air-CO_2

$CO_2 \rightleftharpoons H_2CO_3$

Fig. 2.6. See text at Eq. (2)

3. As a strong acid H_2CO_3 is completely dissociated: first oxydation level:

$$H_2CO_3 \rightleftharpoons HCO_3^- + H^+$$

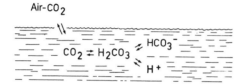

Fig. 2.7. See text at Eq. (3)

The second oxydation level, the dissociation of HCO_3^- into CO_3^{2-} and H^+, can be neglected at a pH of below 8.5 because of its small proportion.

4. When water and carbonate rock come into contact the ions are freed out of the crystal lattice, a physical process:

$$CaCO_3 \rightleftharpoons Ca^{2+} + CO_3^{2-}$$

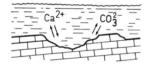

Fig. 2.8. See text at Eq. (4)

5. The newly created CO_3^{2-} associates with the H^+ from step 3:

$$CO_3^{2-} + H^+ \rightleftharpoons HCO_3^-.$$

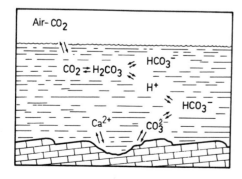

Fig. 2.9. See text at Eq. (5)

Dissolution of Carbonate Rocks

6. The solution along the interface becomes impoverished of CO_3^{2-}. The solution's equilibrium with the solid $CaCO_3$ is thereby disturbed and the ion product no longer corresponds to the solubility product L.

$$(Ca^{2+})(CO_3^{2-}) = L = 4.0 \cdot 10^{-9} \text{ (Picknett, 1973) at } 10°C.$$

Note: round brackets (...) mean the "activity" of their content; square brackets [...] point out the concentration (see Chap. 2.2.4).

The solubility product L has been determined in quite different ways by different authors. Picknett's figures might actually be the most precise.

Table 2.1. L according to Picknett (1973)

	5°C	10°C	15°C	20°C	25°C
L:	$4.1 \cdot 10^{-9}$	$4.0 \cdot 10^{-9}$	$3.9 \cdot 10^{-9}$	$3.8 \cdot 10^{-9}$	$3.7 \cdot 10^{-9}$

In order to replace the expended CO_3^{2-} and thereby bring the ion product again up to the constant L, $CaCO_3$ again dissolves (4th step), whereby $[Ca^{2+}]$ increasingly predominates over $[CO_3^{2-}]$.

The equilibrium between the carbonic acid and its first product of dissociation is disturbed by the combination of H^+ with CO_3^{2-}, and renewed dissociation results. In order to maintain the equilibrium physically dissolved CO_2 is hydrated and thereby the equilibrium with atmospheric CO_2 disturbed, which is the reason why new CO_2 diffuses into the solution. In short: all the steps in the reaction are re-activated by the association of H^+ and CO_3^{2-} until a new equilibrium is reached. Therefore processes of dissolution are not to be comprehended as static only, they must also be considered dynamic.

By the addition of steps 2-5 the well-known and much-used equation for the dissolution of limestone is reached (dissolution equation):

$$CaCO_3 + CO_2 + H_2O \rightleftharpoons Ca^{2+} + 2HCO_3^-.$$
solid limestone   dissolved limestone

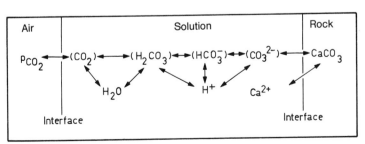

Fig. 2.10. The mutual dependences in the system $CO_2 - H_2O - CaCO_3$; the dissociation of water is not taken into consideration

If Mg^{2+} also takes part in the process of dissolution, the conditions on the interface solution/rock become considerably more complicated. To be precise the dissociation of water should also be included:

$$H_2O \rightleftharpoons H^+ + OH^-.$$

However, under natural conditions this ion portion can be neglected as the extreme pH values necessary for a higher portion are not attained in karst.

2.2.1 CO_2 on Either Side of the Interface Air/Solution

CO_2 is the only component in the system which can vary uninterruptedly within wide margins. The most important source of CO_2 is the air of the atmosphere and of the soil; their CO_2 contents generally make up between 0.00035 and 0.1 of the total pressure (Bögli, 1969a; Miotke, 1972, 1974). Under special conditions the partial pressure p'_{CO_2} sinks below 0.00035, for example with increasing altitude, in forests (assimilation) or in close contact with melting snow, limestone, and air (Ek et al., 1969, Bögli, 1969e). A wind with only 0.016% CO_2 (Bögli, 1970), as an exception 0.010% CO_2, blows at times out of the Schwyzerschacht (Muotatal, Bödmerenalp). At present the mean p'_{CO_2} of atmospheric air lies at sea level at 0.035% but is frequently higher close to the ground.

Table 2.2. p_{CO_2} in dependence on the altitude when the pressure portion of 0.035% remains unchanged

Altitude (m above sea level)	p_H in % of p_0	p_{CO_2} when the pressure portion is 0.035%
0	100	0.035
200	97.7	0.0342
400	95.3	0.0334
600	93.1	0.0326
800	90.9	0.0318
1000	88.7	0.0310
1200	86.6	0.0303
1400	84.5	0.0296
1600	82.4	0.0288
1800	80.4	0.0281
2000	78.4	0.0274
2200	76.5	0.0268
2400	74.6	0.0261
2600	72.8	0.0255
2800	71.0	0.0249
3000	69.2	0.0242

The CO_2 content of the air in the soil exceeds that of the atmosphere which is 0.035%; values between 10 and 100 times that amount are normal. As the upper limit Roques (1962) mentions 10%, i.e., 300 times as much, and Trombe (1952) even 25%, or 700 times as much. In a humus soil at 1650 m above sea level Bögli (1969) measured a p_{CO_2} of 8.3% in relation to 760 mm of Hg; this corresponds to a portion of 10.2% in the soil's air. The CO_2 of the soil's air is biogenic since it is created by processes of decomposition of organic substances by organisms in the soil and by root respiration (Bögli, 1960a, Miotke, 1974).

According to Henry's Law the p_{CO_2} of the atmosphere and the dissolved CO_2 are in the following relationship to one another:

$$p_{CO_2} = D \cdot [CO_2], \quad \text{D: diffusion coefficient dependent on temperature} \tag{2.1}$$

The formula derived from this according to Trombe (1952) is more suitable in practice:

$$CO_2 \text{ g/l} = M \cdot p_{CO_2}. \tag{2.2}$$

$$M = \frac{1}{D} \cdot \frac{1.964}{44};$$

1.964: liter weight of the CO_2; 44: molecular weight of the CO_2

Table 2.3. M as a function of the temperature

	0°C	5°C	10°C	15°C	20°C	25°C	30°C	35°C
M calculated according to Roques (1962)	3.441	2.830	2.343	1.999	1.709	1.473	1.284	1.125
% of M_o	100	82.2	68.1	58.1	49.7	42.8	37.3	32.7
M according to Trombe (1952)	3.364	2.797	2.345	2.001	1.724	–	1.306	–
% of M_o	100	83.1	69.7	59.4	51.3	–	38.8	–

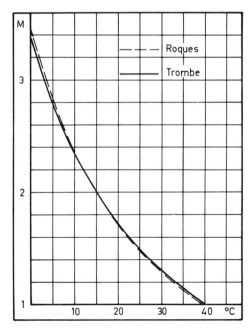

Fig. 2.11. M as a function of the temperature (Roques, 1962; Trombe, 1952)

By means of Eq. (2.2) the CO_2 concentration in the solution is determined when in equilibrium with the p_{CO_2} of the atmosphere. CO_2 is dissolved physically to more than 99%, the remainder, 0.7% at 4°C, is hydrated (Trombe, 1952). When the p_{CO_2} is constant the CO_2 content of the water sinks to 37.3% at 30°C, compared with the content at 0°C. As a result the solubility of the limestone sinks also.

Table 2.4. Amount of the CO_2 dissolved in equilibrium in ppm as a function of p_{CO_2} (%) and temperature T

p_{CO_2}	0°C	5°C	10°C	15°C	20°C	25°C	30°C	35°C
0.01	0.34	0.28	0.23	0.20	0.17	0.15	0.13	0.11
0.03	1.03	0.85	0.70	0.60	0.51	0.44	0.39	0.34
0.1	3.44	2.83	2.34	2.00	1.71	1.47	1.28	1.13
0.25	8.60	7.07	5.86	5.00	4.23	3.68	3.21	2.81
0.5	17.2	14.2	11.72	10.00	8.54	7.37	6.42	5.63
0.75	25.8	21.2	17.6	14.99	12.83	11.05	9.63	8.44
1	34.4	28.3	23.4	20.0	17.1	14.73	12.84	11.25
5	172	141.5	117	100	85.5	73.7	64.2	56.3
10	344	283	234	200	171	147	128	112.5

Little is known concerning the CO_2 hydrates. The existence of a hexahydrate, $CO_2 \cdot 6H_2O$, is certain but according to Roques (1962) nothing can be said about the other hydrates. The kind of hydrate cannot be determined from the dissociation products. Therefore the monohydrate $CO_2 \cdot H_2O$, or H_2CO_3, is taken as a norm.

Between the dissolved CO_2 and the products of the first step of dissociation there exists an equilibrium with the dissociation constant K_1.

$$K_1 = \frac{[H^+][HCO_3^-]}{CO_2^0}.$$

The values for K_1 vary according to the author. Cigna (1972) considers those of Roques (1964) as the best, while those of Harned and Owen (1958) are preferred in the Anglo-Saxon world (Picknett et al., 1976).

Table 2.5. K_1 as a function of the temperature T: (a) according to Roques, 1964 (quoted from Stchouzkoy, 1972, p. 474), (b) according to Harned and Owen (1968) and Harned and Davis (1943)

	0°C	5°C	10°C	15°C	20°C	25°C	30°C	35°C	40°C	45°C	50°C
(a) $-pK_1$	6.646	6.596	6.547	6.500	6.456	6.414	6.372	6.330			
$K_1 \cdot 10^7$	2.260	2.535	2.838	3.126	3.500	3.855	4.246	4.656			
(b) $-pK_1$	6.579	6.517	6.464	6.419	6.381	6.352	6.327	6.309	6.298	6.290	6.285
$K_1 \cdot 10^7$	2.647	3.040	3.430	3.802	4.147	4.452	4.710	4.914	5.058	5.139	5.161

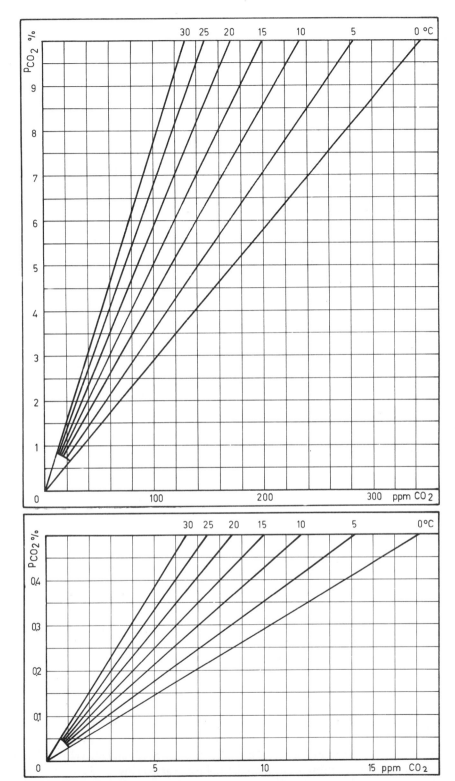

Fig. 2.12. a, b. p_{CO_2} and dissolved CO_2 (ppm) according to the M values of Roques (1962). **a** 0%-9%, **b** 0%-0.5% partial pressure

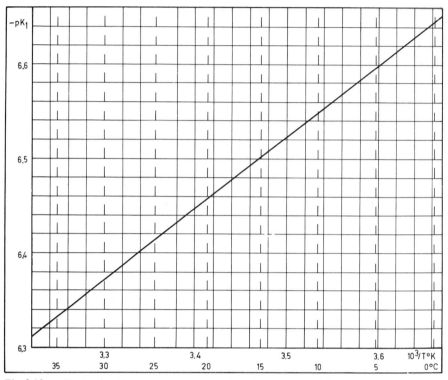

Fig. 2.13. $-pK_1$ as a function of the temperature (acc. to Roques, 1964)

In the second step of dissociation the following is valid:

$$K_2 = \frac{[H^+][CO_3^{2-}]}{[HCO_3^-]}.$$

The values are 10^4 times smaller than is the case with K_1.

Table 2.6. K_2 as a function of the temperature T according to Harned and Scholes (1941) and Maronny (1961, quoted from Roques, 1964, p. 286)

T	0°C	5°C	10°C	15°C	20°C	25°C
$-pK_2$	10.625	10.557	10.49	10.430	10.379	10.329
$K_2 \cdot 10^{11}$	2.37	2.77	3.24	3.71	4.20	4.69
T	30°C	35°C	40°C	45°C	50°C	
$-pK_2$	10.290	10.250	10.220	10.195	10.172	
$K_2 \cdot 10^{11}$	5.13	5.62	6.03	6.38	6.73	

In the solution CO_2, HCO_3^- and CO_3^{2-} are a function of the partial pressure of the atmospheric CO_2. Through dissociation they effect a change of the H^+ content and thereby of the pH value. Between the latter and the p_{CO_2} there is a close relationship during equilibrium which is shown in the following graph (Fig. 2.15).

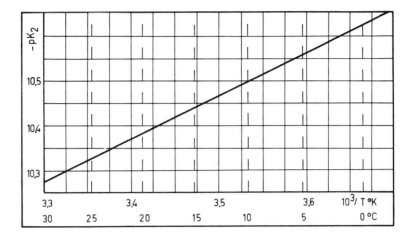

Fig. 2.14. K_2 as a function of the temperature according to Harned and Scholes (1941) and Maronny (1961)

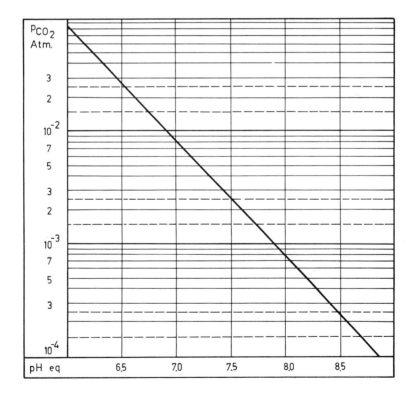

Fig. 2.15. pH as a function of the pCO_2 according to Roques (1964)

During equilibrium the ratio of the amounts of CO_2, HCO_3^- and CO_3^{2-} are always the same at any given pH-value, but they also change when it changes.

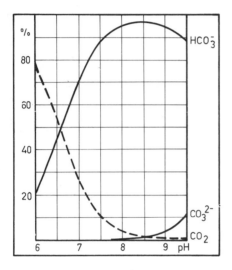

Fig. 2.16. Amounts of CO_2, HCO_3^- and CO_3^{2-} at 10°C as a function of the pH during equilibrium (acc. to Roques, 1954)

It is clear from the graphs that a pH-value of 8.5 CO_2 and CO_3^{2-} each make up 1.2% and that the HCO_3^- reaches its maximum of 97.6%. At a pH below 8.3 the content of CO_3^{2-} can even be neglected in calculations. The same is true of CO_2 at a pH above 8.7. However at 6.6 and below, its concentration even exceeds that of HCO_3^-. In nature, where pH values between 6.25 and 8.75 occur, CO_3^{2-} is quantitatively of no essential significance. In equilibrium karst water generally shows pH values between 7 (corresponding to approx. 330 ppm of dissolved limestone) and 8.5 (50 ppm). HCO_3^- strongly dominates here.

2.2.2 The Kinetics of CO_2

Up to now CO_2 and the other members of the system have been looked at under static conditions, which show definite qualitative and quantitative relationships. If one takes primarily a kinetic point of view — considering the rates of reaction and changes in concentration in general — one must be satisfied with qualitative statements. Either the models used diverge too greatly from the actual realities of nature or the parameters cannot be determined sufficiently precisely. The results, calculated theoretically, are at best approximate values. A few clearly defined cases are an exception, e.g., the changes in concentration during dripping.

The mass transfer from one medium to another always takes place slowly, which is the reason why an existent state adjusts only gradually to a new one. Changes in the p_{CO_2} can, on the other hand, take place abruptly, for example when water moves through a tube without contact with air and then issues into a cavity with an open water surface. In the airless tube the water forms a closed system in which nothing more changes after the equilibrium is reached: if the lime content is low, no limestone is dissolved, if it is high, none is deposited. A theoretical p_{CO_2} prevails which is in equilibrium with the dissolved CO_2. When the pool is reached the system is open and shows a new p_{CO_2} to which it must adjust.

In Fig. 2.17 p_1 is the theoretical partial pressure to the CO_2 in equilibrium (CO_{2eq}, eq = equilibrium) in the closed system, p_2 is that in the air-filled cavity; assumed: $p_1 < p_2$ which for example corresponds to the conditions in bare karst. When the water flows into the air-filled space it is undersaturated in relation to p_2. The graph shows an exponential curve where the partial pressure gradually adjusts to the pressure in equilibrium with the p_2 as atmospheric CO_2 diffuses into the water. τ refers to the time (t) which passes before 90% of the change in concentration necessary for the equilibrium has taken place.

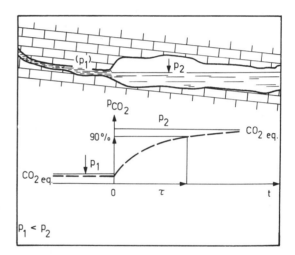

Fig. 2.17. Temporal change in the CO_2 content in underground water when $p_1 < p_2$

In ascending karst springs (vauclusian springs) normally $p_1 > p_2$. In comparison to the air on the exterior (p_2) the water has an excess of CO_2 and therefore gives off CO_2 to the atmosphere (Fig. 2.18).

It has already been mentioned that in both cases the concentrations of the various components are forced to change.

The transition from one medium into the other is determined in the case of quiet water by the diffusion coefficient D of the CO_2 in the air and in the water.

$D_{in\ air}$ = 0.138 cm²s⁻¹ (0°C, 1 atm); 1 cm²s⁻¹ (25°C)
$D_{in\ water}$ = 0.95 · 10⁻⁵ cm²s⁻¹ (0°C).

Because the diffusion coefficient in the water is smaller by the factor 10^{-4}, the rate of diffusion is 10,000 times lower there. The latter therefore determines the amount of the mass transfer through the interface.

Fig. 2.18. Temporal change of the CO_2 content in the water of a vauclusian spring where $p_1 > p_2$

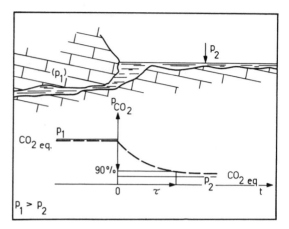

If the two media are motionless, the CO_2 molecules move on only by diffusion. There is no need to speak of a marginal layer thereby since the entire volume reflects its conditions. The adjustment to the equilibrium takes place from the interface and progressively includes parts which are more and more distant. It is obvious that the ratio of the surface to the volume, A/V, is the decisive factor for τ.

When the water is turbulent conditions change fundamentally. CO_2 is carried away from the interface immediately by convection and mixed with the solution, as the water's movement is greater by the power of tens than the rate of diffusion. The whole volume is thereby brought to approximately the same concentration and the difference of the concentration at the interface is always held at the maximum. This greatly accelerates mass transfer. Moreover the water surface is rough and the A/V is increased. This has a similar effect.

One of the best examples that has been investigated is the behavior of CO_2 in a sinter tube. These stalactites are not rare in caves, measuring between 10 and 100 cm in length — as an exception up to 450 cm — at diameters of between 4 and 6 mm. Roques (1964, 1969a) examined the drops which emerge at the lower end of the tube. Its wall can be considered airtight so that no contact exists with the air on the exterior. Thus the water forms a closed system with partial CO_2 pressure $p_1 > p_2$. When CO_2 is given off the p_{CO_2} changes from p_1 to p_2 by way of p', the concentration C_1 thereby changes also by way of C' to C_2, where the equilibrium is reached (Fig. 2.19).

Whether the equilibrium is reached during the formation of the drop is simply a question of time. τ amounts to approx. 4 h. Dixon and Hands (1957, quoted according to Roques, 1964, p. 346/7) investigated the rhythm of drops falling every 0.1-10 s, Roques did the slower ones.

The occurrence of a minimum of CO_2 loss during the formation of a drop seems to be contradictory. If the drop already falls off after 0.1 s, 16% $(C_1 - C_2)$ is given off, after 4 s it is only 9%. When the time of formation is short (0.1 s), the water flowing in quickly, turbulence is great and therefore a maximum difference of concentration to the p_{CO_2} in the air dominates practically constantly. When the period of formation is 4 s the turbulence is so low that a marginal layer can build up in which diffusion determines the rate of discharge of CO_2. Of course the thickness "d" of this marginal layer

Fig. 2.19. Behavior of CO_2 in the water of a sinter tube / straw stalactite

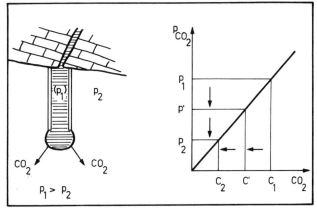

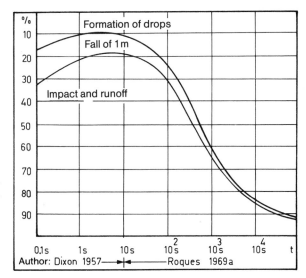

Fig. 2.20. Graph of the losses of CO_2 during the formation of drops, during their fall and while they flow away, in reference to the time of formation t of the drop (according to Roques, 1969a, p. 469).

is still slight so that the concentration gradient C_1/C' and therefore also the rate of diffusion is high even though the latter still lies far below the velocity of convection.

As the length of the period of formation of the drop increases the turbulence continues to decrease. Moreover there is a transition of $(p_1 - p_2)$ to $(p' - p_2)$, decreasing until approximately zero. The discharge of CO_2 sinks thereby, also approaching zero.

τ is reached in approx. 4 h. From this the order of magnitude of τ can be derived for other forms, too. The drop of 0.08 cm^3 shows a A/V of 8.9 cm^{-1}, as long as it hangs on the sinter tube as a hemisphere, and of 11.2 cm when it is released as a sphere. In comparison to this a cylindrical basin with a capacity of 800 cm^3 and with an open surface of 160 cm^2 has a A/V of 0.2 cm^{-1}.

For example that is the case with small camenitsas (see Chap. 3.1.2.1, B.1). If all other conditions are similar, 180 h or 7 1/2 d are the result for τ. This magnitude has

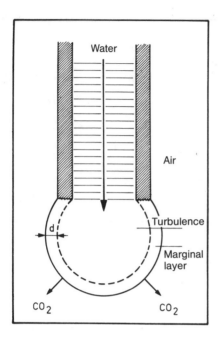

Fig. 2.21. Drops on a sinter tube

been measured approximately several times in nature. Natural processes, wind, evaporation, and temperature changes generally vary this value.

The period of the formation of the drop ends when it falls off. Dixon and Roques both chose a height of fall of 1 m. They observed that as the period of formation of the drop increased, there was a continuous decrease of 16% from the amount $(C_1 - C_2)$ at a rhythm of 0.1 s to 9% at 4 s and to 2% at τ. At the moment when the drop falls off an approximate sphere forms from the hemisphere and A/V increases abruptly from 8.9 cm^{-1} to 11.2 cm^{-1}. Moreover a short turbulence follows after the detachment of the drop falls. That means there are two further constant magnitudes at the same height of fall. There still remains the turbulence taken over from the drop; it results in an additional loss of CO_2 when the intervals between the drops are short. When the drop strikes and subsequently flows away 70%, 82%, and 7% of the $(C_1 - C_2)$ is given off at drip-intervals of 0.1 s, 4 s, and 4 h respectively.

2.2.3 Dissolution of $CaCO_3$

The dissolution of carbonates results in the first place from the release of ions from the ionic lattice, i.e., the passage of the ions through the interface solid/liquid. It does not happen at any arbitrary rate as the ions are surrounded with H_2O molecules (hydration); this is connected with a low energy requirement. In the system $CO_2 - H_2O - CaCO_3$ the dissolution of $CaCO_3$ belongs to the slow reactions.

On the surface of the solid substance a marginal layer forms during dissolution which remains intact even when the turbulence is strong. It is several microns thick. In it the movement of ions takes place only by diffusion, which determines the rate of dissolution by its slowness. During turbulence there exists beyond this layer a balanced concentration C', on the limestone the concentration in equilibrium C_k, which depends on the equilibrium with the H^+ brought in by diffusion. Therefore the concentration gradient in the marginal layer amounts to

$$\frac{C_k - C'}{d}.$$

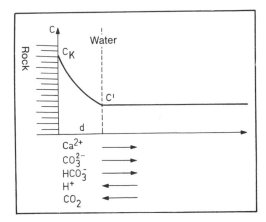

Fig. 2.22. Concentration gradient in the marginal layer

When those ions leave the ionic lattice CO_3^{2-} is transformed by H^+ into HCO_3^-. According to Roques (1964) it is irrelevant whether the H^+ or the hydronium ion is put in. In the marginal layer there are two directions of movement: Ca^{2+}, CO_3^{2-}, and HCO_3^- move towards the zone of turbulence, H^+ and CO_2 toward the rock.

The solubility of $CaCO_3$ is dependent on the kind of ionic lattice built up by the Ca^{2+} and CO_3^{2-}. Under natural conditions rhomboedric calcite is the most stable modification with the lowest degree of solubility. Metastable rhombic aragonite is somewhat more soluble. The other two modifications, hexagonal vaterite and the amorphous $CaCO_3$-gel are unstable in contact with water and are transformed into calcite, and, when the concentration is sufficiently high, into aragonite. Here, too, for the degrees of solubility there are different values given in literature. To be sure Cigna (1975) names those of Roques (1964) and Stchouzkoy-Muxart (1972) as the best; the values of Frear and Johnston (1929), whose figures became classical, are 20% higher, and also those of Tillmans (1932) show similar divergences, apart from the low CO_2 contents below 2 ppm. This causes uncertainty.

Roques and Stchouzkoy worked with the same experimental arrangement. The former dealt with the area of 10° and 15°C for calcite and aragonite, the latter with that of 20° and 30°C for calcite. The results of both complement one another very well, yet those of Roques appear to be somewhat too high; this becomes unpleasantly apparent only when a curve of dependence on temperature is drawn.

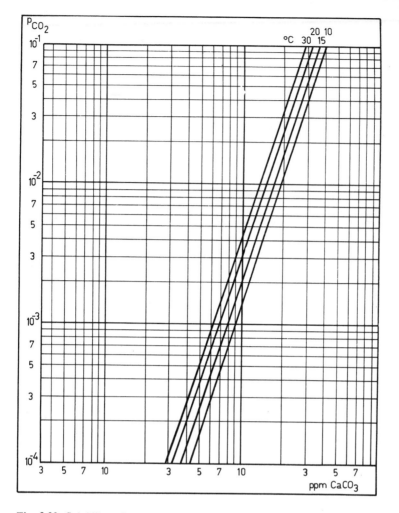

Fig. 2.23. Solubility of calcite and aragonite in dependence on the p_{CO_2} (acc. to the values of Stchouzkoy, 1972, and Roques, 1964)

CO_{2eq} is always associated with dissolved $CaCO_3$. In place of CO_{2eq} the p_{CO_2} resulting from it can also be put in. For a long time Tillmans' empirically gained pairs of values (1932) were valid for calculations. They were subsequently modified by the work of Pia (1953), Roques (1962, 1964) and others and given a theoretical basis, which was, however, similarly based on empirical figures to a large extent.

Ca^{2+}, Mg^{2+}, and HCO_3^- can be determined precisely and expediently today by means of complexometric methods. One can find the pH value with any desired degree of precision. It is closely related to all the components of the system $CO_2 - H_2O - CaCO_3$.

Dissolution of CaCO₃

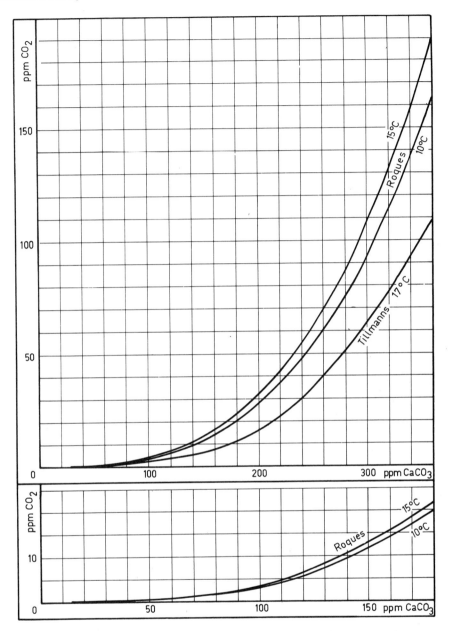

Fig. 2.24. CO_2 eq according to Roques and Tillmans

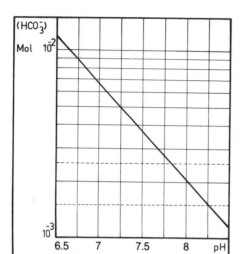

Fig. 2.25. pH and (HCO_3^-) during equilibirum

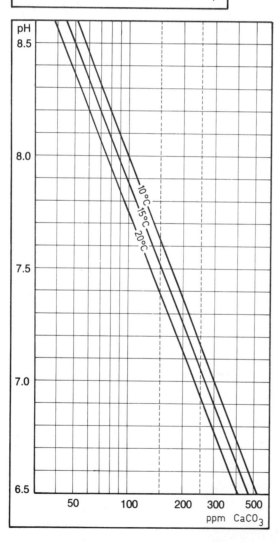

Fig. 2.26. pH and (Ca^{2+}) during equilibrium; Ca^{2+} recalculated into $CaCO_3$

The preceding comments furnish in the following classification of CO_2:

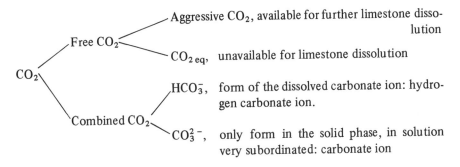

2.2.4 Influence of Other Ions

In natural solutions of the system $CO_2 - H_2O - CaCO_3$ there are almost always other ions to be found, above all Mg^{2+}, more rarely SO_4^{2-}, also alkali ions in the vicinity of the sea and in arid regions with saltwater lakes. In addition ions from mineral springs, weathered rocks and ores, guano, fertilizers, sewage water, and mine water are found locally. Very generally it is true that in the case of a saturated solution of a salt which has a low degree of solubility the addition of a second, more easily soluble salt which has no ion in common with the first increases the solubility of the first. Cigna et al. (1963) proved an increase in the solubility of 10%-25% from the addition of 0.1% NaCl.

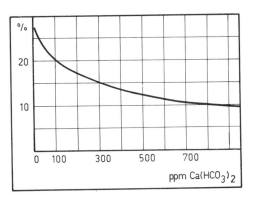

Fig. 2.27. Changes in the solubility in % of the total solubility with the addition of 0.1% NaCl according to Cigna et al. (1963)

Explanation: the activity of the ions is the portion of active ions in the total stock of ions and the activity coefficient f is the factor with which the concentration of ions must be multiplied in order to obtain the activity (Langelier, 1936; Pauling, 1967). According to this the activity of Ca^{2+} is:

$$(Ca^{2+}) = [Ca^{2+}] \cdot f_{Ca^{2+}}.$$

f is always smaller than 1. Everything that hinders the movement of the ions reduces their activity. That is for example the case when the concentration of ions is increased from C to C'. Then the following is true:

$$f_{Ca^{2+}} > f'_{Ca^{2+}}.$$

One must thereby take into consideration the interpretation of the solubility product:

Pure solution: $L = (Ca^{2+})(CO_3^{2-}) = [Ca^{2+}][CO_3^{2-}] f_{Ca^{2+}} \cdot f_{CO_3^{2-}}$.

The addition of other ions reduces the activity from f to f'. The product sinks. In order to reach L an additional dissolution of $CaCO_3$ is required, i.e., the originally saturated solution becomes undersaturated by the addition of other ions. The solubility of $CaCO_3$ thereby increases by the corresponding amount.

Reversely the addition of a substance with one ion the same causes a reduction in the solubility of the limestone because in the solubility product Ca^{2+} or CO_3^{2-} occurs in addition. Therefore the result must be the deposit of limestone in order to retain the constancy of L. Indeed the activity coefficient sinks at the same time, also, whereby part of this effect is again offset. Roques' experiments (1964) show a remarkable increase in solubility with the addition of $MgCl_2$. Mg^{2+} tends to the formation of complexes in a solution with HCO_3^-, i.e., to a fixation in the uncharged $MgCO_3^\circ$ complex and in the $MgHCO_3^+$ complex. The portion of these complexes is of great consequence as opposed to the corresponding Ca complexes which occur only in such slight amounts that they can be neglected. The formation of complexes considerably disturbs the electrical equilibrium between Mg^{2+} and Cl^-. Therefore Ca^{2+} is drawn upon in substitution which leads to a renewed dissolution of limestone quite apart from the change of $f_{Ca^{2+}}$ (see too Garrels et al., 1961).

Gerstenhauer and Pfeffer (1966) found that the primary $MgCO_3$ content of a rock considerably reduces the amount of $CaCO_3$ dissolved in the unit of time. Experiments with artificially produced mixtures of $CaCO_3$ and $MgCO_3$ as sediments show a sinking of the amounts of $CaCO_3$ dissolved in 28 h from 17 ppm in the case of pure $CaCO_3$ to approx. 6 ppm with 2% $MgCO_3$ content and to 3 ppm with 95%. This phenomenon presumably plays an important role in karstification. Closer investigations in the field are still lacking.

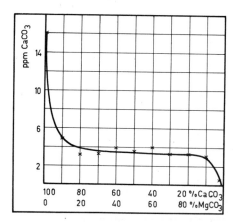 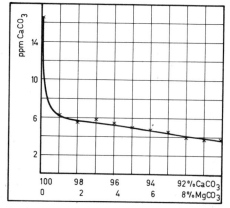

Fig. 2.28. a, b. The "readiness of $CaCO_3$ to dissolve" (Lösungsfreudigkeit) when in mixture with $MgCO_3$ (acc. to Gerstenhauer and Pfeffer, 1966)

The calcium carbonate first precipitates out of solution with Mg^{2+} and Ca^{2+} as sinter tufa, as long as the Mg^{2+} does not dominate (see Chap. 1.3.1). To test this a small stream with a tufa-bed, a medium gradient of 15% and a water flow of 0.4 l/s was examined. There were three measurement stations (1, 2, 3) at distances of 80 m (M_1-M_2) and 160 m (M_2-M_3). The average velocity of flow between M_1 and M_2 was 686 m/h, between M_2 and M_3, 960 m/h and between M_1 and M_3, 847 m/h.

Table 2.7. Ca^{2+} and Mg^{2+} in a brook (Hitzkirch, Ct. Lucerne)

Date, Time, Weather	t_W (°C)		Ca^{2+} ppm	Mg^{2+} ppm	$CaCO_3$ ppm
10.8.73 / 15.30-16.00	M_1	13	124.0	9.7	309
Warm, dry air	M_2	13 1/2	106.2	13.6	265
t_{Air} (°C): 26 1/2	M_3	14 1/2	94.1	15.8	235
10.8.73 / 18.15-18.45	M_1	13	120.2	8.9	300
t_{Air} (°C): 24 1/2	M_2	14	112.2	10.9	280
	M_3	15	102.2	12.2	255
13.8./15.00-15.30	M_1	13 1/2	110.2	12.2	275
Warm, dry air	M_2	14	100.2	13.4	250
t_{Air} (°C): 27	M_3	16	88.2	15.4	220
13.8./19.20-20.00	M_1	14	104.0	14.6	259
t_{Air} (°C): 21	M_2	14	96.2	15.8	240
	M_3	15	90.2	15.8	225
18.8. / 15.50-16.20	M_1	14	112.2	12.2	280
Sultry, misty	M_2	15	106.2	12.2	265
t_{Air} (°C): 29	M_3	16 1/2	96.2	12.2	240
24.8. / 7.40-8.10	M_1	14	116.2	9.7	290
Damp air	M_2	14 1/2	108.2	9.7	270
t_{Air} (°C): 16 1/2	M_3	15	102.2	9.7	255
24.8. / 18.00-18.30	M_1	14	120.2	9.7	300
t_{Air} (°C): 22	M_2	15	112.2	9.7	280
	M_3	16	108.2	9.7	270

From the table it can be seen that the Mg^{2+} content rises when the air is dry and evaporation increases accordingly but that when there is slight evaporation it remains unchanged within the frame of the margin of error. The higher Mg values at stations 2 and 3 are a consequence of the solution's loss of water on days of low humidity.

2.2.5 Mixing Corrosion (Plates 1.3, 1.4)

Buneyew (1932, quoted from Laptev, 1939) recognized the fact that water in equilibrium becomes corrosive again when it mixes with other water with a different concentration of Ca^{2+}. Laptev (1939) pursues this thought. In Germany Klut (1943) mentions this phenomenon, which can lead to very unpleasant corrosion in drinking water and sewage pipes and in concrete constructions in water. This knowledge remained limited to a small group of water supply and sewage engineers. In 1961 Bögli discovered the phenomenon anew and published his results (1963c, 1964a-c), which were completely concerned with karst and speleology. Ernst (1964) established the theoretical basis for them.

Mixing corrosion can be derived from usual corrosion with the aid of the curve of equilibrium. On the convex side lies the water oversaturated with lime, on the concave, the aggressive water with its surplus of CO_2. Any straight line between two arbitrary points of the curve runs through the aggressive zone. Those are the straight lines of mixing in which lie all possible mixtures of the two waters which are in equilibrium, W_1 and W_2. The surplus of CO_2 created by the mixing can be graphically determined with the aid of the curve of equilibrium.

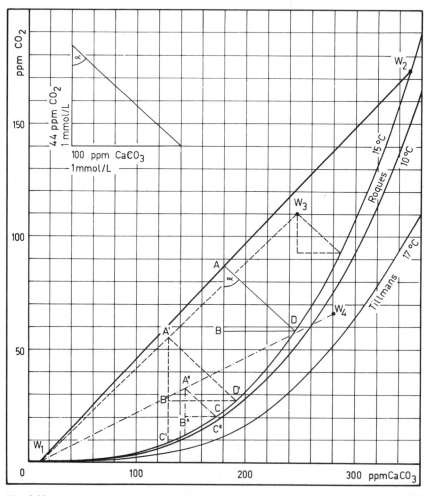

Fig. 2.29. Determination of the aggressiveness of mixed water according to Bögli (1964a, b, c, 1971a). See text

It is assumed that the temperature is 15°C. W_1 is water with 10 ppm of $CaCO_3$ and approx. 0.1 ppm of CO_{2eq}. W_2 is water with 350 ppm of $CaCO_3$ and 174 ppm of CO_{2eq} (according to Tillmans it was only 101 ppm of CO_2). A mixture of such extreme contents of lime occurs only rarely in nature, perhaps where bare karst and soil covering with a high degree of biological activity occur side by side, e.g., in the Mediterranean region. The mixture 1:1 (A) results in a $CaCO_3$ content of 180 ppm with 87 ppm of CO_2. But this lime content demands only 23 ppm of CO_{2eq} (C). Therefore there is a CO_2 surplus of 64 ppm ($\overline{AC}$). The portion $\overline{AB}$ is required to dissolve the additional $CaCO_3$, $\overline{BC}$ is used for the increase of CO_{2eq}. According to the equation of dissolution Eq. (2.2) 1 mol of $CaCO_3$ requires 1 mol of CO_2 for dissolution. Therefore $\overline{AB}$ is to $\overline{BD}$ as 1:1, i.e., 44 ppm of CO_2 to 100 ppm of $CaCO_3$ (1 mmol each), see auxiliary graph (left) above. Its angle α is therefore to be used at A to the construction. $\overline{BD}$ corresponds to the amount of additionally dissolved $CaCO_3$, here 65 ppm – a rare extreme in nature.

The determination of the aggressivity of a mixture can also be extended to primarily aggressive water or to water oversaturated with $CaCO_3$; see W_1 in mixture with W_3 or W_4 respectively. That with W_3 results in an additional dissolution of 62 ppm of lime, yet W_3 shows a primary aggressiveness of 38 ppm so that 24 ppm fall to mixing corrosion. W_4, oversaturated with lime, when in mixture with W_1, still manages an actual, additional dissolution of 28 ppm of $CaCO_3$. In the case of an even greater surplus of lime, as Thrailkill (1968, p. 33) established in the groundwater of Kentucky, it is possible that no additional lime can be dissolved at all – the mixture is then accordingly less oversaturated. In nature usually much smaller differences in the lime content occur. In Hölloch the extremes are 50 ppm and 200 ppm, which results in 14 ppm of $CaCO_3$ as the amount additionally soluble.

Table 2.8. Additional amounts of $CaCO_3$ soluble by mixing corrosion at 15°C calculated according to Roques' values for equilibrium

Initial concentrations		Ratio of the mixing components with the amounts of $CaCO_3$ additionally soluble in ppm		
W_1 ppm $CaCO_3$	W_2 ppm $CaCO_3$	3:1	1:1	1:3
11.5	125.0	4.5	5.0	4.0
11.5	221.6	15.5	24.5	19.5
11.5	329.6	35	58	55
73.9	125.0	1.3	2.5	2.0
73.9	221.6	9.6	14	11
73.9	329.6	25	41	37
125.0	170.5	0.9	1.1	0.8
125.0	272.7	9.5	14	10.8
125.0	358.0	21	34	32

These values are somewhat higher than those calculated by Howard (1966) with the figures of Garrels and Christ (1965). Picknett's investigations (1977) showed that when carbonate waters with different Mg contents mix, aggressiveness also increases: rejuvenated aggressiveness.

2.2.6 Cooling Corrosion and Thermal Mixing Corrosion

With increasing temperature and constant p_{CO_2} the solubility of carbonates of Mg, Ca (Sr, Ba) decreases. On one hand the amount of CO_2 dissolved sinks (Tables 2.3 and 2.4, Fig. 2.12), on the other hand the required CO_{2eq} rises (Fig. 2.24). The values necessary to establish the curves of temperature dependence have not yet been determined experimentally in their entirety. Single values do not offer any common result and even Roques' (1964) and Stchouzkoy's (1972) values are not very suitable for this purpose. Therefore the theoretical deduction of Zehender et al. (1956) will be applied here. The values calculated from it by Bögli (1964a) are correct in their magnitude and above all in the type of the curves, even if numerical deviations are to be expected.

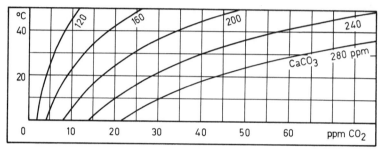

Fig. 2.30. Dependence of the CO_{2eq} on temperature in the case of chosen constant lime contents (acc. to Bögli, 1964a, p. 88)

Whenever the temperature of water in equilibrium sinks CO_2 becomes available for the dissolution of limestone: *cooling corrosion*.

Table 2.9. Additionally soluble amount of $CaCO_3$ during cooling acc. to Bögli (1964a, p. 88). The figures are approximate values in consideration of the respective CO_{2eq}

$CaCO_3$ ppm	Additionally soluble $CaCO_3$ (ppm) during cooling from				
	6° to 0°C	10° to 6°C	15° to 10°C	20° to 15°C	24° to 15°C
120	1.0	0.9	1.2	1.5	2.8
160	2.3	1.9	2.7	3.2	6.3
200	4.2	3.5	5.0	5.9	11.4
240	6.9	5.7	8.1	9.6	19.2
280	10.3	8.5	12.0	14.3	27.1

Cooling corrosion is active in the upper, vadose regions of underground karst. It is connected to zones with strong, daily thermal variations or with clearly marked seasons. In the first case it is active during the day until late in the night, in the second during the warm season. As soon as the temperature above the surface is less than the day's average or the annual average, respectively, cooling corrosion ceases since the temperature at a depth of 10-25 m, depending on the kind of rock, corresponds to the annual mean temperature there. Daily variations are detectable a few decimeters deep in the ground at the most. Below the level of constant temperature the temperature rises anew (geothermal gradient, see Chap. 15.2).

In mature karst circumstances are basically different. Air circulation and flowing water cool the rock off so that even at a depth of several hundred meters nothing of the earth's heat can be detected. In Hölloch (Switzerland) where passages reach as far as 900 m below the earth's surface, a temperature gradient approximately corresponding to that in the open atmosphere ($-1°C/200$ m) dominates in winter.

Cooling corrosion is climamorphologically active. It attains maximum values in continental regions of the moderate and marginally tropical zones where considerable daily variations are added to the annual ones. Compared with mixing corrosion, however, the effectiveness of cooling corrosion is generally less, and, moreover, limited to the uppermost meters of the rock.

Thermal water is a special case. Its capacity to corrode rises to high values, as the amounts of CO_2 which are released by cooling are occasionally enormous. When the pressure decreases the CO_2 can become free spontaneously, especially on emergence at the surface, thereby making the water bubble up: bubbling water. The thermal springs of Budapest created cave systems by cooling corrosion.

Table 2.10. $CO_{2\,eq}$ as a function of t when the $CaCO_3$ content is constant; thermal springs

$CaCO_3$ ppm	$CO_{2\,eq}$ (ppm) 0°C	10°C	20°C	30°C	40°C	50°C
120	1.9	2.8	4.0	5.6	8.1	11.5
160	4.4	6.3	9.0	12.6	18.2	25.9
200	8.2	11.8	16.4	23.7	34.2	48.7
240	13.8	19.7	28.3	39.7	57.2	81.5
280	21.3	30.4	43.8	61.4	88.3	125.9

Table 2.11. The amount of CO_2 released during the cooling of thermal water

$CaCO_3$ ppm	Cooling from 30° to 20°C	40° to 20°C	50° to 20°C	50° to 10°C
120	1.6	4.1	7.5	8.7
160	3.6	9.2	16.9	19.6
200	7.4	17.8	32.3	36.9
240	11.4	27.9	53.2	61.8
280	17.6	44.5	82.1	95.5

Recalculated to $CaCO_3$ (ppm):

$$2.274 \cdot CO_2 \text{ (ppm)} = CaCO_3 \text{ (ppm)}$$

Depending on the lime content of the water between one-quarter of the corrosive CO_2 at 120 ppm of $CaCO_3$ and two-thirds at 280 ppm is required for the additionally necessary CO_{2eq}.

Because of the bend of the curve there is also thermal mixing corrosion (Fig. 2.30). The bend is slight and as a result the amount of additionally soluble $CaCO_3$ is small. At temperatures of 0°C and 24°C it lies between 0.7 ppm for a content of 120 ppm of dissolved $CaCO_3$ and 4.9 ppm for 280 ppm respectively. It can therefore be neglected. Exception: mixing of normal with thermal water.

2.2.7 Pressure Dependence of Limestone Solutions

In the absence of gas bubbles there is no change in solubility with the pressures which occur in karst as the water for these pressures may be termed incompressible. Whether the rate of single reactions depends on pressure remains to be investigated, for example the formation of calcite nuclei in oversaturated solutions.

Liszkowski (1975) mentions unstationary conditions of currents in the phreatic zone which provide pressure shocks. In the case of aggressive water these can considerably accelerate corrosion and thereby the achievement of equilibrium. From this he deducts that there is "accelerated corrosion due to pressure shocks" effective in the case of water in equilibrium also. This he supports with the fact "that the overpressures in the water occuring in the zones of the pressure shocks also lead to an increase in the partial pressure of the CO_2 dissolved in the water, automatically increasing the aggressiveness of the karst water," (p. 196). Now since the sum of the physically dissolved CO_2 molecules per volume unit remains unchanged (incompressibility of H_2O), there is no increase in the partial pressure and therefore also no corrosion in water in equilibrium. On the other hand it is to be expected that such pressure shocks act erosively. The acceleration of corrosion in aggressive water can be accepted as possible, while the hypothesis of an increase in aggressiveness must be rejected as highly improbable.

It is a different matter in the case of enclosed gas bubbles. The following is true:

$$V \cdot p = \text{constant, hence } V_1 : V_2 = p_2 : p_1.$$

At a depth of 10 m in the water there are 2 atm and the volume of the gas bubble decreases to one-half. At the same time the partial pressure p_{CO_2} increases to double. In the phreatic zone several meters of water pressure can occur. Therefore there is additional corrosion in the zone of the air bubbles. It can be calculated according to Table 2.4, Figs. 2.12 and 2.24. In rapidly flowing underground water air is swept along by the water at the transition to the phreatic zone and as the water pressure increases there results a corrosive attack on the cave ceiling and thus the formation of the so-called ceiling half-tube (see Figs. 12.4, 15.5; Plate 8.2; Bögli, 1956a).

2.2.8 Comparison and Evaluation of the Types of Corrosion

The equation of the dissolution of limestone is at the basis of every type of corrosion of carbonate rocks; hence it is clear that the presence of aggressive CO_2 is essential for

the dissolution of limestone. The types of corrosion can be set up according to the kind of this CO_2:
- a) "normal corrosion", called corrosion in short
- b) corrosion by CO_2 as an oxidation product of organic matter in the underground
- c) corrosion by mixing waters
- d) cooling corrosion and thermal mixing corrosion
- e) pressure corrosion
- f) corrosion by volcanic CO_2 or by CO_2 from oil or gas fields

These cases have already been partly discussed comparatively (Chap. 2.2.6, 2.2.7): f) need not be further discussed as its effects can easily be derived from normal corrosion.

Although corrosion by the mixing of water had long received the attention of water-supply and sewage-disposal engineers because of its destruction of water constructions, this mixing effect was unknown to geographers, geomorphologists, and geologists so it was never applied to the dissolution of limestone in nature. Therefore corrosion by means of the intake of CO_2 from the air or else by aggressive water was regarded as the given situation, as normal. Oxidation of organic matter in the underground and the CO_2 resulting from it was frequently applied as a deus ex machina, to make comprehensible those processes of dissolution which could not be explained by "normal corrosion", e.g., in the phreatic zone.

2.2.8.1 Possibilities and Limitations of Normal Corrosion

Normal corrosion depends on a supply of CO_2 from the open atmosphere or from air in the soil and on aggressive water, which amounts to the same thing in the end. Atmospheric CO_2 can only be absorbed by water when the water's CO_2 content has not yet reached equilibrium. On the other hand aggressive water has a surplus of CO_2.

In the bare karst of high mountains where comparatively low temperatures slow down the rates of reaction, water usually seeps away long before it reaches the CO_2 equilibrium with the atmosphere, very especially during heavy rains. During precipitation between 2° and 4°C values around 15-30 ppm were measured, i.e., 20%-40% of the value at equilibrium (Bögli, 1956b, Table 1, 1960a).

At a p_{CO_2} of 0.035% between 0° and 5°C only 1.0 ppm of aggressive CO_2 is soluble in lime-free water; with this only 2.3 ppm of $CaCO_3$ can be dissolved. If the water already contains $CaCO_3$ in solution, however, $CO_{2eq} + CO_{2aggr}$ together make up 1.0 ppm and the amount of $CaCO_3$ which can be dissolved is still smaller. Because of the conditions of equilibrium aggressive water can, by means of its CO_2 content, never reach equilibrium with the atmosphere, which gave rise to its aggressiveness. Therefore the atmosphere must stand at the water's disposal as a source of CO_2 for corrosion until equilibrium is reached.

In closed systems the CO_{2aggr} becomes consumed and an internal equilibrium is created in which the CO_{2eq} corresponds to a fictitious partial pressure in equilibrium with the water. If the water then enters an air-filled, closed cavity, it absorbs CO_2 until equilibrium is reached, thereby dissolving $CaCO_3$ (Fig. 2.17). When the size of the air-filled space is limited its CO_2 content sinks, under certain circumstances even to below the value of the exterior atmosphere. On the other hand, it increases if the water has too high a $CaCO_3$ content and therefore a high content of CO_{2eq}. When the air circulates, however, there is a very rapid adjustment to the exterior atmosphere. The p_{CO_2} in the cave is then usually between 0.03% and 0.05%, which corresponds at 5°C to an equilibrium with approx. 80 ppm of dissolved limestone (Bögli, 1976b).

When water enters narrow interstices, e.g., in the initial stage of underground karst, contact with the atmosphere is lost and a closed chemical system forms which is in equilibrium. In such interstices normal corrosion does not take place. Only changes in temperature can disturb the equilibrium. Roques (1964) showed experimentally that a slight oversaturation of $CaCO_3$ can long be maintained in this situation. When water rises out of greater depths it cools off and can thereby become aggressive. Even if such water adjusts more quickly to the new equilibrium a certain amount of time passes before all the CO_{2aggr} comes into contact with the limestone by convection, dissolving $CaCO_3$. Water issuing from springs is therefore usually not in complete inner equilibrium and is just as little in equilibrium with the exterior atmosphere. In Aire Head Spring (Yorkshire) the water is more aggressive in the cold seasons than in the warm (Sweeting, 1965), which supports this. A general interpretation of the pH/$CaCO_3$ graph does, however, still create difficulties.

In deep caves one can observe again and again that water emerging from joint planes corrodes when it comes into contact with the cave's atmosphere (Plate 1.2). This becomes detectable even without analysis, from the formation of cave grooves, a type of cave karren. In Hölloch in the Rabengang 800 m below the surface, water emerges which has created such cave grooves (Plate 1.1). The p_{CO_2} of the cave's atmosphere there varies between 0.04% and 0.1% (Bögli, 1976b). The necessarily low lime content of the water requires that it seep in at the surface before reaching equilibrium with the exterior air. Moreover this low content must remain intact for more than 1 km in closest contact with the limestone of the joint's walls. This is only possible in a closed system, i.e., in the absence of air.

Conditions in silvan karst (forest karst) and in areas with unbroken soil covering, as found in the Sinkhole Plains of Indiana and Kentucky, essentially diverge from the model of bare karst. The water seeping through the soils is rich in biogenic CO_2 and in dissolved $CaCO_3$, perhaps even in aggressive CO_2. If it then enters the open joints of the limestone, it has, to be sure, not yet quite reached equilibrium with the air of the soil but it is oversaturated with regard to the open atmosphere and to the air circulating in the cave. Any eventual aggressive CO_2 is consumed in the closed system by the dissolution of limestone. When the water reaches air-filled space it gives off CO_2, at the same time depositing calcareous sinter until it reaches equilibrium. The CO_2 content of the air in the enclosed spaces rises thereby. During a phase of air stagnation in Hölloch the CO_2 content rose from 0.04% to 0.12%; this is the partial pressure of the CO_2 in equilibrium with the cave's water with 110 ppm of $CaCO_3$.

In covered karst normal corrosion occurs in underground cavities only as an exception, even when there is a surplus of CO_2 in the atmosphere, e.g., during high water, which always shows a lower lime content, or in winter when there is a minimal amount of biogenic activity in the soils.

Aggressive water which seeps slowly through interstices and narrow joints loses its aggressiveness after a few decimeters or meters as the A/V ratio is very high. Here A is the area of contact water/rock. In large channels with rapidly flowing water remaining aggressiveness can even still be detected after a distance of a few hundred meters: A/V is low, v high.

In conclusion: "normal corrosion" is limited or hindered by

a) the absence of air, e.g., in the entire phreatic zone, under phreatic conditions in the vadose zone (high water, siphons) or in narrow primary interstices which are practically the only water-courses in the initial phase.

b) the achievement of the $CO_2/CaCO_3$ equilibrium or oversaturation with lime.

Consequence: "normal corrosion" is only active as long as the water has not yet reached the CO_2 equilibrium with the atmosphere, and only as long as air, the open atmosphere, air in the soil or the cave's atmosphere stands at its disposal as a source of CO_2.

2.2.8.2 CO_2 from the Oxidation of Organic Substances

From organic matter CO_2 is created by oxidation. When it is not a question of the entrance region this material must be carried in by water. Apart from the soluble portion this presupposes widened interstices with corresponding velocities of flow, i.e., previous widening by corrosion.

Is the widening of a cave by the oxidation of organic material even possible at all? In the endokarst of Central Europe where temperatures are relatively low the amount of CO_2 formed is quite small. In the main system of Hölloch there is an annual mean temperature of 5 1/2°C, in its upper system 4°C. Older and younger cave clays — Old Pleistocene and recent — differ in their organic content by approx. 1%. It is to be assumed that in warmer caves (tropics, subtropics) where temperatures are between 15° and 25°C oxidation progresses much more rapidly so that a high portion of autochthonous CO_2 can be reckoned with. Investigations of this situation have not yet been made.

In natural shafts without any air circulation the CO_2 content can reach dangerous concentrations as a result of decaying plant remains and cadavers. Those are exceptional cases and they do not happen deep inside the earth but rather in the vicinity of the surface or in the entrance region and when the temperature is temporarily increased.

The magnitude of the removal by oxidation of organic matter is calculated as follows, based on the conditions in Hölloch.

It is assumed that a cave clay has lost 1% of its dry weight through the oxidation of organic matter since the Günz-Mindel Interglacial. Three-quarters of this loss is carbon, C. One C dissolves one $CaCO_3$ by way of the intermediary product CO_2. Therefore when recalculated, 37.04 cm³ of $CaCO_3$ (P = 2.7 kg/dm³) correspond to 1 mol of C.

1 m³ of dry clay weighs 1.5 t, and the C content of 0.75% amounts to 11.25 kg, which dissolve 34.7 dm³ of limestone.

In general:

$$K = \frac{100 \cdot D_L}{12 \cdot D_K} V_L \cdot 0.75 \, x.$$

K : vol. of the limestone dissolved
100 : 1 mol of $CaCO_3$
12 : 1 mol of C
D_L : density of dry clay
D_K : density of the limestone
V_L : vol. of the dry clay
0.75 : C portion of the organic matter
x : content of organic matter in the dry clay

If one calculates the volume of limestone dissolved as the average amount removed d from a limestone surface A_K then:

$$d = \frac{K}{A_K}.$$

This will be applied to the cross-section of a passage which for the sake of simplicity is assumed to be 1 m square. The cave clay fills the cross-section of 1 m². A_K is then 4 m² per meter, the volume 1 m³. A depth of removal of 0.0087 m per 1% of oxidated organic material results from this, i.e., 8.7 mm. One must take into consideration that the sediments which come into question — they are cave clays — make up only a small part of the total cavity in the rock; thus one can conclude:

Any additional hollowing performed by the decaying of organic matter in underground cavities can be neglected.

2.2.8.3 Possibilities of Mixing Corrosion

Much in speleogenetics had to be laid aside as inexplicable before the introduction of mixing corrosion. Above all, this was true of the corrosive widening of primary interstices at the beginning of underground karstification, so that O. Lehmann (1932) believed the creation of caves to be possible only from larger than capillary open joints. Corrosion in the phreatic zone was interpreted as the result of the water's own aggressiveness or of the effect of biogenic CO_2, or the passages below the actual karst water surface were explained as submerged parts of the vadose zone.

Mixing corrosion occurs everywhere that two different kinds of water mix. Characteristic of the karstified area is its water circulation in a network of more or less independent water-courses. Thus mixing is possible everywhere. Mixing corrosion therefore occurs throughout the entirety of karst, on the surface just as well as in the deepest parts of the phreatic zone — that is the great difference from normal corrosion. Jimenez (1976) mentions oil-drilling in northern Cuba on the Hicacos Peninsula which struck a cave at a depth of 2952 m. A water influx of the magnitude of 1 cm² cross-sectional surface suffices to form such a cavity. Such forms were mentioned by Bögli (1964a, b) and described by Fink (1968) from the many quarries in the vicinity of Vienna.

The fact that even water with the highest concentration of lime becomes corrosive when it mixes, breaks down another important barrier of normal corrosion. It is a "paradox of mixing corrosion" that the more the concentration of the lime-rich components increases, the more additional limestone is dissolved. This explains the previously puzzling fact that large caves occur in green karst where all water shows a high $CaCO_3$ concentration. Among these is the longest cave in the world, Flint Mammoth Cave in Kentucky (Well and Desmarais, 1973) with 320 km (1977), but also Greenbrier Caverns in West Virginia with 51.4 km (1977), Domica Baradla Cave in ČSSR/Hungary and many others. Many of the large caves are also to be found where covered and bare karst surfaces come together, i.e., where maximum differences in concentration, up to 200 ppm and more, are possible. To this group belongs the second longest cave in the world, Hölloch (central Switzerland), with its 140 km (Jan. 1980), Eisriesenwelt with 42 km, Carlsbad Caverns in New Mexico (33.2 km) and Postojnska Jama (Adelsberger Grotten), to name only a few.

The question also arises as to the origin and the possibilities of the occurrence of different lime contents. The following examples are given in answer:

a) The origin of water from various points of seepage is the normal case. Investigations concerning the p_{CO_2} in the ground show considerable variation even at intervals of less than 1 m, similarly in the vertical profile. This leads to different lime contents in the seeping water, especially where the water originates partly from bare karst, partly from covered (Bögli, 1969a).

b) In the course of 24 h the biological activity in the ground changes and with it the p_{CO_2} as well as the $CaCO_3$ content of the water seeping away (Gerstenhauer, 1972). These variations lead in time to temporal differences in concentration in one and the same channel. The water in an open joint flows fastest in the thread of flow and the more slowly the closer it is to the wall. Therefore slight mixing corrosion takes place (Franke, 1965a).

c) Also water which has an uniform origin can lead to differences in concentration. If water out of a moraine with 300 ppm of $CaCO_3$ seeps into rock moving downward through a narrow interstice in a closed system, the concentration does not change. On the other hand if a part of the same water flows in an open channel, it strives to reach equilibrium with the cave's atmosphere, giving off some of its CO_2. If this water thereby reaches a concentration of, for example, 100 ppm, the mixture with the first fraction 1:1 dissolves an additional 23 ppm of $CaCO_3$ (10°C).

Corrosion by mixing waters is, apart from exceptional cases, considerably less effective per unit of volume than normal corrosion. However, it reacts under conditions under which the other forms of corrosion no longer have any possibility of effectiveness, especially in primary interstices and in the phreatic zone.

2.2.9 Further Possibilities of Corrosion in the Phreatic Zone

Besides "corrosion accelerated by pressure shocks" Liszkowski (1975) mentions the occurrence of CO_2 created mechanochemically. It has been experimentally proved that when carbonates are pulverized CO_2 is created (Peters, 1962).

$$CaCO_3 + \text{pulverizing energy} \rightarrow CaO + CO_2 + \text{heat}$$

From 5 g of pure calcite at 25°C Peters obtained approx. 2 cm³ of CO_2, from 5 g of ankerite ($FeCO_3$) even 185 cm³ (p. 86). Liszkowski assumes that the movements of the earth's crust "whether they are epirogenic movements due to the effects of the earth's tides, or else isostatic rising and falling, orogenic tangential movements... or exogenous mass movements (p. 194) enough mechanochamical CO_2 is supplied" that corrosion can proceed partly according to the equation of dissolution even in the deepest parts of the phreatic zone" (p. 197).

Alpine caves had their origin in the Quaternary and Upper Pliocene, i.e., at a time when mountain formation had more or less ceased and the production of mechanochemical CO_2, if there was any, must have been at a minimum for the first time. And what happened to the CaO? It automatically became $CaCO_3$ again. And the swelling of the CaO when in contact with water? There is nothing known about the extent of $CaCO_3$ thus transformed nor about the production of mechanochemical CO_2 in nature. If such a production really takes place then it would be reasonable to assume a reaction with the CaO created. Such a process might be expected in lime mylonites in any case — but it does not result in the creation of caves.

2.3 Karst Denudation

The total corrosive removal in karst comprises the amount of dissolved rock which is carried away out of the karst area. It is the loss of rock in m^3/km^2 per year or the removal in mm/1000 y.

The total amount K_G/y is calculated out of the water discharge (surface A in km^2 times outflow H in m) and the $CaCO_3$ content of the water in ppm or else g/m^3:

$$K_G = A \cdot H \cdot G \; [km^2, m/y, g/m^3] = A \cdot H \cdot G \; [10^6 \; g/y] \text{ i.e. } [t/y].$$

Recalculated to the volume:

$$K_V = A \cdot G \cdot H \frac{1}{\rho} \; [m^3/y] \text{ and per } km^2 \; K_V' = G \cdot H \frac{1}{\rho} \; [m^3/y \cdot km^2].$$

To simplify matters Corbel (1959a) chose the value of 2.5 kg/dm^3 for the density of calcite which is 7 1/2% lower than its maximum of 2.7. This is justified because it lies within the margin of variation of ρ of the limestone:

$$K_V' = \frac{1}{2.5} H \cdot G \; [m^3/y \cdot km^2] = 0.4 \; H \cdot G \; [m^3/y \cdot km^2].$$

The absolute value corresponds to the removal of rock in mm per millenium.

$$K \; [mm/1000 \; y] = 0.4 \; H \cdot G.$$

Corbel suggests incorporating the portion of the carbonate rock occurring in the region and its alluvials in the calculation. However, that is not easy for on one hand impermeable rocks can also contain limestone, e.g., marly schists, or the limestone may be buried under several meters of residual clays and red soils; this is usually not in evidence in geological maps. Limestones and dolomites can also crop out under permeable rocks free of limestone like the Carboniferous limestone under Big Clifty Sandstone in Mammoth Cave National Park. Water can also flow many hundreds of kilometers through limestone which reaches the earth's surface in only a few places. The vauclusian springs of central Florida, e.g., Silversprings with a discharge of 30,000 l/s are fed from southern Georgia according to Jordan (1950).

Besides the above formula Groom and Williams' (1965) is used in England; it is identical to the first if K_G is expressed in 10^6 g/y:

$$K\ [mm/1000\ y] = \frac{K_G}{\rho\ A\ 10^6}\ .$$

Since the amounts of limestone carried away out of karstified regions are proportional to the discharge and thereby dependent on precipitation and evapotranspiration, a strong dependence on climate results. According to Wüst (1922) the amount of evaporation in the tundra is 120 mm/y, in the cool-temperate zone 300 mm/y, in the subtropics 450 mm/y and in tropical rain forest even 1200 mm/y. There are deviations from this, among others those in karst, because underground drainage noticeably reduces evaporation. In bare karst there is the additional lack of transpiration.

Mature karst landscapes are drained 100% in the underground. In bare karst the water disappears in the underground long before it reaches equilibrium. In addition there is corrosion by mixing water in the underground. Therefore limestone is dissolved on the surface as well as in the underground. The proportions of the two areas of removal cannot be determined from the lime content of the water. The rate of reaction increases to double with a rise in temperature of 10°C. For this reason the dissolution of limestone is accelerated in warm zones, and similarly the adjustment to the lower p_{CO_2} of the open air, also. Oertli (1953) measured for the Pivka (Slovenia) at 16°C as far as Postojna, a distance of 12 km, a drop of 25% in the lime content and in the Postojnska Jama a drop of 9% after 6.5 km (see also Gams, 1962, and Table 2.7). Thus measurements in the tropics must be made within the karst region itself or else at the springs.

Karst denudation is that amount of limestone that is actually dissolved and carried away. Within the area investigated lime can be redeposited as calcareous sinter, calcareous tufa or calcareous mud. The fact of the previous removal of limestone is not thereby nullified. If the lime content of the water is measured on the margin of the karst area it is not the amount of limestone removed which is measured but the balance of limestone, i.e., the difference between the removal and the deposit of limestone. In cold regions the limestone balance is almost identical with its removal because of the low rates of reaction, in the tropics on the other hand it is considerably smaller. This gives the impression that there is only slight removal of limestone in the tropics and must lead to false conclusions. Corbel (1958) assumed the removal of limestone in the subarctic zone to be ten times greater than in the tropics. Sweeting (1964) showed that lime contents also in regions with scarcely diverging climatic conditions are very different. It is a dependence on ground vegetation in the first place, in the second on lithology (chalk or crystalline limestone) that is responsible for this. Pitty (1966) established for his part that the dominant factor in the dissolution of limestone is biogenic CO_2 (see Table 2.12 and Bögli, 1960a).

Table 2.12. The lime contents of karst water in Great Britain (partly according to Sweeting, 1964)

	$CaCO_3$ (ppm)
Durness, N. Scotland, cryst. limestone, partially covered	60- 80
South Wales, Carboniferous limestone, partially covered	80-100
N. Pennines, Carboniferous limestone, almost completely covered	140-180
Mendips, Carboniferous limestone, completely covered with vegetation	220-240
S. England, Cretaceous chalk, complete cover of vegetation	280-300

Similarly Bögli (1971b) proved the dependence of denudation on ground vegetation in the alpine karst of the Muotatal (central Switzerland). In high places over 2000 m above sea level 15% more precipitation falls than in the zone around 1300 m above sea level, which receives 2100 mm. In spite of the greater run-off and a temperature lower by 3°C, karst denudation in the high mountains amounts to only 78% of that in the lower-lying silvan karst. This shows the decisive significance of ground vegetation and the soil covering of rendzina related to it. Moreover it was established that in covered karst 89% of removal results on the earth's surface, in bare karst only 20% (see Table 2.13).

Table 2.13. Karst denudation in the silvan karst (800-1600 m above sea level) and in the bare karst (over 1800 m above sea level) of the Muotatal according to Bögli (1971b)

	Silvan Karst S	Bare Karst B	(S) : (B)
Total removal in 1000 years	91 mm	71 mm	1 : 0.78
thereof:			
A: above ground	81 mm	14 mm[a]	1 : 0.17
U: underground	10 mm	57 mm	1 : 5.7
A : U	8 : 1	1 : 4	

[a] This value was determined morphometrically, with the aid of pedestals of karren tables among other things (see Chap. 3.1.2).

When precipitation is more or less comparable the value of the lime balance decreases from the humid tropics with approx. 50 mm per millenium towards the subtropics to below 30 mm; it strives toward the maximum of approx. 100 mm in the temperate zone of deciduous and coniferous forests, in order to sink again in the subarctic. In Table 2.14 the use of extreme climatic examples was avoided.

Table 2.14. Karst denudation in various climatic zones (Bögli, 1971b)

	Precipitation mm	Removal of limestone mm/1000y	Authors
Warm humid climate			
Jamaica, Rio Bueno	2000	51	Birot et al. (1958)
Jamaica, Martha Brae	1800	54	Birot et al. (1958)
Puerto Rico, Rio Camuy	1650	40	Birot et al. (1958)
Cuba, Rio San Vicente	1600	46[a]	Lehmann, H., et al. (1956)
Florida, Kissimmee River	1200	27	Douglas[c] (1962)
Temperate climate			
Kentucky, continental	1100	64	Douglas[c] (1962)
Lee, Essex, GB	700	63	Douglas[c] (1962)
Malham Tarn, N'Pennines, GB	1100	58[a]	Sweeting (1964)
Areuse, Swiss Jura	1700	89	Burger (1959)
Muotatal, silvan karst, Switzerland	2200	91	Bögli (1971a)
Birsigtal, Swiss Jura	1100	50	Barsch (1969)
Subarctic to arctic climate			
Muotatal, bare karst, Switzerland	2400	71	Bögli (1971a)
Spitsbergen	280	17[b]	Nagel

[a] Calculated by the author from literature
[b] The same on the basis of personal information
[c] Unpublished diss. by M. Sweeting

Not only corrosion but also erosion plays a part in removal. It is dependent on the climate, the orography, and the rock's resistance to weathering. Erosion is especially effective in regions where frost is frequent, especially in the zone of frost weathering (see Table 2.15).

Table 2.15. Erosion in limestone according to Corbel (1959b)

	mm/1000 y
Gold Creek, SE Alaska, mountainous, subarctic	100
Svartisen, Norway, mountainous, oceanic-subarctic	100
Rio Usumacinto, Yucatan, humid marginal tropics	15
Sligo, Ireland, hilly, temperate, oceanic	4
Kissimmee River, Florida, USA, plain, humid subtropics	0.5

3 A General View of Exokarst

Superficial karst phenomena, exokarst, are created by the dissolving effects of precipitations. They are often related to karst hydrology. For this reason the reader will be given a review of such surface forms. The manifestations of exokarst are by definition corrosion forms on soluble rocks at the earth's surface. Also belonging to this category are forms which are only indirectly the result of corrosion, e.g., collapsed dolines, dry valleys, stepped pavement karst, and poljes. The forms are classed in the following categories:

3.1 Karren, small solution features measuring a few millimeters to a few meters, and their higher classification karren fields.
3.2 Small, closed karst depressions measuring a few meters to 1 km: dolines (sinkholes), uvalas (compound dolines), cenotes, cockpits.
3.3 Corrosion plains.
3.4 Fluviokarst: karst valleys, dry valleys.
3.5 Glaciokarst: stepped pavement karst.
3.6 Large, closed karst basins, measuring 500 m to many km: poljes.

3.1 Karren, the Small Solution Feature

Karren is derived according to Eckert (1895, 1902) from the old High German word char = kar which means a bowl. Karren are indeed usually named according to their hollow form: solution groove, undercut karren, wall karren. But there are elevated forms, too, such as karren spines, pinnacles, and clints. It is more probable that the word is derived from the Slavic Kras or Karst, which has a paleoindoeuropean root meaning a stony, bare rock surface.

Bare limestone surfaces such as have been created by climatological processes and anthropological influences show a great variety of forms (Cvijić, 1924), of which only a few can be mentioned here. These are joined by further karren forms under a partial or complete covering of soil.

3.1.1 Introduction

The forces which create these forms are determined, all other conditions being the same, by the process of dissolution and the manner of drainage. Bögli (1960a) classified the processes of limestone dissolution into three morphogenic types of corrosion effect.

Introduction

The *first type* results in a physical dissolution of $CaCO_3$, upon which the association of CO_3^{2-} with H^+ follows directly to form HCO_3^-. H^+ originates from the dissociation of H_2CO_3. Thereby the product of the concentrations of Ca^{2+} and CO_3^{2-} falls below the value of the product of solubility. For this reason limestone is again dissolved until this value is reached anew. H^+ has thereby disappeared from the equilibrium of dissociation of carbonic acid. It must regenerate from the physically dissolved CO_2 until the concentration of H^+ reaches the value at which K_1 is re-established. With that the *second morphogenic type of corrosion effect* has already been introduced. The transformation of CO_2 into H_2CO_3 now controls the system's velocity of reaction. Since the concentration of CO_2 in the solution diminishes, its equilibrium with atmospheric CO_2 is disturbed. The latter begins to diffuse into the solution. As soon as the rate of diffusion of CO_2 from the atmosphere into the water determines the velocity of the dissolution processes the *third morphogenic type of corrosion effect* begins.

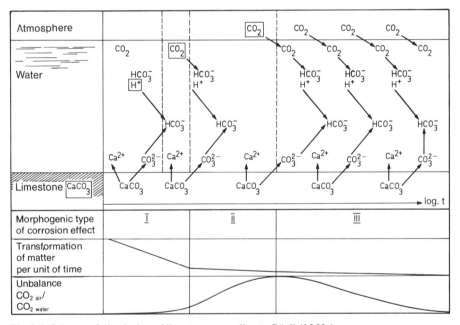

Fig. 3.1. Scheme of dissolution of limestone according to Bögli (1960a)

To summarize, the following can be said about the three morphogenic types of corrosion effect:

First type of corrosion effect: relatively high velocity of reaction, increasing with the temperature, limited by the rate of dissolution of $CaCO_3$. This type of action is to be found on bare limestone surfaces which are exposed to precipitation. Because of the dependence on temperature the correlated forms, solution flutes, are especially long in the tropics (H. Lehmann, 1953).

Second type of corrosion effect: the velocity of reaction is lower but also increases with the temperature. On the other hand the quantity of CO_2 dissolved at $0°C$ is three times larger than at $35°C$. For this reason this type is favored in cold climates. Correlated forms are the large solution flutes of the second order, three to six times wider than the those of the first order. Trittkarren (heelprint karren) appear on less sloped surfaces. The hydration of CO_2 and the dissolution of the calcite are the limiting factors.

The third type of corrosion effect: the velocity of reaction is low and is controlled by the rate of diffusion of CO_2 from the atmosphere into the water. For this reason the exchange of materials is 10^2-10^3 times lower than is the case with the other two types. The third type of action is directly aimed at attaining the equilibrium $H_2O - CaCO_3 - CO_2$. All karren forms not yet named fall into this group.

Karren forms are strongly dependent on the manner of flow, whether the water runs over bare limestone surfaces or seeps through very permeable or less permeable soil. Wherever water flows freely over limestone, it stays on the surface only a short time, disappearing into the underground long before it reaches the balance. In the limestone Alps of central Switzerland 27 ppm of $CaCO_3$ were measured at an equilibrium of more than 80 ppm in a 13-m-long solution groove (Bögli, 1960a). During a heavy rainfall the values sink to below 20 ppm. The influence of area-wide sprinkling is reflected in the low limestone content of this water.

All three types of corrosion effects are temporarily prolonged by rain, since precipitation delays the succession of one by the other. There are morphological consequences for the forms of the first two types of action. Rillenkarren grow somewhat longer and the surface part of the heel-print karren increases greatly. The lime content of the film of water varies frequently during heavy rainfall between the second and third type of action. It must be pointed out, however, that the occurrence of rillenkarren and heel-print karren still depends on lithology, on whether the amount of bare limestone surface is sufficient, and on the slope of the surface. Therefore the lack of these forms is not proof of the lack of the first two types of action. No detailed investigations have yet been made of this problem.

Water behaves quite differently in the soil. For one thing the air in the soil has a characteristic content of biogenic CO_2 which varies greatly but always exceeds the CO_2 content of the atmosphere. In rendzinas, a humus-carbonate soil of the forest and meadow level in the limestone Alps of central Switzerland CO_2 contents of 0.04%-9% have been measured. Moreover water drainage is hindered by the soil. Thus the contact between the CO_2 in the soil's air, the water and the limestone lasts much longer than is the case on bare karst where the precipitating water disappears into the grikes after only a few minutes and before it reaches the $CaCO_3/CO_2$ equilibrium. Water seeping away through the soil is, on the contrary, so rich in lime that it is oversaturated with respect to the atmosphere.

Water from very permeable soils is lower in lime than water from marly soils. In Bisistal (Muota Valley, central Switzerland) spring water from bare karst contains on average 70 ppm of $CaCO_3$, that from karst with humus patches contains 85 ppm, that from covered karren fields with clayey soils inserted in rendzinas contains 130 ppm, and that from below clayey soils even reaches 190 ppm at temperatures around $7°C$.

Table 3.1. Lime content of various waters from karst regions (Bögli, 1960a)

Place	Author	Kind of surface, remarks	Lime content (ppm)
Bisistal (Muotatal)	Bögli	Bare karrenfields low in humus	70
Bisistal (Muotatal)	Bögli	Karren fields covered with vegetation	85
Muotatal	Bögli	Covered karren fields	130
Muotatal	Bögli	Forest and meadow in karst springs	105
Hölloch (Muotatal)	Bögli	High water	56- 80
		Low water, autumn	120
		Joint water, winter	180
		id	200
Lauiloch (Muotatal)	Bögli	Surface with humus soil	105-140
Littau, (Ct. Lucerne)	Adam[a]	Water out of calcareous gravel	220-250
Hitzkirch (Ct. Lucerne)	Bögli	Springs from calcareous moraine	300-390
Pivka (Yugoslavia)	Oertli	Slovenian karst, cave river	132-187
Unica (Yugoslavia)	Oertli	Slovenian karst, cave river	
		Beginning of April	137-210
		Beginning of June	157-265
Cuba	H. Lehmann	Joint water	97.5
		Joint course 4 m, in cave	150
		Joint course 10 m, in cave	152
		Rio San Vicente, cave river	140-150

[a] personal communication

Bögli (1951, 1960) was the first to approach the karren problem by means of measurements and water analyses. Then followed Bauer (1964) and Miotke (1968), and in England Sweeting (1966) and Jones (1966).

3.1.2 The Genetic System of Karren Forms

The multiplicity of possible karren forms makes a morphological system endless, while a genetic one allows a meaningful collection. The manner in which they are moistened and the conditions of drainage of the corroding water are fundamental to it. Therefore a differentiation is made between karren forms which are created by free, unhindered water flowing off over bare limestone surfaces, those which are caused by a patchy covering of soil, and those which develop beneath a closed covering of soil. These are the basic forms. They change in appearance when conditions change, e.g., when overgrown: subsequent forms. In addition to single forms, form complexes and groups of form complexes can be differentiated.

3.1.2.1 Single Forms

A. Free Karren
Bare karst; the water flows unhindered over the limestone surface.

A.1 Basic Forms in the Case of Area-Wide Moistening by Precipitation
When water runs off over the limestone surface, the concentration of lime increases and

corrosion decreases. Therefore the karren should flatten out toward the lower end. This is, however, contrary to all observations. Area-wide precipitation adds new, unspent water which counteracts the slowing-down of the chemical reaction. In fact, due to the increasing quantity of water moving toward the lower areas, corrosion is even intensified; the karren become deeper. The gradual singling-out of individual streams leads to the generation of grooves.

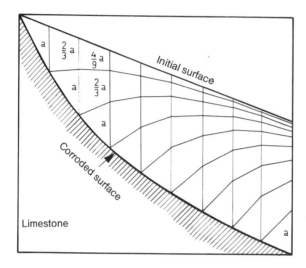

Fig. 3.2. The amount of superficial deepening by an exponential decrease in the water's capacity for corrosion (see text)

If the moistened surface is separated into horizontal strips, the amount of superficial deepening by the water of each strip is added in the direction of the flow. The final shape consists of a steeper slope in the upper part and a parallel deepening of the lower part (Fig. 3.2.).

Solution flutes, (Germ.: *Kannelierungen, Rillenkarren,* Austria: *Firstkarren*), (Table 2.1), occur only in places where fresh unspent precipitation is active and they end where the water attains too high a content of lime or where spent water is added. Solution flutes are the module of the first morphogenic type of corrosion effect (Bögli, 1960a). However, Gerstenauer and Pfeffer (1966) contradict this hypothesis. Yet Pfeffer (1976) observes: "Hence on a basis of findings in the field, and contrary to earlier supposition (Pfeffer, 1967), a specific type of action must be accepted for the genesis of solution flutes," (p. 11). In a later publication Heinemann et al. (1977), p. 59, state more precisely: "that the explanation of this type of action as offered by Bögli cannot, on the other hand, be supported. The explanation of the phenomenon by the dynamics of flow and the corrosion connected with them (Bauer, 1963; Miotke, 1968, 1972)... is not sufficient to explain the phenomenon of solution flutes, either," and p. 78 "An explanation of the findings and therewith a new theory concerning the origin of solution flutes is not possible at the present time." Accordingly the question is still open today.

The length increases with the slope, the temperature, and the amount of rainfall, reaching 1 m and more in the tropics (Cuba: H. Lehmann, 1953), up to 50 cm, and even as an exception 100 cm in the Alps. In the Mediterranean region a length of 100 cm is common if the rock allows it. The width varies between 1 and 3 cm. They are together in rows with no space in between, with sharp intermediary ridges, which hardly reach 1 cm in height. They increase at all freely exposed peaks and ridges where fresh rainwater alone is at work. In Fig. 3.3 the grooves of the rillenkarren (R) gradually flatten out to a smooth surface (A).

In the USA this karren type is rare. It is most readily found in the Rockies, in the Canadian Rockies (Sweeting 1972), and on the edges of Kamenica, e.g., on the Kaibab Plateau near Kaibab Lodge (AZ) and on the Mesa Verde (CO).

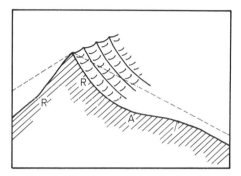

Fig. 3.3. Solution flutes (R) with adjacent flat surface (A) (Germ.: Ausgleichsfläche)

Trittkarren (heel-print karren) (Table 2.3) are almost connected with subhorizontal, adjacent, flat plains. On the surface they create a form which resembles the imprint of a heel (Bögli, 1960a). They move back into the slope, cut like steps by retrogressive corrosion. The semi-circular form is preserved by the "horseshoe falls effect" which concentrates the main amount of water on the innermost part of the heel-print. At the upper rim the water gains speed. The thickness of the film of water is indirectly proportional to the speed of flow. On one hand a higher speed causes a greater effectiveness of fresh precipitation added to the flow on the ground, on the other hand it causes the diffusion of atmospheric CO_2 and heavier corrosion.

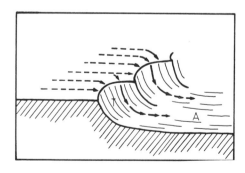

Fig. 3.4. Trittkarren (T) with even surface (A)

Many trittkarren have originated from the rim of the grike lying below and have moved upward into the surface through retrogressive corrosion. Bögli (1951) and Haserodt (1965) showed that at the foot of rather steep surfaces where snow collects deep, almost funnel-shaped trittkarren appear, which are of subnival origin. Trittkarren are common in the Alps. Sweeting (1972) mentions them as a peculiarity of the coast of Burren in County Clare, Ireland.

On almost vertical walls solution flutes form which are of the width of 3 to 6 rillenkarren and can become as long as 3 m; these are *solution flutes of the second order* (Table 2.2). The latter two karren forms named belong to the second morphogenic type of corrosion effect; the following belong to the third.

Rinnenkarren (solution grooves) form where runoff water is collected in streams. When the whole surface is moistened, the amount of water increases downwards with the result that the grooves are widened and deepened at the bottom. This distinguishes them from other similar forms. When the slope is slight they are coiled, but become straighter with increasing inclination. A cross-section is round to flat at the bottom; the transition to the surface of the rock is not usually sharp but rounded. The more water runs over the rim into the groove, the more rounded the rim becomes. Rinnenkarren are sometimes interpreted as being subcutane forms grown below the soil cover. This manner of origin is not to be excluded but it is probably relatively rare. Solution grooves can be found in all climates; in arid zones they are present as relics of the past when the climate was damper (pulvial time).

A.2 Basic forms of karren with water flowing off without area-wide moistening

As a rule in this group of forms the water flowing off arises from snow melting or seeps out from humus patches and rarely from bedding interstices or joints. The capacity to corrode decreases downward, the higher the temperature, the faster.

Meandering karren (Table 2.4) are small grooves cut directly into the surface, a few centimeters wide and deep; they remain the same size or decrease down the slope and usually have small meanders which show typical, undercut slopes and slip-off slopes. They frequently appear in the bottom of larger grooves, e.g., rinnenkarren.

Wall karren (Table 3.1) form on vertical walls. Occasionally area-wide sprinkling plays a role in their creation. Their cross-section forms a semi- to a three-quarter circle. If the wall is high enough the wall karren diminish downward.

If the water originates in humus covering, meander and wall karren have a different characteristic: they become *humus-water grooves.* When leaving the humus covering, the water contains an excess of CO_2 and, therefore aggressive, it dissolves limestone. For this reason humus-water grooves are deep at the beginning. The biogenic CO_2 which it carries is used up after 2-3 m, the grooves flatten out and continue as normal meanders or wall karren (Fig. 3.5). On rare occasions the water is so rich in lime at the beginning that, when the $CO_{2\ aggr.}$ is used up, it is oversaturated with lime and so deposits $CaCO_3$, e.g., Eggstock (Karrenalp, Ct. Schwyz).

B. Half-Exposed Karren

On otherwise bare limestone surfaces there are patches of soil which attack the rock by means of biogenic CO_2.

Single Forms

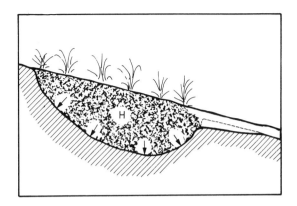

Fig. 3.5. Humus patch (*H*) and water seeping out of it. The water, rich in CO_2, creates corrosion troughs and humus-water grooves. (↓: direction of the corrosion by biogenic CO_2)

B.1 Basic Forms
Solution pans (Slovenic kamenica; Table 3.2) are basin-shaped, closed depressions in limestone. The initial form is a closed hollow created by a humus patch. At its best the solution pan has over-hanging side walls and a flat floor covered by algae and small pieces of broken rock; diameter 0.1-1 m, depth rarely more than 15 cm. After filling the hollow, the rainwater, which is low in lime content, absorbs atmospheric CO_2 through its surface. This results in corrosion which acts mainly sideward. The mud on the floor, moreover, protects the bottom of the basin from corrosion.

Solution notches form wherever humic soil borders on a very steep or vertical limestone surface. The rock becomes undercut by water rich in biogenic CO_2. In the cone karst of the humid tropics foot caves occur which are over-sized enlargements of solution notches.

B.2 Subsequent Forms
Undercut karren (Table 3.3) are rinnenkarren (solution grooves) which have been transformed by humus filling, their side walls have been hollowed under by biogenic CO_2.

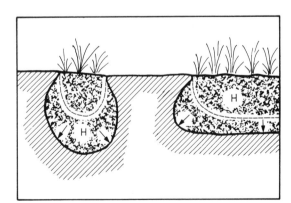

Fig. 3.6. Cross-section of undercut karren (*left*) and solution notch (*right*), *broken lines* indicate the initial state; *H* humic soil

C. Covered Karren
Soil covers the whole limestone surface; the draining water is oversaturated with CO_2 as compared with the atmosphere; corrosion is intensive (Gams, 1959; Bögli, 1960a).

C.1 Basic Forms

These are easily overlooked. In our climate wavy surfaces like corrugated tin are occasionally formed: *wave karren*. When denuded they are a disposition for the formation of rinnenkarren. Beneath compact soils roots etch into the limestone: *root karren* (Table 3.4). They are so small and flat that, once laid bare, they are quickly removed by surface corrosion.

To the basic forms belong also the *cavernous karren* of the recent and Tertiary red-earth regions and of the humid tropics. They consist of pitted limestone. However, this form has nothing to do with similar-looking, weathered dolomite — but there are numerous transitions to the latter.

Finally the *geological organ* must be mentioned which leads to forms which have nothing more to do with karstification. They are found as karst forms in the soft, Pliocene limestones of Apulia, in the soft chalk of northern France and southern England and quite frequently in gypsum.

C.2 Subsequent Forms

These are widespread, especially *round karren* (Table 4.1). In the soil the water does not flow freely; it corrodes away all edges and points. The small karren forms disappear, grooves and grikes are widened and deepened.

One to two centuries after being laid bare, the earlier rounded edge is only just recognizable. Therefore round karren and their remains give evidence of an earlier soil covering.

D. Grikes (Table 4.4)

Grikes are very widespread and are hardly ever missing. They form both in bare karst and under a soil covering. In bare karst in the Swiss Alps they become 20-30 cm wide, under the soil up to 50 cm. Their depth varies, but is difficult to determine because of the narrowness of the deeper parts. In bare karst they reach several meters in depth, in silvan karst (forest karst) up to 10 m and more. Bögli (1960a, 1964d) assumes that the disposition for many grikes existed earlier than the last Pleistocene glaciation (Prewürm, Prewisconsin); Pigott (1965) agrees while Haserodt (1965) sees them in their entirety as postglacial. However, these diverging views may be less the result of a difference in substance than of a lack in the definition which gives no indication of the width at which a corrosively opened joint is to be called a grike. A widening to 2 mm complies with the facts of the formation of a grike and such should also be called a grike. Such a widening requires a great deal more time than the widening from 2 to 200 mm.

E. Karren Tables, Tables of Corrosion (Table 5.2)

Area-wide moistening lowers a limestone surface by means of corrosion. Boulders protect the rock below so that in time a pedestal forms; its height on a horizontal plane indicates the extent of limestone denudation since the boulder was deposited there. Boulder and pedestal together form the karren table (table of corrosion according to Moser, 1956, limestone table according to Moser, 1967). In the Märenberge (2250 m above sea level, central Switzerland) there are erratic boulders from the Daun (10,000 years ago) lying on the Malm limestone with a pedestal of a maximum of 15 cm in height (Bögli, 1961b). Moser (1956, 1967) found similar values for the Dachstein (Austria). Corbel (1957) mentions the like in Ireland with pedestals of one-half-meter height.

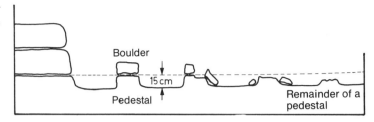

Fig. 3.7. Karren tables (acc. to Bögli, 1961a)

F. Surf Karren

Surf karren form along marine limestone and dolomite coasts where the surf sprays water onto abrasion surfaces which lie a little above normal sea level. It is an effect of corrosion by the mixing of sea- and rainwater. They do not exist under the surface of the sea — seawater is not limestone-corrosive. Beyond the splashwater zone the karren are much less sharp: rain karren (Mensching, 1965). Mensching (1965) describes them as found on the Cantabric coast, Trudgill (1976) describes the same on the coral limestone of the Aldabra-Atolls. They are often found on the coasts south of Bari and Tarent. Sweeting (1972) mentions them in many places along the British and Irish coasts.

3.1.2.2 Complex Karren Forms

Complex karren forms are created by the combination of several single forms, especially of grikes with other karren, e.g., pinnacles, clints, and pavements.

Pinnacles (Bögli, 1951, 1960a, Plate 4.2) are mature forms of karren. The side walls of grikes and rinnenkarren cut across one another forming sharp edges and peaks which can reach several meters in height. In limestones with a thin sheet structure these soon fall into fragments which form *debris karren*. Pinnacles need a long period of formation. The post-glacial period is obviously not sufficient. Therefore they are extremely rare in the Alps, only to be found in especially protected areas which moving ice does not reach, e.g., the peak of the Hengst (Schrattenfluh, Switzerland) and Misthaufen (Ct. Schwyz). However, they are frequently found in extraglacial regions, e.g., in the Mediterranean region and in many such places they dominate. They are common in the tropics and attain great sizes, e.g., in New Guinea or northern Australia.

Wherever glaciers have broken off karren pinnacles, the remains of bedding planes delimited by grikes, are left behind; these are *clints* (Bögli, 1951, 1960a). They can spread over several square meters and are classed as the smallest unit of glacial karst (Plate 4.3).

3.1.2.3 Groups of Complex Forms

Groups of complex forms are large forms of karst with karren as the physiognomically dominant form. They are types of karst.

Karren fields appear as bare karst and consist of the sum of exposed and half-exposed karren, occasionally also of covered karren which have become exposed. They attain the size of a few hectares to a few hundred square kilometers.

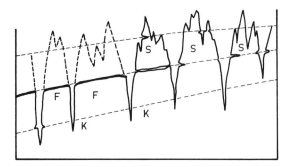

Fig. 3.8. Pinnacles (*S*) and clints (*F*)

3.1.3 Karren and Karst Hydrology

The importance of karren consists in the generation of a surface drainage pattern with extremely short branches and closely situated absorption points. The distance of surface flow is limited to a few decimeters up to 20 m, rarely more. Grikes and pits serve to absorb the water.

3.1.4 Pseudokarren (see Chap. 1.1)

Pseudokarren develop on insoluble, mostly silicate rocks by means of weathering processes. They show a rounded type of rinnenkarren, less frequently the form of atypical solution pans (Fränzle, 1971). Klaer (1956) describes well-developed rinnenkarren on Corsican granite. Morawitz (quoted by Maull, 1958) names such forms out of the tonalite of the Hochgall group (South Tyrol) and in the gneiss of the Hochalm peak (Austria). The best-developed forms were discovered in the tropics, for example, on Itatiaja (Maull, 1930) on Syenit, in the Seychelle on granite (Bauer, 1898) and in many other places.

3.2 Small, Closed Hollows in Karst

Closed hollows in karst are created by karst processes, whether directly by corrosion or indirectly by subsidence and collapse into underground cavities created by dissolution. The main forms are dolines and uvalas, cenotes, and cockpits. There are also numerous atypical karst basins of various sizes which cannot be classified in any one of these categories.

3.2.1 Dolines

Dolines are the most wide-spread karst form. They are to be found in every karst landscape, whether they appear as funnel-shaped, bowl-shaped, or as flat, dish- and trough-shaped dolines. In the Slavic languages dolina means a valley, but also small, closed karst hollows. The term was introduced into geomorphology by Austrian geologists. Dolines are defined as "simple, funnel-, bowl- or kettle-shaped, closed karst cavities with underground drainage and a diameter which is greater than their depth" (Fink, 1973). Trough-shaped should be added to this definition. To avoid confusion with dolina = valley, vrtača has been used exclusively for doline in Slovenic karst literature for several decades.

The diameter of the doline varies between a few meters and 1000 m, with a depth of up to 100 m (Cvijić, 1893) and the surface between 0.17 m^2 and 159,200 m^2 (Cramer, 1941). They can appear singly but they can also form doline fields (Sinkhole Plains, IN, KY, USA) and long-drawn-out rows on rock boundaries and on joints and faults. Concerning the density of dolines Cramer (1941) names values between 0.57 dolines per km^2 and 2450 d/km^2. Sweeting (1972) states an average of 100-200 d./km^2. Morawetz (1965) found 400 d/km^2 in Istria and Mallot (1945) found the same number in the Mitchell Plains (IN, USA). The following may pass as extremes: the 51 small solution dolines per hectare (5100 d/km^2) on the Glattalp in the limestone Alps of central Switzerland (Bögli, unpub. investigation, 1952-1962) and the 20-30 subsidence dolines per acre (5000-7500 d/km^2) in the Millstone Grit of South Wales (Thomas, 1954). Cramer (1941) names a maximum doline surface of 300,000 m^2/km^2, and Gams (1969) calculates that doline surface makes up 24% of the karst in Yugoslavia, even 64% in Montenegro.

A genetic classification differentiates between solution dolines, solution subsidence dolines, and collapse dolines.

Solution dolines are formed when the limestone is dissolved away from under a soil covering by a widening of the interstices (Terzaghi, 1913). Nicod (1967) found in the Provence that dolines in bare rock are not deepening any more, but that those under soil are today still continuing to develop. Rain is the most important factor but drifting snow effects additional corrosion until late in the spring, in the high mountains until summer. The deepening takes place because of, among other things, the widening of the interstices in the rock with resulting settling (see also Palmer et al., 1975). Blocks and soil formations sink, too. Fine material, later also sand and stones, are washed away through the widened interstices. If the drainage ways are widened enough in the overlying, loose material, inner erosion results and funnels form on the surface. Williams (1969) calls this type an *alluvial doline.* Further developments in the underground bring about a settling of the rock and the loose material slides after. Thus the alluvial doline is genetically an intermediary type between the pure solution doline and the subsidence doline.

The *subsidence doline* is created by a slow downward movement of a mass while the *collapse doline* is formed by one rapid, usually single occurrence, caused by a cavity which lies near the surface. The collapse of the cavity's roof can open an entrance to a cave, e.g., at the well-known cave of St. Kanzian (Škocjanske Jame, Slovenia) through the Small and the Large Doline (Mala and Velika Dolina), and at the Grotta di Castellana south of Bari (Italy). Deeply situated cavities gradually break through until finally the roof collapses (see Chap. 12.2.4). The famous collapse dolines of Imotski belong to this type. The Modro Jezero (Blue Lake) lies in a funnel doline which is 245 m deep with steep, in places even vertical, walls. Weathering has made a funnel out of the origi-

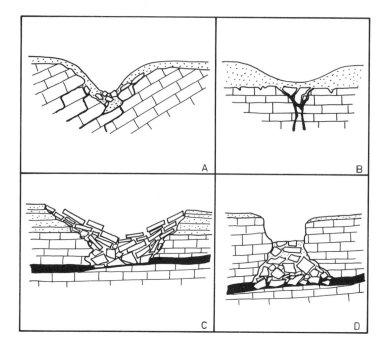

Fig. 3.9. Solution doline (*A*), alluvial doline (*B*), subsidence doline (*C*), collapse doline (*D*)

nal kettle form. The Crveno Jezero (Red Lake) lies in a 481-m-deep shaft which is only 200 m in diameter (Guidebook, 1965). The facts that the ratio of its width to its depth is 2:5 and that the lake is filled by underground karst water hardly allow this form to be counted among normal collapse dolines. The Red Lake is a cenote.

3.2.2 On the Morphology of Dolines

The morphological classification is still today essentially according to Cvijić (1893): funnel-shaped dolines, bowl-shaped dolines, trough-shaped dolines, kettle-shaped (shaft) dolines. Well-developed, funnel-shaped dolines (Plate 5.3) form in loose material or in slightly weathered rock, e.g., in thinly layered limestones. Bowl-shaped dolines show a flat floor, trough-shaped dolines have a concave one (O. Lehmann, 1931). Usually secondary dolines form in them; these are generally funnel-shaped with rocky outcrops or large stones visible.

In thickly layered limestone denudation is slight; it is also only slightly susceptible to frost weathering; corrosion works vertically into the depths, and only a little horizontally. The result is forms with vertical walls, and irregular in their cross-section, also rounded: karren dolines (Plate 5.1), kettle-shaped dolines (Kotliči, see Kunaver, 1965). They are not collapse dolines. Already early there was a difference made between *asymmetrical* and *symmetrical* dolines, a morphogenetic division. *Symmetrical* dolines are round to roundedly elliptical; *asymmetrical* dolines are lopsidedly extended with variously inclined slopes.

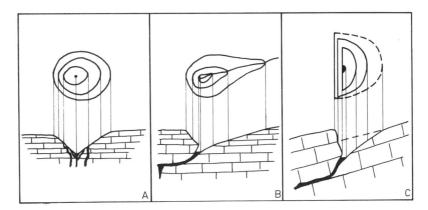

Fig. 3.10. Ground plans and cross-sections of a symmetrical solution doline (*A*), an asymmetrical doline with a runnel (*B*) and a structurally asymmetrical doline (*C*)

Asymmetry is caused by superficial water courses which end primarily in a swallow-hole and develop into a form that can be morphographically called a doline. Watercourses can, however, rip up an existing doline. Structurally asymmetrical dolines form on inclined limestone banks.

There can also be climatic causes of asymmetry, e.g., when rain or snow are regularly brought in from one certain direction of the compass; e.g., in Central Europe from the west. Chabot (1927) mentions finding in the French Jura asymmetrical dolines with steep inclines toward the north and east, which he assumes are caused by drifting snow. The south-exposed side of dolines with soils rich in humus is biologically more active because of the greater warmth; this causes a flattening process. Haserodt (1965) observed in the case of large dolines on the borders of frost-weathered zones that the steepest and frequently bare rock slopes are exposed to the east and northeast. Sweeting (1972, p. 51) mentions similar forms in the Permian limestone of Vaughan (NM, USA).

There is little to be found in literature concerning the inner structure of dolines. Aubert (1966) made a cross-section of an asymmetrical solution doline in the Vallée de Joux (Swiss Jura), which is very complicated in its structure; it contains loose material from the Prewürm Age up to today and shows an open canal, 80 cm in diameter beneath. The author investigated numerous funnel-shaped dolines on the Glattalp (central Switzerland); they were usually small dolines in marly limestones of the Upper Malm, 2-10 m in diameter and 0.5-2 m in depth. Here before the Würm Period underground water courses developed with dimensions of up to 200 cm x 20 cm. Digging in a broad trough-shaped doline in the silvan karst over Hölloch (central Switzerland), we found a small system of caves which led down into the depths; it could not be explored to the end because the passage became too narrow. The single branches of the system began in secondary dolines. Clay with limestone fragments formed the ceiling.

3.2.3 Cenotes

Cenotes are collapse dolines above a high karst water surface and reaching into it; they show a diameter/depth relationship under 1. Their walls are vertical, occasionally overhanging. Neighboring cenotes as a rule show the same high water level. This type is frequent in Yukatan but by no means limited to that peninsula. Jennings (1971) refers to cenotes in Turkey, and the author's opinion concerning Crveno Jezero was already dealt with at the end of Chapter 3.2.1.

Gerstenhauer (1968) explains the origin of the cenote by corrosion by mixing waters; water seeping in mixes with the karst water-body in the underground. Incasion (breakdown) can cause the bottom of the cenote to extend beyond the water level: the "dry cenote".

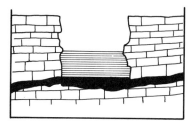

Fig. 3.11. Cenote

3.2.4 Karst Window and Karst Gulfs

Two other hollows offer a glimpse of underground karst water, on one hand the karst window, on the other the karst gulf. These forms were first described and named by Mallot (1932) when he found them on the Mitchell Plains (IN, USA). A *karst window* is a large funnel-shaped doline, in the depths of which a short stretch of a cave river is visible. This type is found in Spring Mill State Park (IN, USA). This karst window has a diameter of 130 m and a depth of 17 m. The cave river is open for an extent of 70 m, but an alluvial surface is missing (cf. Osinksi, 1935).

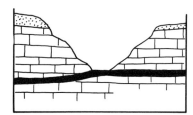

Fig. 3.12. Cross-section of a karst window

The *karst gulf* is created by the collapse of a broad cavity which is situated close to the surface. The fragments are transported away by corrosion, less by erosion. At the bottom of the gulf sandy-clayey sediments are deposited. Later extensions follow when corrosion cuts back under the overhang and lateral breakdown occurs. There is no relation to an uvala or a small polje. The type's locality is the Wesley Chapel Gulf (IN, USA; Mallot, 1932, 1945). It comprises 3.4 ha, of which 2.5 ha alone are covered with deposits (loam and sand). When the water level is high the underground Lost River overflows, covering the whole area and depositing further fine material.

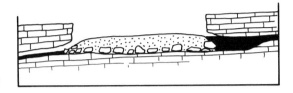

Fig. 3.13. Cross-section of a karst gulf

3.2.5 Cockpits

Cockpits are a hollow form in the cone karst of the humid tropics. This climate does not exclude the possibility of the occurrence of normal dolines, too, when lithological conditions are suitable. In Jamaica cockpits and karst cones develop in the pure crystalline White Limestone (Middle Eocene to Lower Eocene) but only normal dolines in the marly Yellow Limestone (Lower Middle Eocene), (Sweeting, 1958). The contours of the cockpit are not rounded as is the case with dolines, but star-shaped with indented sides, which indicates that cone formation is the decisive factor.

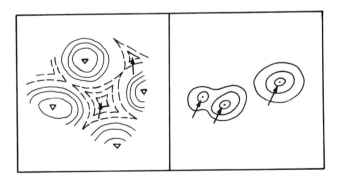

Fig. 3.14. Cockpit and doline (*triangle* peak of the karst cone)

3.2.6 Uvalas (Slovenic)

Uvalas, also called karst troughs by Cvijić (1901), are the most controversial of the karst hollows. Cvijić (1901, p. 77) defines them as "larger, broad-bottomed karst hollows with broken ground" which show "no even areas on the bottom". H. Lehmann (1970, p. 783) defines uvalas as "elongated, sometimes twisted like a valley, but generally bowl-shaped basins in karst", but wrongly neglects to mention the uneven ground which often shows dolines. The ISU (International Speleological Union, Fink, 1973) took this into consideration in its definition, but otherwise holds mainly to the definition of Lehmann. Jennings (1971) and Sweeting (1972) give too narrow a definition of the uvala as a hollow form with an uneven floor, which is the product of two or more dolines that have grown together (compound dolines); this is the best definition of uvalas in plateau regions (England, USA) but does not do justice to all those in folded areas, e.g., in the Dinaric region.

3.3 Corrosion Plains

Corrosion plains are widespread in Dinaric karst, occur, however, also on the edge of and inside cone-karst regions. They are found in most poljes, but they are also not infrequently outside them: on the lower course of the Cetina and of the Krka (Yugoslavia). They are usually horizontal; older ones are later also either tectonically bent, lifted, or cut up by later processes of corrasion. They spread out right across the tectonic structures in the limestone. Kayser (1934, 1955, 1973) differentiates between marginal karst plains and polje floors. Here the question will be left open as to whether other corrosion plains within the karst should be explained as marginal karst plains — because of their wide expanse they also encounter unkarstifiable rocks — or as parts of a polje which was later split open, or even as parts of a river polje in the sense of Kayser (1934).
  Gams (1965) explains the creation of such plains by accelerated corrosion in flooded areas, whereby the soils with their biogenic CO_2 supply the water with the necessary CO_2. Roglić (1957) explains the extension of the polje by a growth of the corrosion plain from the boundary of the rock into the limestone; this is due to the action of the polje waters which on the one hand make the permeable underground watertight, on the other hand, however, dissolve the limestone laterally around the bordering ponors. This, however, does not explain the striking constancy of height over relatively large distances. According to Rathjens (1954) Dinaric corrosion plains are the forms of an earlier age when the climate was warmer and wet seasons alternated with dry. According to Kayser (1934) they should continue to develop today; this can be observed on the lower Neretva and on the Skadersko Jezero (Skutari Lake). Pfeffer (1973, 1975) points out that in humid climates corrosion plains develop only at the level of inundation, but that the morphodynamics of dry climates also cause such plains in karst.

3.4 Fluviokarst Forms: Karst Valleys, Dry Valleys

Underground drainage is typical of karst; therefore valleys, which can only have been formed by surface waters, are felt to be foreign forms. Today they are ordinarily no longer active and their creation must therefore date back to an earlier period. One differentiates between the surface rivers, which are still active today, and inactive dry valleys. Karst rivers are generally allochthonous and flow from an unkarstified area into the karst, the whole extent of which they cross overground, e.g., the Tarn (France), or else the river flows alternatingly on the surface and in the underground as is the case with the Pivka between the basin of Postojna and Ljubljana.

There are several reasons for the existence of karst rivers on the surface:

1. Karstification is not so advanced that the underground cavities (e.g., open joints) are able to swallow the water. During its course the river loses water into the underground karst, in dry periods so much that the river may even dry up.

2. The water has cut in to meet the body of karst water. A reciprocal action begins between the river and the karst water. The river becomes the local base level (deep karst).

3. The river has reached the underlying, impermeable layers (shallow karst). Karst springs supply it with water.

4. The riverbed was subsequently made watertight so that the river flows undisturbed through permeable rock, as is the case with many polje rivers (Lika in the Likapolje, Yugoslavia).

5. Underground drainage can be hindered by permafrost (Pleistocene) and the result is renewed surface drainage: blocked karst. On the one hand the old, superficial water branchwork is reactivated, on the other hand new valleys form which are relatively flat. After the permafrost disappears the young generation become entirely dry valleys, the old ones at least partially, e.g., in the dry-valley karst of the Swabian Alb (Friese, 1933; Dongus, 1963; Villinger, 1969). In the north of France and in southern England (Warwick, 1964) they are not rare.

The characteristic valley form of a permanent river in karst is the *canyon* (Cvijić, 1893; Grund 1903). How steep the sides of the valley are depends on the susceptibility of the rock to weathering. In dense limestone with thick layers they are very steep to vertical. The creation of canyons is controversial. They are explained partly by fluvial erosion, and corrosion, partly by collapsed underground river courses. A general theory does not exist; one must decide from case to case (cf. pocket valleys).

Frequently rivers flow only a little way into the karst and end: *blind valleys*. In Indiana and Kentucky they have significant names like Lost River and Sinking Creek; their valleys are only few meters deep, and dry valleys interlaced with dolines form their continuation (Mallot, 1939, 1945; Powell, 1965). Also autochthonous karst valleys which still end in the karst count as blind valleys. Dry valleys, which lead on from blind valleys to a higher level, bear witness to an earlier active phase and to successive underground tapping. *Pocket valleys* lead the water of larger karst springs (river source) away out of the karst. They are often called bout du monde. They are frequently explained as collapsed underground water-courses, thus Malham Cove in the North Pennines (GB, Hudson et al., 1933).

Further details concerning karst valleys can be found in Cvijić (1893), Scheu (1918), Corbel (1957), Maull (1958), Roglić (1960, 1964b), Jennings (1971), Sweeting (1972), and Pfeffer (1975).

It must be pointed out that there are also nonkarst dry valleys in humid climates, e.g., where the water seeps into permeable alluvial deposits and where Pleistocene valleys which once carried the water from melted snow have become inactive. In arid or semi-arid climates dry valleys are a normal occurrence (Pfeffer, 1975).

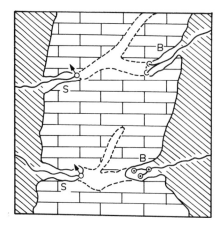

Fig. 3.15. Blind valley (*B*), pocket valley (*S*) and the dry valleys belonging to them (*broken lines*)

3.5 Glaciokarst

Stepped pavement karst and *cuesta-like karst* (Plates 5.1., 5.4.) together form glaciokarst, a karst glacial form complex (Bögli, 1964d). The upper bedding-planes are widened by corrosion and the cohesion of the limestone layers is loosened. During the Pleistocene glaciation the loosened limestone banks were pushed down. They separated from the part which stayed behind along joints or grikes. A stepped surface was thereby produced; when the steps lie contiguously horizontal, they are like a staircase, when inclined, like a cuesta.

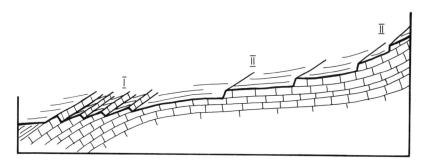

Fig. 3.16. Cuesta-like karst (*I*) and stepped-pavement karst (*II*)

3.6 Poljes

The largest karst hollows, the poljes, were first described by Franz von Steinberg (1761) (Polje of Zirknitz). Since then the term polje has been used in geomorphology for the entirety of these karst hollows. "Polje" originally had nothing to do with karst. The term first rather comprised a piece of arable land in the midst of uncultivable land, whether the latter was a stony surface, as in the case of karst where the green fields are especially striking, or swampy meadows as in the Sava lowlands. Only a few karst morphologists, e.g., Roglić, are aware of the ambivalence of the expression and thereby influenced in their manner of thought. For most the original term is no longer anything but an interesting reminiscence, so that it can today be morphologically clearly defined. The attempt to define the term polje genetically is thwarted by the multiplicity of possibilities and opinions, as H. Lehmann (1959) makes clear. Therefore the polje is today defined morphologically. The ISU (Fink, 1973) suggests: "Polje, an extensive hollow form in karst, closed on all sides, for the most part with an even floor, with a steep border in places and a clear angle between slope and polje bottom. The polje has underground drainage. It can be dry the whole year, or have water flowing through it at times, or be inundated." H. Lehmann (1970, p. 868) describes the form as follows: "Poljes are hollow forms of various sizes, but are generally several kilometers long and wide; they are partly basin-shaped, partly twisted like a valley; they are closed all around and have a flat floor which is covered with fluvial deposits of residue from weathering (terra rossa, gravels)." The polje's floor is generally impermeable. Occasionally isolated, pointed, cone-shaped hills have been left over after the formation of poljes, the so-called hum (pl. humi); these are characteristic of many a polje.

The size varies greatly. The Likapolje in Croatia, the largest of all, has a surface of 700 km^2. According to an inventory made by Ballif (1896) of Bosnia and Herzegovina, the polje of Livno comprises about 405 km^2; the smallest, the Blatcapolje, which is southwest of the mouth of the Neretva, however, only 0.3 km^2. There are smaller ones outside of this region.

No other form of exokarst is so closely bound up with karst hydrology as is the polje. Cvijić (1893, 1960) already pointed this out, and O. Lehmann (1932) gave the reason for it. He recognized that during floods poljes can be filled not only from the sides by inflowing water, but also from the underground by water being pushed up out of the underground karst tubes which are under overpressure. This becomes visible when water discharges from the original swallow-holes, the ponors, so that they function as springs. Such alternate karst openings (estavelle) are characteristic of poljes which are periodically or episodically flooded. Beside the water spouts there are swallow holes at the same level at the same time. Therefore Grund's explanation (1903) of such flooding proves to be false, for he believed it to be caused by rising ground water. O. Lehmann also rejects this theory categorically.

Poljes with perennial lakes are rare. The Lake of Zirknitz dries up in the course of the summer (Löhnberg, 1934; Breznik, 1962). The present-day perennial lake has artificially been made watertight; yet new swallow-holes have formed. The two open poljes — poljes where the basin has been broken open on the side — one shortly before the mouth of the Neretva and the other the Skadersko Jezero (Skutari Lake) — show perennial lakes. It seems that the corrosion plains are developing further at the margins; the steep slope of the shores support this assumption.

Many poljes contain periodical lakes, which form at the beginning of the rainy season and become dry again at its end. But numerous poljes remain dry all year. In Bosnia-Herzegovina Ballif (1896) counted 32 dry poljes; among them were the large Nevesinjsko and the Duvansko Polje; but he counted only 17 periodically flooded poljes; to these the Livanjsko Polje also belongs. The new hydroelectric installations have changed the ecological balance of the water system to such an extent that many of the periodically inundated poljes have today dried up completely, among them the Livanjsko Polje and the Popovo Polje.

The genesis of the polje is still controversial today. Cvijić (1893, 1901) assumed that poljes were formed by uvalas which grew together so that consequently there was one genetic sequence of forms: the doline — uvala — polje. This theory has now been completely discarded, it is a thing of the past. There are no common causes of origin if one disregards the fact that tectonics influence the creation of all three forms. In any case Cvijić (1893) saw the relationship of the polje to geological structure; he assessed it as a tectonically favorable condition. This is emphasized by the facts that poljes are lacking in plateau regions and that they occur only in folded regions. Grund (1903) gave the extreme explanation, however: "The karst polje is a karstified tectonic subsidence field. Only by virtue of its underground drainage does it become a component of the karst phenomenon." (quoted from H. Lehmann, 1959, p. 260). Here the hollow form itself is attributed to tectonic processes. J. Martin (1965) described such poljes in the Middle Atlas of Morocco. But otherwise this opinion is not held by many others since this type of tectonic polje seems to be very rare. It is, however, a generally accepted fact today that tectonic factors such as folds, synclines as well as anticlines, large faults and the boundaries of rocks are important for the formation of poljes (Nicod, 1969). The formative, corrosive processes start from these structures. Roglić (1939, 1957) showed how water from impermeable rock gradually works its way into the limestone by means of lateral corrosion. There are always new swallow-holes being created at the front, while the corrosion plain which is forming gradually becomes watertight from the sediments carried in fluvially. Poljes have therefore been created by exogenous factors, by the erosion of insoluble rock, and by the corrosion of carbonate rock (Kayser, 1934, 1973; Louis, 1956; H. Lehmann, 1959; Cvijić, 1960; Roglić, 1964a; see Chap. 3.3).

Points of departure for the formation of poljes can be highly diverse. While Cvijić firmly rejected their development from valleys, precisely this manner of formation has become a certainty in many cases today, as in the case of Popovopolje, to name one of many. Characteristic of this type of polje are a long, drawn-out form, drainage along the great axis, and the threshold at the lower end. Still others illustrate the significance of the rock's margin for the formation of poljes by lateral drainage from impermeable into permeable — and also soluble — rock. Most of such cases began precisely at the border of rock.

The disposition for poljes was created for the most part in the late Tertiary and they were also hollowed out then. Tertiary sediments in many polje floors bear witness to this. With the change in climatic factors in the Pleistocene Period their lateral growth ceased. Moreover in many polje surfaces situated at higher altitudes in the Dinaric area (Cvijić, 1918; Kayser, 1934; Rathjens, 1960; Liedtke, 1962) and in Greece (Defaure, 1966) as well as in the Apennines (Rühl, 1911; H. Lehmann, 1959; Demangeot, 1965; Pfeffer, 1967) and in the Venetian Prealps (Fuchs, 1970) fluvial and glaciofluvial deposits were made and in places moraines and moraine ramparts were formed. This led to changes in the physiognomic character of the floors of the poljes, without the polje losing its function as such because of that.

A geographical division shows on one hand a zone of poljes in subtropical climates with alternating wet and dry seasons, on the other hand such a zone in the humid tropics where the poljes are mainly bound to cone-karst regions. Thus they show a considerable dependence on climate (climate-morphology). It is, however, striking that in warmly temperate North America there exists till now but one known polje, Grassy Cove, which was first taken to be a huge uvala (Lane, 1952). It lies on the subtropical, slightly folded Cumberland Plateau in Tennessee. Poljes are amassed around the Mediterranean Sea, with concentrations of them in the Dinaric, Greek region and in the mid-Appennines, as well as in the Taurus Mountains (Louis, 1956) in Anatolia. In southern France there are a few examples, e.g., the Plan de Canjeurs (Nicod, 1969); in Spain they are even rarer; they can be found occasionally in Morocco, in the Mid-Atlas, e.g., the Daja Chiker (Ek and Mathieu, 1964) and the Polje of Aguelmane Azigza on the Causse d'Ajdir (Martin, 1965).

The poljes in the cone-karst regions of the humid tropics, the interior valleys enjoy a special position. They frequently occur as marginal poljes, e.g., in Cuba on the Sierra de los Organos (H. Lehmann et al., 1956) or in Puerto Ricco (Monroe, 1960, 1976). In Jamaica a rare form lies in the midst of the cockpit karst, the tectonic polje of Lludas Vale (Sweeting, 1958). In the cone-karst regions of southeastern Asia they are relatively rare (Balazs, 1968; von Wissmann, 1954; H. Lehmann, 1936).

Classification of poljes according to H. Lehmann (1959) (simplified)

1. High-surface poljes, sunk into an elevated flat relief and without a system of valleys as forerunner.
1.1 Even-floored basin-poljes (Dinaric type) with Pleistocene filling in the basin.
1.2 Trough poljes, without noticeable filling in the basin.
2. Valley poljes, sunk into older systems of valleys.
2.1 Even-floored, rubble, valley poljes with a sharp angle between floor and slope.
2.2 Trough-shaped valley poljes without the sharp angle.
3. Semi-poljes, physiognomically a genuine polje, but bordered on one side by a non-karstifiable rock.

Lehmann meant this classification to be a suggestion and not generally valid, yet it has since been frequently used.

Classification of poljes according to Gams (1973)

1. Semi-poljes: impermeable layers reach far into the polje on at least one side of it (complex semi-polje according to H. Lehmann, 1959).
2. Marginal polje (the same as a semi-polje according to H. Lehmann, 1959).
3. Over-flow polje (according to Gams, 1969): a polje with karst springs on one side and ponors on the other; because less permeable layers within the polje force the water to rise on the side of its influx and to flow over at the surface.
4. Storage polje: polje floor on the level of the piezometric fluctuations of the water level, because the karst water outside the polje is dammed up by impermeable layers.

5. Parapolje: polje-shaped forms without even floors but clearly prominent fluvial, glacial, or periglacial features.
6. Polymorphous (polygenetic) poljes: they have the characteristics of more than one of the above types.

The advantage of this classification lies in the consideration it gives to karst hydrology. To what degree it will be accepted remains to be seen.

4 Endokarst and Karst Hydrology

4.1 Introduction

Endokarst is underground karst. It is not a primary phenomenon because sedimentation takes place without forming syngenetic cavities. Simultaneously, however, the disposition for bedding interstices (banking) is created. Reef limestones are an exception to this rule because primary cavities are formed in them during the growth of the biological reef components (bryozoa, sponges, corals). Such cavities do not belong to karst because they are not created by processes of karstification, e.g., corrosion. They can, however, be incorporated into the process of karstification and promote it (Bretz, 1960).

Underground karstification can set in at the earliest when consolidation has begun, as Jennings' (1968) investigations of the limestone dune sands of SW and S Australia, including Tasmania, established. He calls this type *syngenetic karst* which in this case is to be understood as syngenetic with diagenesis (consolidation of rock), not with sedimentation as is usual. This is, however, an established term. Although processes of dissolution occur in sands, as corroded grains prove, the mobility of the material prevents any creation of karst forms. Karst corrosion cannot be assumed to take place below the surface of the sea yet the widespread occurrence of beach-rock indicates a tendency to limestone precipitation.

Karst is controlled by the dissolution of limestone and therefore by the presence of water. The underground drainage which is necessary for the development of endokarst presumes a network of subterranean water courses; these consist primarily of joints and bedding planes and/or sufficient porosity of the rock. Such permeable networks do not occur only in karstifiable, but also in many insoluble and brittle rocks. To these belong, among others, the already-mentioned quartzites and quartz sandstones of Venezuela (see Chap. 1.1 and White, 1960; White et al., 1966; Urbani and Szczerban, 1974) as also volcanites, especially volcanic tuffs. Lavas tend to considerable jointing as a result of fast cooling; the tuffs are very porous; these two properties are the basis of underground drainage. In the humid tropics the weathering of porous, volcanic ashes proceeds extremely rapidly, when seen from a geological point of view. The clays created thereby seal the interstices and pores in the course of only a few years or decades, so that there is a noticeable decrease in the seepage of surface water. It is characteristic that most volcanic pseudokarst areas lie in semi-arid and arid regions where it is impossible to distinguish between karst and nonkarst areas on the basis of underground drainage. Such landscapes should not be called karst, although this is sometimes done in consideration of the partial underground drainage through solid rock. In the interest of a clear definition of karst one should not speak of volcanic karst. When, however, forms occur on the surface which resemble karst features e.g., dolines, pseudokarren, and nonkarstic caves (e.g., lava tunnels), the correct term to use is pseudokarst.

It must be pointed out that underground drainage is not completely identical with karst in any case since occasionally streams ooze away into the valley fill.

A rock complex which is capable of karstification is at first exposed to karstification only on the surface. Processes of corrosion set in at the outcrop of the capillary interstices of bedding planes and joints, effecting their enlargement. On the surface small V-shaped karren are thereby formed, a few millimeters wide and deep, features of exokarst. Corrosion which reaches deeper is scarcely recognizable. This must have been partly the reason why O. Lehmann (1932) assumed as a prerequisite of cave formation the existence of "initial cavities" (Germ.: Urhohlräume) with a minimal width of 2 mm. Experience has shown that such open joints are so rare in the Alps that they could never have been the cause of the formation of the giant caves.

The condition which sets underground karstification in motion is a difference in height between the limestone surface and the local base level (Vorflut). As soon as this condition is met, there is theoretically no limit set to the depth karstification can reach, as long as a head keeps the water moving. Especially Russian and North American researchers have pointed out that a previous valley formation is necessary for endokarst (Popov et al., 1972; Davies and Legrand, 1972). In the initial stage, when phreatic conditions prevail, besides aggressive water, also mixing corrosion works within the rock. The process is steadily intensified, for the more the cavities extend and permeability increases, the greater is the potential for corrosion until at last a limit is set by the quantity of water and its capacity for corrosion. Considering the small amount of water moving through and the slow speed with which it moves, the initial phase lasts very long, in the order of tens of thousands, even hundreds of thousands of years. In order to set water in motion at all, at least a primary perviousness (Germ.: hydrographische Wegsamkeit) is necessary (O. Lehmann, 1932) for the water must not only penetrate into the rock, it must also be able to seep out again. With time the interstices of a primary permeability become karst-hydrologically-active open joints. The initial phase is then at an end. This is noticeable on the surface by the gradually increasing loss of water. Endokarst is made accessible only when the underground cavities further develop into caves. According to the internationally valid definition a cave is a "natural underground cavity which has more than the size of a human being," (Fink, 1973), or according to another definition an underground cavity large enough "to allow a human being access to it" (Bögli, 1976a, p. 6). This specification of size is irrelevant for karst hydrology but it does represent the threshold across which the diversity of the subterranean world reveals itself directly to the seeker.

4.2 The Origin of the Water in Endokarst

Karst groundwater is fed by seeping water from precipitation on bare limestone (N), by water from soils (B), by water flowing in from superficial streams (G) and by condensation water (K) (Fig. 4.1). The various components mix quite fortuitously in the underground and finally collect in the phreatic zone (*hatched*) or above the impermeable stratum, along which they flow to the karst spring.

The Origin of the Water in Endokarst

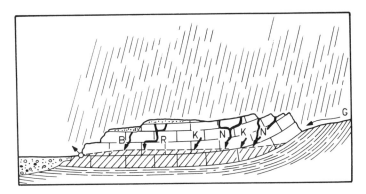

Fig. 4.1. Underground karst water (see text for explanation)

During seepage the different types of water show different behavior and different lime contents. Seen statistically, precipitation seeps into the underground in a diffused fashion, although making use of well-defined swallow-points; it shows a low lime content. $CO_{2\,aggr}$ is found in quantities below 1 ppm or is completely lacking. Because of its low lime content, the water becomes corrosive when it comes into contact with the CO_2 of the cave's air. Corrosive water obtains the CO_2 necessary for the dissolution of limestone either directly from the cave's atmosphere or by mixing, as opposed to aggressive water, which carries the CO_2 in solution from the surface or releases it when cooled. In general it is true that with increasing precipitation the lime content of seeping water decreases. There are exceptions to this rule which indicate conditions which should be investigated more closely.

From a statistical point of view the water from soils also seeps in diffusedly. As explained earlier, it is rich in dissolved carbonates and is, moreover, often aggressive. But this aggressiveness is quickly lost when contact is made with the rock, since the A/V is large. In contact with the cave's atmosphere this water is normally oversaturated and can deposit calcareous sinter. Prolonged precipitation causes the lime content to decrease.

Lake- and river-water is frequently in balance with the CO_2 of the atmosphere and therefore shows a lower lime content than water from soils. The lime content of water in little streams is dependent on the springs which feed them, but also on the time the water has been flowing since it left the spring. If water from lakes, rivers, or streams flows into the underground, it forms cave rivers in the vadose zone. In the phreatic zone these rivers can either continue to exist independently or their water is dispersed in the existing network, losing its independence.

The formation of underground condensation water is undisputed but its quantitative importance is all the more disputable. Aristotle already observed it in the Thrazian Mountains and Seneca partly explained underground water in this way. Volger (1877) began with the observation that there was a difference between the large outflow from karst springs and the aridity of the catchment area. From this he concluded that "all groundwater should form from the condensation of the water-vapor in the air in the underground" (quoted from Ule, 1925, p. 6). He thereby rated the importance of condensation for the occurrence of karst water very high, much too high; it resulted in the rejection of his theory. It was therefore almost lost. In the first two German text books on speleology by Kraus (1894) and by v. Knebel (1906) condensation was not even mentioned. Kyrle (1923) writes: "Much more modest in their amount and therefore also in their effect are condensation waters... Apart from a certain influence on the temperature of underground wind they are of no importance for caves," (p. 163 f). Gèze (1947) took up Volger's idea, however, and emphasized the importance of this water. His point of departure was southern France, where important differences of temperature are found in the Mediterranean climate. The following numerical example illustrates to what extent this is true: when air measuring 30°C with a humidity of 70% (21 g water/m^3 air) is cooled to 10°C (100 % humidity: 9.4 g water/m^3 air), 11.6 g of water is condensed per m^3 of air. In caves there is often a considerable current of air. The result is many liters of condensed water per second in an underground cave system. Nevertheless a quantity of water of the size of m^3/s seems unlikely. Trombe (1952, p. 40 ff) deals with this problem more closely. He mentions that Roman Theodosia on the Crimea met its water requirements from 13 large heaps of limestone blocks which provided up to 720 m^3 on warm days. Chaptal (1932) illustrated with experiments that a mass of rock of 4 m^3 near Montpellier can provide over 2 l/d.

Condensed water is free of limestone, and therefore obtains its entire requirement of limestone from the cave. This amount exceeds by many times the amount per liter of water of the other types of cave water. In places of condensation small dents, the size of a large drop of condensed water, can usually be found so that earlier points of condensation can be reconstructed from them. Such forms are described by Mais (1975) in Schlenkendurchgangshöhle in Salzburg, Austria (see also Andrieux, 1970).

The difference in the carbonate contents of the water types results, when they are mixed, in mixing corrosion. Some zones are especially favored for this. In shallow karst it occurs most frequently on the impermeable underlying bed where descending water merges with a sideways flow. In deep karst this mixing occurs in the zone of the permanent karst water level, which leads to the creation of cave levels of the piezometric type (Chap. 8.2).

5 Physical Behavior of Karst Water

Hydrostatics and hydrodynamics determine the flow-behavior of subterranean water. Since the shape of caves and of rock surfaces defies quantitative description in physical or geometrical terms, hydrodynamics may only be summarily considered (for more detail see Prandtl, 1969).

5.1 Hydrological Perviousness – Karst-Hydrological Activity – Velocity of Flow

O. Lehmann (1932) established three types of joint: hydrologically impervious, hydrologically pervious, and karst-hydrologically active. If a joint is closed in only one place, then it is *hydrologically impervious*. The water course is blocked and the joint is of no account for karstification.

Hydrological perviousness implies a continuous connection between a point where the water seeps in and the spring/water outlet. It is unimportant whether the underground water passages end at the surface of the earth or in a cave. In hydrologically pervious joints the "water may ooze so slowly that it needs more than a year for its passage and in no way responds to precipitation with the strong short-term fluctuations of karst water" (O. Lehmann, 1932, p. 12). The main velocity of flow may amount to only a few centimeters or meters per day. Hydrological perviousness is hindered but not prevented by long distances. A few dozen meters are as much within the normal frame of reference as a few dozen kilometers.

Hydrologically pervious interstices in limestone may be widened by corrosion and thereby become *karst-hydrologically* active. Lehmann draws the line between the two at an open-joint width of 2-3 mm. Stini (1933) distinguishes narrow and broad watercourses and sets the limit between the two at 0.2 mm. In karst-hydrologically active cavities water flows within hours or days toward the spring, or as Lehmann defines it "in much less than one year." Heim (1919, p. 702 ff) assumes that the thermal springs at Baden near Zurich originate in the autochthon of the Aar massive, which is a distance of at least 80 km (Engelberg-Baden), their water flowing at depths exceeding 1500 m. The time needed would, therefore, amount to roughly 20 months, which would correspond to an average velocity of flow of 5.5 m/h, i.e., to that of lesser karst-hydrological activity. This activity expresses itself in the potentially high velocity with which water can flow through an underground cavity, rather than in the velocity with which a spring's discharge reacts to precipitation – there are some karst springs which react very slowly.

Apart from a few exceptions the velocity of flow cannot be directly followed in the phreatic zone; it can only be indirectly calculated — in favorable situations, for instance by interpreting the granulometric composition of mechanical sediments in cave passages (see Hjulström Curve Fig. 13.1) and extrapolating for other sizes of cross-sections. The average linear velocity of flow v_1 can be determined by observing the time marked water takes to flow. In karst this average is between a few meters and 0.5 km/h (Tables 5.1 and 5.2). The time the water takes to flow is related to the distance a between the swallow point S and the karst spring K (Fig. 5.1). The way the water really flows is always longer because it runs in all three dimensions and it is an exception when it runs in a straight line for a longer distance. v_1 must therefore be multiplied by the lengthening factor $\frac{L}{a}$ (L: real length), in order that the average real velocity v_e be obtained.

Fig. 5.1. Scheme of a possible water-course between S and K (see text)

The distances between the Danube's sinks and the Aach Spring measure 11.7 km from Brühl and 18.3 km from the loop of Frieding. In the Swiss limestone Alps by means of water-tracing a connection over 20.8 km long was proved to exist between Schrattenfluh (Ctn. Lucerne) and the Lake of Thun not far from Interlaken. The v_1 amounted to 540 m/h at a gradient of 4.7% (Δh: 980 m; Knuchel, 1972). In investigations for the utilization of karst water in Yugoslavia water was marked hundreds of times. Many of these investigations showed a water-course of over 20 km. In Taurus it was proved of the connection Homat/Homa that a = 134, that the flow took about 370 d and that v_1 = 15.2 m/h (Bakalowicz, 1973).

Table 5.1. v_1 between the Danube's sinks and the Aach Spring (Batsche et al., 1970)

Sinks	a (km)	Δh (m)	gradient (%)	v_1 (m/h) [a]
Danube sinks:				
South of Immendingen	13.3	179.9	1.35	364
Brühl	11.7	178.6	1.53	450
Friedinger Schleife/Loop of Frieding	18.3	142.5	0.78	332
Others:				
Doline E Neuhausen	15.4	260	1.69	226
Drainage shaft E Liptingen	11.8	245	2.08	190
Quarry S Buchheim	19.8	310	1.57	24

[a] in reference to the first appearance of the tracer.

It is clear from Table 5.1 that no direct relationship exists between the linear velocity of flow and the gradient. Yet it has been shown that lower v_1 are found with smaller quantities of water, such as in the quarry S of Buchheim.

Investigations made in the upper Muota Valley (Central Switzerland) provided the figures shown in Table 5.2.

Table 5.2. v_1, a, Gradients in the upper Muotatal according to Bögli (1960b, 1970). Concerning tracing points and resurgence of tracer see Fig. 5.2

	1.	2.	3.	4.	5.
6. v_1 m/h	150	86	40	112	131
a m	2700	2480	3360	3580	1530
Gradient (Grad.) %	17	18	13	12	30
7. v_1 m/h	222	148	60	185	189
a m	3560	3410	4230	4440	2460
Grad. %	14	15	11	11	20
8. v_1 m/h	231	155	62	190	201
a m	3560	3570	4370	4580	2620
Grad. %	13	14	11	11	20
9. v_1 m/h	288	148	63	200	213
a m	3750	3700	4400	4600	2770
Grad. %	14	14	11	11	20
10. v_1 m/h			38	129	
a m	Unrecorded		5230	5420	Unrecorded
Grad. %			12	12	
11. v_1 m/h	363	221	94	327	368
a m	6540	6640	7060	7200	5900
Grad. %	16	16	15	15	18
Date:	15.7.1947	23.9.1946	9.2.1958	27.8.1947	8.7.1954

Points of tracing:
1. Glattalpsee NW
2. Glattalpsee SW
3. Glattalpsee S
4. Glattalpsee SE
5. Schafpferchboden

Resurgences:
6. Taschibach
7. Eigeli
8. Feldmoos
9. Höchweidquellen
10. Richliswaldbach
11. Brünnen/Hinter Seeberg

The mean linear velocities of flow show scarcely any noticeable dependence on the gradient, but they show a marked dependence on the amount of water flowing. In July when the melting of the snow is coming to an end, the water level is high in the Glattalp Lake and there is a large amount of precipitation, the maximum velocities are found; in February when there is a minimum of water the velocity of flow is also at the minimum.

5.2 Catchment Area – Local Base Level

The local base level is the stream which gathers and carries away the water of a catchment area, and which thereby becomes the hydrological base. It is in accordance with

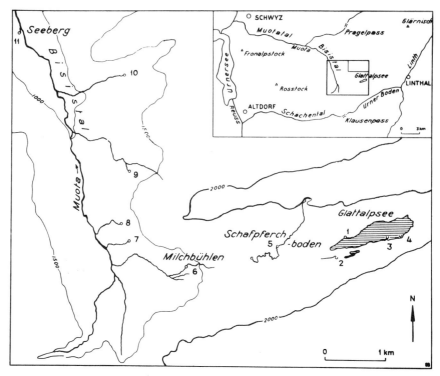

Fig. 5.2. Points of tracing and resurgence in the upper Muotatal according to Bögli (1960b)

the character of underground karst that catchment areas of karst springs can overlap, especially in young karst which is not yet fully developed. These various local base levels can also have parts of the catchment area in common, as tracing has proved conclusively (Fig. 9.1). The absolute hydrological base level of a karst area lies on its edge. However, this edge by no means always coincides with the limit of the karst visible on the surface. Permeable layers can dip under impermeable, unkarstifiable ones: "covered zones" according to Villinger (1972), "underground karst phenomenon" according to Penck (1924). Penck's term should definitely be avoided as it takes in quite general occurrences of underground karst such as underground water-course and caves. The water can rise to the surface through a fault. The water level in the spring then becomes the reference level, the local base level for the karst (Fig. 5.3).

Tectonics and the stage of development of the karst determine the occurrence and the number of springs which are karst-hydrologically independent of one another. In young underground (endogenic) karst which is only slightly developed the interstices are still slightly widened; for this reason the passage of water requires higher pressures. The piezometric surface lies correspondingly high. The number of active water passages is large, consequently also the number of karst springs. Fully developed karst shows only few, although large, active underground water passages, excluding those which bring the water in. Analogous to this, there are also only few, though large, karst springs (river sources). The area of Hölloch, 22 km², is drained by the Schleichenden Brunnen alone.

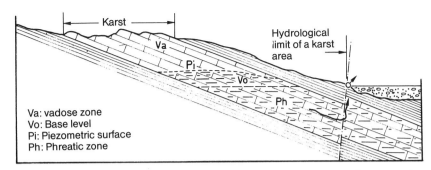

Fig. 5.3. Hydrological limit of a karst area. *Broken lines* piezometric surface or surface separating phreatic/vadose respectively

5.3 Shallow and Deep Karst

In *shallow karst* the impervious rock lies at the same height or higher than the local base level — the phreatic zone is missing. Villinger (1972) observes that in this case karst springs "emerge at the base of their karst water reservoir" (p. 158). Actually the classification of shallow and deep karst should not be dependent on the spring but on whether only the vadose zone occurs (shallow karst) or the vadose and the phreatic zones together (deep karst), Groschopf (1969) agrees with this definition also.

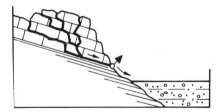

Fig. 5.4. Shallow karst

In shallow karst the hydrological conditions cannot be influenced essentially by the local base level, i.e., by the spring. A rise in the base level (high water) only creates backwater behind the spring.

In *deep karst* according to Villinger (1972) "large parts of the karst water reservoir ... lie under the water level of the spring" (p. 159). The water level of the spring determines the level of the subterranean karst water table. The permanent karst water table is a fictitious surface deducted from the single piezometric surfaces of the lowest water level. It can rise considerably at high water. In the case of fully developed karst hydrology and slight discharge the velocity of flow is so low in the karst tubes of the phreatic zone that this surface runs almost horizontally, practically corresponding to the surface on which water would be at rest. It is unimportant whether the point at which the karst water discharges from the karstified rock is at the base level or deeper in the valley fill or subaqueous; decisive is always the water table in the spring, the local base level.

Mixed forms of shallow and deep karst are frequent. The higher part of the Swabian Alb shows shallow karst. However, in folded areas the order can be different, e.g., by perched aquifers (Fig. 5.5). Therefore the local base level does not determine the permanent karst water table in all cases of deep karst.

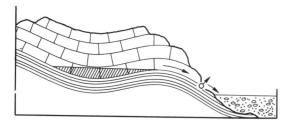

Fig. 5.5. Mixed form of shallow and deep karst with a perched aquifer

5.4 Pressure Flow – Gravitational Flow; the Cave River

Underground water is in motion as a rule and takes part in water circulation. On the other hand, water that is almost stagnant or which moves only by convection currents does not renew itself. In this case it can be fossil or else the expression of immediate climatic conditions (arid climate). In Australia and under the Sahara fossil groundwater has been bored; it dates from the last Pluvial Age. Jäckli (1970) calls it "groundwater with periodic renewal." In the joints of the syncline of Coulon in the catchment area of the Fontaine de Vaucluse water was struck during boring operations; its age was determined to be 3400 ± 400 years (Renault, 1970, p. 27). Such water, which undergoes extremely slow renewal, is to be classed hydrologically as almost stagnant. Geyh (1969) divides karst groundwater into four categories according to age. In the fourth he places "old water of covered karst; it is several thousands, or even tens of thousands of years old and contains no recent components. ...As far as can be seen, this water does not seem to stagnate for it is always in motion, though extremely slow. First measurements taken near Ingolstadt ... showed velocities of flow of about 1 m per century."

Flowing underground water occurs:
a)   as seeping water
b)   in gravitational flow with a free surface
c)   in a pressure flow where it fills up the whole space
d)   as karst water-bodies which embrace the whole phreatic zone.

In a gravitational flow the water does not move much differently than it does in streams on the ground's surface and it creates the same forms, too, by corrosion and erosion. Linear deepening is characteristic. Canyons form, a key-form of the vadose zone. Cave rivers frequently end in a siphon, a water-filled depression in the course of the passage. The roof dips into the water, which is the origin of the French expression voûte mouillante. Because of the possibility of a change from free to pressure flow, the term cave river is used to embrace the whole flow, whether it has a free water surface or whether it flows under phreatic conditions.

Introduction

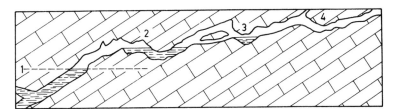

Fig. 5.6. Siphons. *1* karst water-table; *2* closed siphon; *3* open siphon; *4* dry siphon

In *Höhlenkunde* (Speleology) by Kraus (1894), in keeping with the practically undisputed professional views of that time, the cave river was not yet treated in a chapter of its own. It was a different matter with Knebel (1906) who had a remarkably independent opinion since the groundwater theory of Grund (1903) had supplanted all previous theories almost completely. It was his opinion that the water of the Danube flows as a cave river to the Aach Spring and does not discharge into the groundwater. Katzer (1909) and Bock (1913a), however, helped the theory of the cave river so thoroughly to victory that since then many German-speaking speleologists and karst researchers have flatly rejected the term groundwater as applied to karst — and still do so today. This is justified in so far as it refers to Grund's conceptions but not if it refers to the present-day picture of a phreatic zone or of a karst water-body where the term karst groundwater may appropriately be used. This difficulty occurs only in the German language; in English groundwater is all subsurface water below the permanent water table (Schieferdecker, 1959).

Pressure flow is the only form of flow in the phreatic zone; it does, however, also occur in the vadose zone, as in siphons. In this case an excess of pressure, and thereby a head, is created toward the air-filled lower passages or towards the karst water table. Interstices which lead downward are therefore used and widened. There is an increased loss of water in the siphon. Finally the interstices are widened to cave passages and carry the cave river down into the depths with a gradient in one direction. While young, vadose cave rivers in a two-cycle cavern which have formed first in the phreatic zone often show siphons (cf. Chap. 14.4), a fully developed, vadose cave river shows a gradient in one direction. The same is true of young cave rivers of purely vadose origin.

5.5 Piezometric Surface

5.5.1 Introduction

Karst water tables in passages which rise out of the phreatic zone (piezometric tubes) are the surfaces to which water from a given aquifer will rise under its head. The pressure at this surface is equal to that of the atmosphere; it is a *level of equilibrium*. Not only the individual water tables in the cavities are understood under the term piezometric surface, but also the fictitious table deducted from it (Warwick, 1960). Roglić (1965) refers to this fictitious table when he writes: "In compact limestones, no groundwater table can be formed".

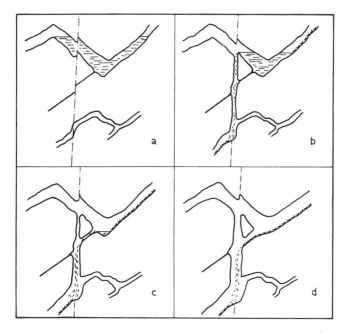

Fig. 5.7 a-d. Development from a young to a mature cave river in a two-cycle cavern

The basic component of the head in a karst water-body is the hydrostatic pressure, which is given by the local base level. Over the hydrostatic pressure positive and negative hydrodynamic pressure components are superimposed which help to determine the height of the individual piezometric levels.

In addition, at high water the feeders carry a large amount of water and the velocities of flow are greatly increased. The difference in elevation between the high-water level and the low-water level in Hölloch can be 100 m in the front part and 180 m in the back. In Flint Mammoth Cave (KY, USA) it is only a matter of a few meters.

Table 5.3. The altitude of the piezometric surfaces in Hölloch at low water; $Q = 0.3$ m³/s

Place	altitude m above sea-level	Distance from local base level	Gradient from place to place (%)	Gradient to the local base level
Local base level: Schleichender Brunnen	637			
Zurichsee	639	400	0.5	0.5
Orkus	643	2100	0.24	0.29
Fjord	643	2250	0	0.27
Passage E Anubissee	648	2800	0.9	0.4
Rätselsiphon	654	3300	1.2	0.5
Donnertal Siphon	665	5000	0.65	0.56

5.5.2 Static Karst Water-Body, Local Base Level

If hydrostatic conditions alone prevail, the individual local piezometric water tables and the base level are at the same level; the water does not move: a static karst water-body.

In the humid climate of Western and Central Europe the discharge of water is only episodically interrupted – if ever. Prolonged dry periods are rare; underground channels are too slow to react to dry weather and therefore hardly ever dry up. Weidenbach (1960) describes a static karst water-body in the Brenz region. The fossil water already mentioned in the syncline of Coulon (see Chap. 5.4) is only conceivable in a quasi-static karst water-body.

In the Mediterranean climate temporarily static karst water-bodies are more frequent due to the months without rain, however underground condensation compensates somewhat for this. Only in fully arid climates can permanently static karst water-bodies be found. In imitation of "groundwater with periodic renewal" (Jäckli, 1970) it could be called groundwater with episodic renewal.

Usually every body of karst water has an inflow and an outflow. The movements of water thus caused are concentrated in individual streams with temporally and locally varying velocities, even though there exists a spatial connection within the phreatic zone. The result is the occurrence of various high piezometric surfaces in ascending karst tubes. As opposed to the static, this is a dynamic body of karst water.

5.5.3 The Bernoulli Effect (Equation of Continuity)

With a given discharge Q the velocity of flow is determined by the law of continuity. According to this the same volume V or the same mass m of an imcompressible medium flows through every cross-section A of a tube in the interval of time t. The stream of water Q is:

$$Q_V = \frac{\Delta V}{t} \quad \text{or} \quad Q_m = \frac{\Delta m}{t}.$$

Fig. 5.8. Flow in a tube (pressure flow)

From Fig. 5.8 can be concluded: $Q_V = A \cdot v$ and $Q_1 = Q_2$
and $A_1 \cdot v_1 = A_2 \cdot v_2$
or
$$\frac{v_1}{v_2} = \frac{A_2}{A_1} \tag{5.1}$$

For stationary streams which are free of friction the Bernoulli equation is valid with its various forms of equation for pressure, energy, and head.

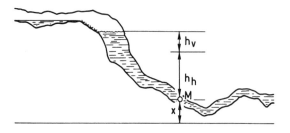

Fig. 5.9. Pressure fractions in a karst tube with flowing water (see text)

In Fig. 5.9 $h_v = \frac{v^2}{2g}$ is the drop in the water table caused by the velocity of flow v, h_h is the head which exerts hydrostatic pressure on the point of measurement M, and which causes the pressure p,

x is the elevation of the point of measurement above the level of reference.

Equation for pressure according to Bernoulli

$$p + \rho g x + \frac{\rho}{2} v^2 = \text{constant} \qquad (5.2)$$
$$\text{(a)} \quad \text{(b)} \quad \text{(c)}$$

a) actual static pressure at point of measurement
b) pressure from the elevation of the point of measurement above the level of reference
c) decrease in pressure which the hydrostatic head undergoes due to the velocity of flow (impact pressure; ρ = density).

By introducing the element of volume V of a section of water into the equation one arrives at the energy equation:

$$p \cdot \Delta V \quad + \rho g x \cdot \Delta V \quad + \frac{\rho}{2} v^2 \cdot \Delta V \quad = \text{constant}$$
$$(p = \rho g h; \rho g \Delta V = \Delta G, \text{ G: weight})$$
$$\Delta G h \quad + \quad \Delta G x \quad + \quad \frac{\Delta m}{2} v^2 \quad = \text{constant} \qquad (5.3)$$

Energy from pressure + energy from position of M
$$\qquad \qquad E_P \qquad \qquad + \qquad E_K \qquad = \text{constant}$$

The sum of all energies is the same at every point of an unbranched tube.
If we divide the equation of pressure by ρg, the result in the equation of the head:

$$h' + x + \frac{v^2}{2g} = \text{constant} \qquad (5.4)$$

The practical facts in a network of underground cavities often show results of the Bernoulli equation which appear unusual. Thus water rising through a piezometric tube out of an enlargement of the passage can flow backward over the main stream and into another tube which begins at a narrow place in the same passage.

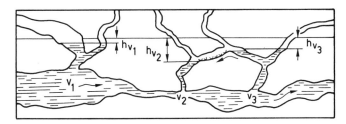

Fig. 5.10. Piezometric surfaces in a natural system of tubes

5.5.4 Torricelli's Theory (Law of Outflow)

Up to here the air pressure has not been mathematically included since its amounts are more or less balanced. It will, however, be taken into account when the Bernoulli equation is applied to a receptacle with an outflow (Fig. 5.11). The outflow will be the level of reference.

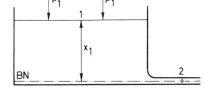

Fig. 5.11. Pressure conditions when the law of outflow is applied. *BN* level of reference

The following equation is valid (pressure equation):

$$p_1 + \rho g x_1 + \frac{\rho}{2} v_1^2 = p_2 + \rho g x_2 + \frac{\rho}{2} v_2^2.$$

The air pressure p_1 is hardly different from p_2, consequently $p_1 = p_2$. Moreover the water surface in the receptacle is very much larger than the area of the outflow's cross-section; therefore v_1 can be disregarded. Point 2 lies at the level of reference which is why x_2 is also zero. It follows that:

$$\rho g x_1 = \rho \frac{v_2^2}{2}.$$

x_1 is the height of the column of water which creates the hydrostatic pressure; $x_1 = h_h$.

$$h_h = \frac{v^2}{2g} \tag{5.5}$$

or

$$v = \sqrt{2 g h_h}. \tag{5.5a}$$

Equation (5.5a) is Torricelli's theorem according to which water flows out of a wide receptacle with a velocity as if its parts had fallen down from the height h of the water surface to the level of outflow free of losses.

Table 5.4. Height of pressure h and velocity of flow v according to the Law of Outflow

h (m)	v (m/s)	h (m)	v (m/s)
0	0	0.46	3.00
0.000127	0.05	0.82	4.00
0.002	0.20	1.27	5.00
0.0127	0.50	1.83	6.00
0.05	1.00	2.50	7.00
0.115	1.50	3.26	8.00
0.20	2.00	4.13	9.00

At a low v, up to 0.2 m/s, friction plays a small role and can be disregarded. Above this amount an increasing loss of pressure occurs (see Chap. 5.5.5). The energy leaving the system warms the water and v sinks with increasing h further and further below the value calculated.

Nothing is changed in Eq. (5.5) and (5.5a) as long as the water flows out through a tube under conditions of no friction. Practically, this is realized when the tube is wide enough for v to amount to close to zero. The tube for the outflow then belongs to the receptacle. Such conditions are not rare in nature in the case of the rising karst spring, the so-called vauclusian spring. Such is the Blautopf near Ulm.

The Blautopf shows an A_2 of 2.8 m² in the narrow pass at the base of the spring's basin, while A_1 in the inflow passage is between 10 and 30 times greater (Keller, 1963, p. 227 f). At low water (0.5 m³/s) the following measurements were taken in the narrow pass: v_{max} = 0.25 m/s and v_m = 0.178 m/s. In the passage behind the pass v_m is 0.006-0.018 m/s; thus the losses of pressure are unnoticeably small.

Opposed to these figures for low water are others for greater discharges:

Average over years: Q = 2.16 m³/s, v_m = 0.77 m/s
                                     v_{max} = 1.1 m/s
Maximum discharge: Q = 26.2 m³/s v_m = 9.35 m/s.

Loss of Pressure in Flowing Water

In the passages behind the narrow pass a v_m of 0.026-0.08 m/s is valid for mean water, for high water, however, 0.31-0.93 m/s.

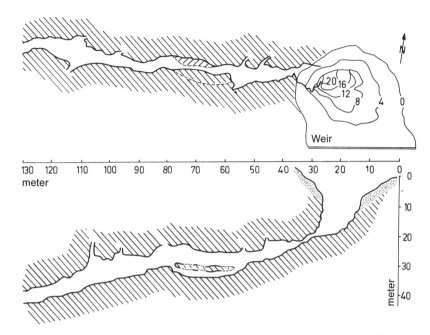

Fig. 5.12. Longitudinal section of the Blautopf near Ulm (according to Keller, 1963, p. 222).

5.5.5 Loss of Pressure in Flowing Water

As soon as water flows, viscosity (internal friction) and external friction cause losses of pressure. This can be presented graphically by means of curves of pressure.

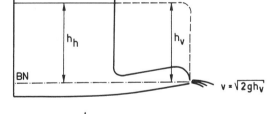

Fig. 5.13. Pressure curve (*dashed line*) when friction is lacking (narrow outflow). *BN* level of reference

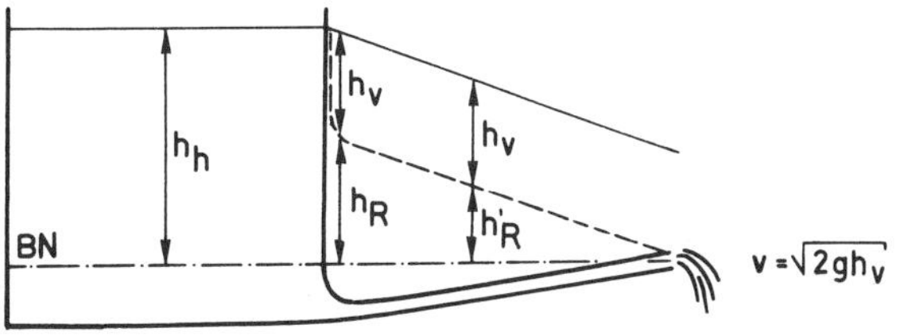

Fig. 5.14. Curve of pressure (*dashed line*) under the influence of friction. *BN* level of reference

h_v height required to accelerate the water to v, more precisely to v_m.

h_R is the height necessary to counteract the friction. h_v remains constant for the whole length of the tube – provided the cross-sectional area stays constant – h_R diminishes throughout the length of the outflow tube to zero and thereby shows the losses of pressure. According to O. Lehmann (1932, p. 52) the portion h_R of the total pressure increases with an increasing tube length so that for a length of 1500 tube diameters the value for h_v is close to zero.

Table 5.5. v_m and h_R dependent on the relative length l/d according to O. Lehmann (1932, p. 52)

l/d	v_m [m/s]	(h_R + 0.09 h_h) [m]	h_v [m]
0	14.0	0	10
25	11.25	3	7
84	8.86	6	4
466	4.43	9	1

At the beginning of the outflow tube the water is accelerated to v. At this point the velocity diagram across the tube shows almost the same velocities. The parts on the sides, however, are braked by wall friction, while the water rushes ahead in the thread of maximum velocity. Because water cannot be compressed:

v_m must remain the same if A of the tube is unchanged.
$v_{max} = 1.3\ v_m$ (Prandtl, quoted from O. Lehmann, 1932, p. 48).

The factor 1.3 is an approximate value which can also be applied approximately to underground water-courses which are geometrically incomprehensible, as was shown in the Blautopf (see Chap. 5.5.4).

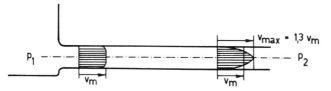

Fig. 5.15. Diagram of velocities in an outflow tube at turbulent flow

An acceleration from v_m to v_{max} requires 0.09 h_h according to Prandtl (quoted from O. Lehmann, 1932, p. 48).

5.5.6 Losses Through Friction, Losses of Pressure

There are various kinds of pressure loss: friction from the walls (due to roughness), changes in the cross-section and changes in direction. Each must be calculated individually. Since the situation of the piezometric surface is of special interest in karst hydrology, losses of pressure are expressed as changes in the pressure height.

a) For a flow between *rough walls* when the surface of the cross-section remains the same the following equation is valid:

$$v_m = \sqrt{\frac{2g}{\lambda} r_h \cdot \frac{\Delta h}{\Delta l}}. \tag{5.6}$$

r_h is the hydraulic radius which represents the relationship of the area of the cross-section A of the water to the moistened periphery U, $\Delta h/\Delta l$ is the gradient and λ a dimensionless coefficient. From Eq. (5.6) follows:

$$\Delta h = \frac{v^2 \lambda \Delta l}{2g\, r_h}. \tag{5.7}$$

In hydraulic engineering the Manning-Strickler equation is often applied:

$$v = k\, r_h^{2/3} \left(\frac{\Delta h}{\Delta l}\right)^{1/2} \quad \text{(quoted from Müller, 1958, p. 524)}. \tag{5.8}$$

If v is the same in Eq. (5.6) and (5.8), then

$$k = \left(\frac{2g}{\lambda} r_h^{-1/3}\right)^{1/2} \quad [\text{cm}^{1/3}\, \text{s}^{-1}] \tag{5.9a}$$

or

$$\lambda = \frac{2g}{k^2} r_h^{-1/3}. \quad \text{respectively.} \tag{5.9b}$$

If λ is put into Eq. (5.7) and h is taken as the loss in the pressure height h_r, then it follows that

$$h_r = \frac{v^2}{2g} \frac{2g\, \Delta l}{k^2\, r_h^{4/3}}. \tag{5.10a}$$

with $\xi_r = \frac{2g\, \Delta l}{k^2\, r_h^{4/3}}$ follows

$$h_r = \xi_r \frac{v^2}{2g}. \tag{5.10b}$$

Table 5.6. Coefficients of roughness k ($m^{1/3}\, s^{-1}$) for surfaces important to karst hydrology according to Manning-Strickler (Press, 1969, p. 695)

Smooth masonry	50-60
Smooth paving	40-50
Coarse paving, smooth rock protrusion	30-40
Wall of coarse boulders	20-30
Very coarse rock protrusion	15-20

In underground water-courses the kinds of surface are most varied. Small, phreatic karst tubes have usually been formed by water alone and are therefore relatively smooth, frequently covered with scallops (k: 30-50). In passages there are additional obstacles, sharp edges, boulders (k: 15-30). The wider the passage is, the more boulders are created (k: 8-20).

b) *Changes in the cross-section* are, as opposed to the case in technical constructions, very frequent. They cause losses of pressure which are heavily dependent on geometry. Constant and slow changes in the cross-section cause scarcely any loss of pressure. If the inlet is angular, contraction of the inflowing water and a loss of pressure result:

$$h_e = \xi_e \cdot \frac{v_e^2 - v_w^2}{2g} . \qquad (5.11)$$

h_e is the loss of pressure height when the water flows into a narrower section of tube. When the entrance is rounded it is close to 0-0.1, when it has sharp corners it is between 0.5 and 1.3 (acc. to Press, 1969, p. 807).

In karst hydrology rounded inlets are unusual. They show coefficients of around 0.3, the others show coefficients of 0.5 and 0.9, with a maximum of 1.1. The losses reach their maximum only when the transition occurs abruptly.

When there is a transition from a narrow pass to a wider cross-section considerable losses occur when there are but small deviations from the ideal form which would cause no loss. Therefore the coefficient is higher, around 0.8-1.2 according to Gibson (quoted from Müller, 1958, p. 538). The walls are frequently hollowed out on the sides behind narrow passes:

$$h_w = \xi_w \frac{v_e^2 - v_w^2}{2g}. \qquad (5.12)$$

c) *Losses due to curvature* are dependent on the angle of deflection $A°$ ($\xi_A°$) and on the relationship radius of curve/diameter of tube ($r_k/d; \xi_{kr}$).

$$h_k = \xi_k \frac{v^2}{2g} = \xi_{kr} \xi_A° \frac{v^2}{2g} . \qquad (5.13)$$

Table 5.7. Coefficients of the losses due to curvature according to Press (1969, p. 809)

r_k/d	ξ_{kr}	$A°$	$\xi_A°$
0	1.3	30	0.15
1	0.7	60	0.40
2	0.33	90	1.00
3	0.25	120	1.60
4	0.25	150	1.80

When the angle of deflection is 60° or smaller and the r_k/d is around 1, the simplified approximate formula can also be used according to Müller (1958, p. 533):

$$h_k = \frac{v^2}{2g} 0{,}0034 \, A°.$$

In karst hydrology all $A°$ up to 180° — as spirals more than that — and all r_k/d are possible. Straight sections of passage are frequent in joint caves, but they are rather rare in caves of phreatic origin. A distance of vision of a few meters is the rule in smaller passages (H < 1.5 m, B < 2 m); in larger ones it can measure 20-40 m and as an exception it can extend to the limits of the range of sight (Hölloch up to 90 m).

A cave map does not always show the true deflections, as these can be situated anywhere in the space. Vertical changes of direction are not recognizable on the map. Small changes in direction, e.g., around one line of sight, often are sacrificed to a generalization of the map.

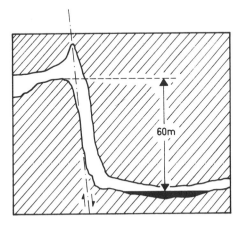

Fig. 5.16. Vertical changes in direction in the Orgelwand, Hölloch (Bögli, 1973a)

d) All *losses of pressure height* caused by friction can be summed up in:

$$h_R = \frac{v^2}{2g} \Sigma \xi,$$

$$\Sigma \xi = \xi_r + \xi_e + \xi_w + \xi_k.$$

5.5.7 Analysis of an Inaccessible Water-Course

The following example is meant to test the applicability of the equations gained from technology for the analysis of inaccessible phreatic water-courses.

Example: Hölloch (Muotatal, central Switzerland) has an inaccessible phreatic collector into which all feeders discharge; above this lies an accessible high-water zone (100-180 m thick) with a 9-km-long high-water course, and the inactive zone. The section Orkus-Schleichender Brunnen was chosen from the whole for treatment because there is a large quantity of reliable data on this section.

The distance a: Orkus – Schleichender Brunnen (emergence of Hölloch) measures 2100 m. An experiment with a tracer showed at low water a linear v_m of 0.01 m/s. Conclusions drawn about the inaccessible water-courses of the phreatic zone were based on the situation in the accessible water-courses of the high-water zone. The resulting values can therefore only be approximate. The extension factor L/a lies at 1.25. This results in a true length L of 2625 m, an average passage cross-section A_m of 24 m² and a v_m of 0.0125 m/s.

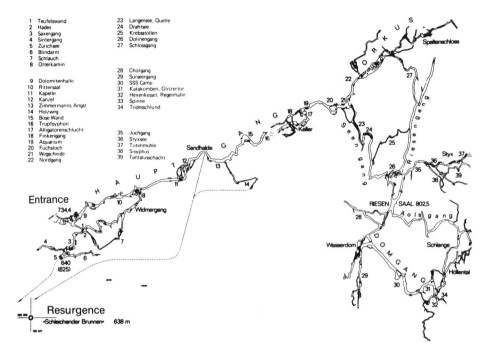

Fig. 5.17. Excerpt from map Orkus – Schleichender Brunnen (AGH, 1970; Bögli, 1970)

Table 5.8. Basic values for the calculations (low water)

	Proportion (%)	L (m)	A_m (m²)	v_m (m/s)	r_h	k
a) Narrow sections	10	265	2.5	0.120	0.36	30
b) Smaller passages	15	395	7.5	0.040	0.65	20
c) Medium value	65	1705	24	0.0125	1.2	15
d) Big rooms	7.5	197	94	0.0032	2.6	10
e) Breakdown	2.3	60	1.0	0.300	0.13	10
f) Narrowest pass	0.2	3	0.3	1.000	0.11	10

1. Losses due to roughness: k according to Manning-Strickler is to be found in Table 5.8. Section (E), which is especially narrow, can only have been created by boulders' being swept into an already narrow spot.

$$h_r = \frac{v^2 \, L}{k^2 \, r_h^{4/3}} \, .$$

Results:

a) $h_r = 0.017$ m

b) $h_r = 0.003$ m

c) $h_r = 0.001$ m

d) $h_r = 0.000\,005$ m

e) $h_r = 0.820$ m

f) $h_r = 0.569$ m.

$\Sigma \, h_r = 1.410$ m

Analysis of an Inaccessible Water-Course

2, 3: Losses from changes in the cross-section: of the numerous changes in the cross-section of the inaccessible part, two became manifest when the water swelled up into the accessible part. For analogous reasons one can assume that there are at least 18 larger pairs of narrow stretches – the many small ones are not of importance.

Table 5.9. Basic values for calculation

		Number of pairs	$\frac{v_m^2}{2g}$	ξ_e	ξ_w
1. c/f	f/c	1	0.0510	1.1	1.1
2. e/d	d/e	3	0.0046	1.1	1.1
3. e/c	e/c	3	0.0046	1.1	1.1
4. c/a	a/c	6	0.0007	0.6	1.0
5. b/a	a/b	5	0.0007	0.4	1.0

Calculations:

$$h_e = \xi_e \frac{v^2}{2g} \qquad h_w = \xi_w \frac{v^2}{2g}.$$

Results:

1. $h_e = 0.056$ m $\qquad h_w = 0.056$ m
2. $h_e = 0.015$ m $\qquad h_w = 0.015$ m
3. $h_e = 0.015$ m $\qquad h_w = 0.015$ m
4. $h_e = 0.003$ m $\qquad h_w = 0.004$ m
5. $h_e = 0.001$ m $\qquad h_w = 0.003$ m

$\Sigma h_e = 0.090$ m $\qquad \Sigma h_w = 0.093$ m

4. Losses due to curvatures: in the high-water course which was investigated there are close to 80 curves in the stretch Orkus – Lake of Zürich with radii r_k of 0.5-1.5d and angles of deflection A° between 30°-150°. Not included are those changes in direction where the side of the angle is so short that it does not cut the line of sight of the passage. The stretch Lake of Zürich – Schleichender Brunnen is not accessible. By means of extrapolation 20 further curves can be counted. There are close to 100 curves variously situated in the cavity, which should be classified in different v_m-categories.

Table 5.10. Number of curves dependent on A°, on ξ_{kr} and on $\xi_A°$

A°	30°	60°	90°	120°	150°
$\xi_A°$	0.15	0.40	1.0	1.6	1.8
Number of curves	21	31	32	10	6
Of them $\xi_{kr} = 0.7$	13	18	20	5	3
Of them $\xi_{kr} = 1.0$	8	13	12	5	3

Table 5.11. Distribution of the curve radii in the v_m-categories

v_m (m/s)	30°		60°		90°		120°		150°	
ζ_{kr} 0.7/1.0	0.7/1.0		0.7/1.0		0.7/1.0		0.7/1.0			
a) 0.120	2	2	4	2	5	2	3	2	–	1
b) 0.040	7	4	7	6	9	7	1	2	2	1
c) 0.0125	3	1	5	4	5	2	1	1	1	1
d) curves not able to be surveyed										
e) 0.3	1	1	1	1	1	–	–	–	–	–

The calculations result according to Eq. (5.3):

$$h_k = \zeta_A \circ \zeta_{kr} \frac{v^2}{2g}.$$

There are 34 individual results but the portion of the losses due to curvature is less than 2% of the total losses. Therefore it is not necessary to reproduce them here.

$$\Sigma h_k = 0.028 \text{ m}.$$

Total losses due to friction:

$$h_R = h_r + h_e + h_w + h_k$$
$$h_R = 1.410 \text{ m} + 0.090 \text{ m} + 0.093 \text{ m} + 0.028 \text{ m} = 1.621 \text{ m}.$$

Discussions of the Results

It must be stated first that the value found is deceptive because the decimal places give the impression of greater precision than is actually the case, as will be seen in the following discussion.

The losses due to roughness make up the major part with 87.1%, losses due to changes in the cross-section make up 11.2% and those due to curvature 1.7%. Since results from comparable investigations do not exist it cannot be determined if these values are generally valid. However, it can be assumed that they are approximately correct for caves of a larger cross-section which are of phreatic origin.

A comparison of the measured sizes with the calculated ones offers some insight. At low water there is the measured 6-m loss of height of pressure opposed to the calculated 1.621 m. The extent of divergence can be explained easily by the fact that the water table in Orkus, which is accessible with difficulty, does not yet belong to the karst water surface; it lies higher. This can be tested during high water since then the pool overflows and is thereby eliminated.

When there is slight high water with $Q \approx 2$ m³/s, the water rises high through Orkus reaching the Seengang where it forms a piezometric surface. The same thing occurs in the Sandhalde where the water table lies 28 m deeper (Bögli, 1966). The drop in pressure occurs along the 1400 m from Orkus to Sandhalde; that is 53% of the total distance. The narrow spot (E) already mentioned borders on this stretch; the latter's loss due to friction is passed on to the next stretch. The loss of pressure amounts to 53% of 1.050 m, i.e., (h_r − 0.569 m) = 0.557 m at low water. This value can be recalculated for slight high water, when the amount of water and thereby also the v is 6.7 times greater. Then the losses due to friction increase to 6.7^2 times and are 45 times greater. According to this the calculated difference in height of the water table amounts to 25.1 m or 90% of the measured value of 28 m. This result lies within the margin of error.

When there is mean high water the discharge in the Schleichenden Brunnen reaches 3 m^3/s, and the water surface lies immediately behind the entrance, 96 m above the water table of the spring. Since there are no sizeable losses of water between the entrance and Sandhalde, a distance of 700 m, and the water comes to rest after the cavities are filled, the surface of the water in the entrance of the cave is the piezometric surface to Sandhalde. The 96-m elevation of the water table would, according to the Torricelli theorem, result in a v_m of 43.2 m/s in the narrow pass when the flow is without friction. This is unrealistic for a v_m of 10 m/s must be considered the upper limit under karst-hydrological conditions (see Blautopf Chap. 5.5.4). This presupposes a cross-section of 0.3 m^2. The loss of pressure in the narrow pass then is 56.9 m = 100 · 0.569 m, or 100 times the loss of pressure at low water, since v_m is 10 times greater. To this must be added the 47%-portion of the whole distance of the other losses which amounts to 49.4 m. The calculation shows a resulting total pressure loss of 106.3 m, a value which is 10.7% above the measured value of 96 m. That is within the margin of error.

This discussion of the values shows that the margin of error must be wide, about ± 15%; that the method of calculation is correct in principle, and practically applicable if there are sufficient and adequate measurements at one's disposal.

5.5.8 Cavitation

Cavitation occurs when as a result of the velocity of flow the pressure of water flowing through a tube sinks below vapor pressure. Vapor bubbles are formed which again break when the pressure increases. The water is thereby accelerated toward the concentration of the bubbles. Water particles collide creating very high pressures; Prandtl (1969, p. 475) mentions 4100 atm. The destructive effect of this process is feared in technology. Cavitation should be expected from the locally high velocities in karst. Yet we do not know of any forms which would indicate such an occurrence. Verdeil (1961) states that cavitation must evoke significant changes rapidly, but that this phenomenon is exceptionally rare, "sera excessivement rare" (p. 48). There have been no investigations of it. To be sure Hjulström (1935) assumes that scallops are an indication of cavitation but this has not yet been confirmed. Allen (1971) for his part observes: "The process has otherwise been ignored in the earth sciences." (p. 211).

5.6 Poljes as Karst-Hydrological Regulating Factors

In Chapter 3.6 it was stated that poljes are filled partly from above by water flowing in from the ground surface, partly from below by water bubbling up from independent karst water-courses under excess pressure. Swallow-holes thereby become springs, estavelles (O. Lehmann, 1932). There is a third possibility where poljes and the bordering karst show an especially close relationship and are karst-hydrologically coupled. Karst which is situated higher and just provides the polje with water but receives none, or karst which is situated lower so that its karst water-body never reaches high enough to have an influence on the polje water do not count as karst-hydrologically coupled (Bögli, 1973b).

In the catchment area of the Areuse (Jura Neuchâtelois, Switzerland) the volume of the underground hollows makes up 0.45% of the rock (oral communication from Burger), that of the Hölloch makes up 0.3% of the Schrattenkalk. In the Dinaric zone, which has been karstified for longer, or in Kentucky (Mammoth Cave National Park) it may be around 0.5%. Weidenbach (1954) determined 1.8% for the catchment area of the Dischinger Springs (Swabian Jura), but this value seemed to him to be "somewhat too high" (p. 66).

When it rains the water table rises in the rock by an amount which is equal to the amount of precipitation, divided by the portion of the volume of the hollows in the rock c. For example if c is 0.005 (0.5%) and 10 mm of rain falls then the karst water table would have to rise by 2 m — provided there is equal in- and outflow on the sides. In a polje coupled with this karst a water layer of only 1 cm would correspond to these 2 m.

The equivalent height of water W in the rock is the thickness of an underground layer of water in m, which shows a volume which is equal to that of 1 m in height in the polje. It is indirectly proportional to c and proportional to the relationship of the polje surface A_P to the surface of the karst-hydrologically coupled karst area A_K.

$$W = \frac{1}{c} \cdot \frac{A_P}{A_K}$$

W = 50 m means that water which stands (50 + 1) m high in hydrologically coupled karst fills up the horizontal polje bottom 1 m deep when it runs out.

During a rainfall the karst water table rises and the water begins to flow out of the rock as soon as the water level is higher than the water level of the polje bottom. After the precipitation is over the karst water table sinks again in the rock since water is supplied to the polje until the water levels are again similar. If it sinks further, the polje water runs out, feeding the karst water-body and keeping the karst water table on the same level until the polje is emptied. Consequently poljes have a karst-hydrological balancing effect; they are important regulating factors (Bögli, 1973b). The karst water table is thereby fixed to the polje water level for a longer or shorter period. The arrangement of horizontal caves or even of whole systems of cavities on one such level indicate such a function.

6 Karst Hydrological Zones

Underground cavities with karst-hydrological activity pervade the karstified region in a three-dimensional network. The movements of the water, especially the slow ones, are hindered only slightly so that there is a division in space, the accumulations of water below, the water-impoverished areas above. The very dense network of interstices and narrow open joints shows a behavior of its own and will only be included in this division to a limited degree.

6.1 Introduction

The classical work of Cvijić (1918) in which he drew up the karst-hydrological zones was partly the result of the controversy between Grund's groundwater theory and Katzer's cave-river theory (1903, 1909, resp.). Swinnerton (1932, 1942) reached similar results 14 years later. O. Lehmann (1932) on the other hand repudiated this theory concerning the behavior of water — wrongly, as was shown. Practical work in extensive cave systems has proved that this division into three zones is justified, but that in small caves or in caves which are situated high above the valley bottom (Eisriesenwelt, Dachstein Caves, Tantal Cave) not all three zones occur or not all three are accessible.

Cvijić (1918) divides karstified areas into

a) a dry and inactive zone with seeping water, crossed by feeders
b) the periodically flooded, upper active zone and
c) the lower active, continuously water-filled zone.

This is only valid for the complete series in deep karst (see Chap. 5.3). In shallow karst the lower active zone and in places also the occasionally flooded, upper zone are lacking today — it could be that these were present earlier and that the lack of them is a late stage of development.

Swinnerton (1932, 1942) also derived a division into three zones from the facts of the frequent and easily accessible karst regions in the table lands between the Appalachians and the Rocky Mountains in America:

a) Vadose zone, air-filled and with vertical drainage — vertical is to be understood cum grano salis because it does not mean the vertical movement of water but the downward flowing of water toward the phreatic zone.

b) High-water zone, alternatively air- and water-filled — in the presence of air, vadose conditions hold sway, when the water is high, phreatic conditions do.

c) Phreatic zone with sideways drainage and all cavities continuously water-filled, whether they are corrosively widened karst cavities or primary interstices and open joints.

Swinnerton's terms vadose (Lat. vadosus = shallow) and phreatic (Gr. phrear = well), introduced by Meinzer in 1923 correspond to the widespread English geological terms (Trowbridge, 1962). Today vadose and phreatic are used by German karst experts with these meanings, too. (German-speaking geologists call the water from precipitation in the underground vadose and use phreatic only in connection with the overheating of vadose water in contact with the magma; Murawski, 1963). On the other hand, the French school rejected the two expressions because they were out of keeping with the original meanings (Mangin, 1975).

Corrosion is of central significance for karst and karstification. From this point of view, also, the division into three is completely justified. Under vadose conditions the system $CaCO_3 - CO_2 - H_2O$ is open and striving toward equilibrium with the CO_2 of the cave's atmosphere, in the phreatic zone the system is closed because the atmosphere is lacking, while in the high-water zone the two types alternate according to the discharge.

Cvijić introduced the criterium of activity or inactivity, further characterizing the three zones thereby. A synthesis of this and Swinnerton's conception establishes a classification which does justice to both authors:
a) vadose, inactive zone
b) vadose, active or high-water zone
c) phreatic, active zone.

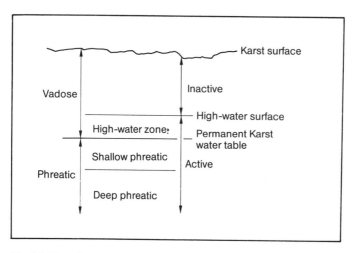

Fig. 6.1. Karst hydrological zones

When Katzer (1909), Martel (1921), O. Lehmann (1932) and others reject a karst water table they are right if they thereby understand a surface in the sense of a water table in the valley fill. The essentials concerning this are dealt with in Chapter 5.5.1. The piezometric surface consists of individual pressure tables in the piezometric tubes and they separate the phreatic and the vadose zones. These pressure tables are situated at various heights arranged around an average value. Depending on the tectonic facts, the distances between the individual surfaces are greater or smaller and in porous limestones they are so small that the surface of the phreatic zone is similar to a groundwater table in gravels and sands. Jennings (1971) reports a subhorizontal water surface in the porous limestone of the Nullarbor Plain in Australia.

6.2 Vadose Zone

The vadose zone embraces on one hand the inactive region which is crossed by feeders, and on the other hand the high-water zone. Its characteristic is the permanent or temporary presence of atmosphere. Its CO_2 content determines the equilibrial content of $CaCO_3$ in the water. When the $CaCO_3$ content is too low, CO_2 is absorbed and limestone dissolved; when there is an excess of limestone, CO_2 is released, whereby $CaCO_3$ is deposited (calc-sinter and calc-tufa).

6.2.1 Inactive Vadose Zone

Inactivity in the vadose zone means that a cavity is not extended either by erosion or by corrosion. However, this does not mean that nothing changes any more in inactive zones. The forms from the active period can be partly or entirely destroyed by breakdown (see Chap. 11), or buried under its sediments, a chaos of boulders. Seeping water which is rich in $CaCO_3$ deposits cave formations (speleothems) e.g., stalactites, stalagmites, flowstone, rimstone (see Chap. 13.3). Biogenic sediments like bat excrement, saltpeter beds, phosphates in primary beds, form only in the inactive zone (see Chap. 13.2). Where water which is low in lime content runs out, solution pockets can develop on overhanging walls and on the ceiling (Bögli, 1963a; Franke, 1963a). Condensation water creates little marks on the ceilings and when it drops off, it creates cave karren (see Chap. 12.3.1). On the rims of cave lakes which are low in lime content occasionally solution notches are formed, e.g., the dry lake in Tantal Cave (Hagengebirge, Salzburg, Austria).

6.2.2 Feeders

Feeders form active strips across the inactive zone. Their water flows downwards towards the karst water-body or towards a collecting channel (shallow karst): vertical drainage (see Chap. 5.4). At high water they are active mainly by erosion; when the amount of water carried is normal they are now more erosively active, now rather corrosively, depending on conditions. Water which is rich in lime cuts in only by erosion, water low in lime cuts in practically only by corrosion when the incline is slight. Between the two extremes there are all kinds of transitions.

6.2.3 High-Water Zone

Whenever there is air in the cavities in the high-water zone vadose conditions are at work. During high water, when the whole cross-section of the passage is filled with water, only phreatic conditions are active. When the water is low, vadose forms arise which may be from the inactive zone or from the feeder zone where canyons and other forms of gorge-like passages give evidence of vadose origins (see Chap. 12.2.3). When the water is high it corrodes and erodes in the high-water zone just as it does in the phreatic zone. The rock's surface shows solution forms; on the floor of the passage scallops and on the ceiling flat hollows occur which are characteristic of a water-filled passage. Pockets caused by corrosion by mixing water form at the exit of interstices from which water emerges; such hollows can be a few centimeters to many decimeters or meters wide and deep. In the same pockets stalactites are occasionally deposited when, at low water, air has taken the place of the water. In ascendent piezometric tubes fine-grained sediments are deposited in the almost still water (see Chap. 13.1). Vadose and phreatic forms mix, overlay one another. Stalagmites stand on clay and may be partially covered again by fluvial sediments or may also be tipped over by the undercutting action of running water at their base. An analysis of the forms offers a relative chronology which often extends from the final phase of formation of the cave to its ultimate inactivation (see Chap. 13.4 and Fig. 13.10). The high-water and the following shallow phreatic zones are the areas of the most intensive cavity-forming processes.

6.3 Phreatic Zone

Only in deep karst are the karst-hydrological zones complete and consequentially the phreatic zone occurs only there. It is frequently lacking in mountainous landscapes — or is not always recognized.

According to Grund (1903) the water-filled zone contains groundwater which has the sea as its hydrological and erosive base level. From here the groundwater surface gradually climbs toward the interior of the land. The groundwater scarcely moves or it stagnates completely. Above it lies a zone of karst water in which the water moves sideways to the Vorflut, in this case to the sea. Already in 1894 Martel was an exponent of the theory of cave rivers according to which water in the depths also moves forward as a cave river. As proof of this theory there is a submarine karst spring at Cape St. Martin. But he did not know of the eustatic movements of the sea-level. However, Katzer (1909) is recognized as the representative of the cave-river theory. His conception of the phreatic zone is completely different from that of Grund.

The cave-river theory so completely supplanted the groundwater theory among German-speaking karst morphologists that use of the term groundwater for karst has been taboo ever since; today it is still partly so. In other languages all phreatic water is called groundwater (see Chap. 7).

In contrast to the multiplicity of European karst theories there is a unity of American theory which is based completely on groundwater.

Fig. 6.2 a, b. A comparison of the groundwater (a) and the cave-river (b) theories. *Upper dashed line* (a) karst water-level, (b) high-water level. *Lower dashed line* (a) groundwater-surface, (b) limit of the phreatic zone

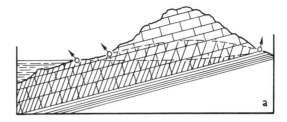
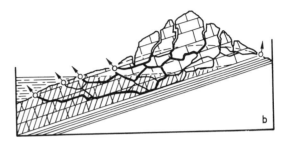

However, it is not the scarcely moving groundwater envisaged by Grund, but a body of karst water in motion; it moves partly as a whole, yet partly in single streams as described by Katzer.

Davis (1930) advanced by deduction his classical theory of cave formation based on his wealth of geomorphological experience and on the application of his theory of cycles. It was the beginning of actual scientific speleogenesis. When the *Origin of Limestone Caverns* appeared, Davis, at 80 years of age, was too old to be able to verify his deductions in nature, or to let them be substantiated in controversial discussions. Therefore this publication found a weaker echo than it deserved. In this treatise groundwater is the central problem. From cave development Davis concluded that groundwater moved slowly, circulating right down into the far depths. The farther the point where the water seeps in from the place where it emerges, the deeper the water-courses run. Piper (1932) came to the conclusion that the water's movement through the depths must happen very quickly if the water was not to lose its aggressiveness already in the uppermost phreatic zone — corrosion by mixing water was not yet known at that time. Oil-drilling in Florida ran into large caves 2000 m below the piezometric surface (Jordan, 1950), one of the many proofs of the correctness of Davis' theory.

Swinnerton (1932) leaves many ways open for the water to move from the place of seepage to the point where it emerges. The one along the shallow phreatic zone offers the shortest way with the least resistance and is therefore preferred. This hypothesis gives a simple explanation of the frequent occurrence of horizontal passages of caves in the Paleozoic plateau of the USA. It should, however, also explain the rarity of deeply situated cavities. How correct this explanation is must remain to be seen, for their rarity could well be only the result of inaccessibility, since drilling activities resemble but a few pricks of a needle over a huge surface. This is the reason why this hypothesis has been contradicted. Dealing with Swinnerton's hypothesis, Hubbert (1940) fell back on Davis' theory when he came to the conclusion that the water must move more deeply through the phreatic zone. But it was Bretz (1942) who made Davis' theory really popular.

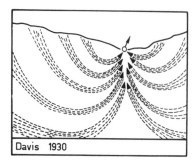 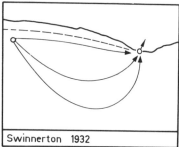

Fig. 6.3. Comparison of the movements of water according to the theories of Davis (1930) and of Swinnerton (1932)

In 1968 Thrailkill took up the problem of groundwater and compared the behavior of water in granularly porous aquifer, e.g., gravel, and in compact limestone which is full of closed and open joints. In the gravel the groundwater is limited by the barrier of permeability, whether it is a piezometric surface or an impermeable stratum. The cross-section of the pores and holes is smaller than that of the exit of the spring. It is different in limestone: the cross-sections of karst-hydrologically active cavities are as a rule larger than the points where the water emerges.

Thrailkill (1968) questions what is to be termed aquifer in karstifiable rock. He sees three possibilities:

1. It is common to regard the entire phreatic zone as an aquifer. Of course it is characteristic that one hole drilled in karst will strike water while another only a few meters away will remain dry.

2. An aquifer could, however, be defined only by the water-filled cavities which run through the rock. To consider it from this point of view is only useful for calculations when the outlining forms are very simple and unbranched. Otherwise the arrangement of the equipotential surfaces becomes complicated and difficult to comprehend (Fig. 6.4a).

3. The circumferential surface of the total of the karst-hydrologically active cavities can, however, also be seen as an envelope enclosing this region like a permeability barrier. Then, when the water flows very slowly, the coarser rock pores are also included. The water-carrier is thereby similar to a very coarse gravel. The equipotential surfaces turn out to be correspondingly simple, and the karst water-body can more easily be approached by calculations (Fig. 6.4b).

In order to clarify the controversy between the theories of Davis and Swinnerton concerning the paths of flow of the water in the phreatic zone, Thrailkill (1968) calculated a model. It consisted of three pipes lying one above the other and connecting the point of inflow with the point of outflow (Fig. 6.5). That would correspond to a natural system of cavities where the cross-section and the average head-loss were the same. The flow volume through each branch would then be dependent on the length of the pipe. The depth T and the distance between the swallow-point S and the karst spring K result in the length of flow f in a square cavity network:

$$f = 2T + \overline{SK}.$$

Phreatic Zone

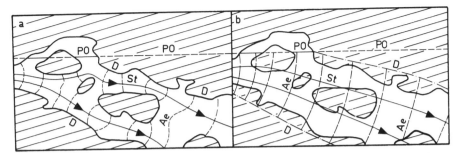

Fig. 6.4 a, b. The aquifer in karstified rock according to Thrailkill (1968); *Ae* equipotential surfaces; *D* permeability barrier; *PO* piezometric surface; *St* flowline. **a** the cavities as aquifer. **b** the zone of the cavities as aquifer

Examples according to Thrailkill (p. 26 f).

a

	Laminar	
↓ 1.0000	→ 0.3388	1.0000 ↑
↓ 0.6611	→ 0.3322	1.6611 ↑
↓ 0.3289	→ 0.3289	0.3289 ↑

	Turbulent	
↓ 1.0000	→ 0.3382	1.0000 ↑
↓ 0.6618	→ 0.3317	0.6618 ↑
↓ 0.3300	→ 0.3300	0.3300 ↑

Fig. 6.5 a (10x vertically exaggerated). When T = 0.005 $\overline{SK}$, 0.010 $\overline{SK}$ and 0.015 $\overline{SK}$ is f = 1.01 $\overline{SK}$, 1.02 $\overline{SK}$ and 1.03 $\overline{SK}$. The shares the three tubes have of the amount flowing through are 33.9, 33.2, and 32.9% when the flow is laminar and 33.8, 33.2, and 33% when the flow is turbulent. **b** When T = 1/2 $\overline{SK}$, 1 $\overline{SK}$ and 1 1/2 $\overline{SK}$ is f = 2 $\overline{SK}$, 3 $\overline{SK}$ and 4 $\overline{SK}$. The lengths of flow are 150%-200% in the shortest connection. The proportions of the flowing water carried are 62.5%, 25% and 12.5% for a laminar, 54%, 27% and 19% for a turbulent flow

b

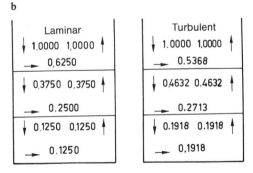

From these two calculations it becomes clear that when the input is close to the karst spring the water favors the use of the shallow paths; when the input is far from the karst spring, however, there is no significantly greater discharge to be expected in the shallow phreatic zone than in the deep phreatic zone. This corresponds completely with the theories of Davis (1930), of Hubbert (1940) as well as of Rhoades and Sinacori (1941).

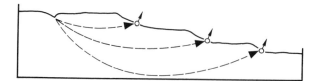

Fig. 6.6. Water-courses from one swallow-hole to various springs

The further development of underground water-courses must, however, also be taken into consideration. The heavier the flow, the more intensively the cavities are widened. Thus when the distances $\overline{SK}$ are short, those lying shallow are preferred, attracting more and more water. From this a manner of flow results which approaches Swinnerton's hypothesis (1932). When the distance $\overline{SK}$ is large this effect is not to be expected since the slight differences in the proportion of the flow are greatly outweighed by the effects of lithological and tectonic differences. But there is still another effect. The water-courses which develop in the shallow phreatic zone are favored by corrosion (aggressive water, corrosion by mixing water) and widened to a greater degree than those in the deep phreatic zone. It can even come to the development of underground karst levels (see Chap. 8). The prerequisites for the calculation, namely equality of all conditions with the exception of the length, thus prove to be valid for only a short time. Thrailkill nevertheless showed that water-courses which reach down into the depths are also entirely possible and probable, particularly since corrosion by mixing water is possible at any, even at a very great depth.

A deduction of the discharge of water-courses of various lengths is applicable horizontally as well as vertically. Since the same additional conditions of corrosion are valid here for all paths of flow, only the differences in length become effective, all other conditions being equal. Length factors of 1.01 or 1.03 have scarcely any effect. On the other hand, values between 1.1 and up to more than 2.0 are not rare on the horizontal plane (see Chap. 5.5.7). Local variations, narrow passes caused by incasion or sediments for example, are decidedly noticeable, so that the dominance of one single connection, which is not always the shortest, can result.

The American school's knowledge and experience can be applied in Europe to areas of deep karst with a comparable geomorphological and tectonic basis, that is, to peneplains with approximately horizontal bedding, like certain parts of Ireland, southern England, or France (Parisian Basin, northern side of the Aquitainian Basin). In the deep karst of the Swabian Jura German hydrogeologists recognized the existence of karst groundwater already at the beginning of this century. In 1960 Weidenbach gathered together his experience and wrote: "that there are areas in the Jurassic of the Swabian Alb in which there are no karst cavities but at best open joints in the limestone rock ... There are then no continuously connected karst water reservoirs consisting of a network of karst caves; rather there are several karst water reservoirs beside one another, each with its own karst mechanism" (p. 176 f.). But it would be incorrect to leave out the following sentence: "On the other hand there is a karst water table which is continuous throughout the whole Swabian Alb" (p. 177). The apparent contradiction in these sentences is resolved as soon as it is recognized that the separating limestone areas, which are nevertheless crossed by open joints, are difficult for flowing water to pass through, but provide no hindrance at all to hydrostatic pressure.

In orogenic areas other conditions hold sway. Karstified regions are cut by deep valleys or border on such. These also become the base level wherever the limestone plunges beneath the valley bottom. Moreover in synclines and faults bodies of karst water occur in elevated positions. The piezometric surfaces lie deep under the earth's surface, their situation is decidedly the opposite of the shallow position of the piezometric surfaces beneath the peneplains of North America.

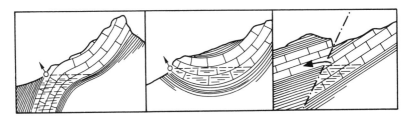

Fig. 6.7. Three types of karst water reservoir in the Alps

There are also large karst water-bodies in the Alps, as proved by the investigations of Bauer et al. (1958), of Maurin and of Zötl in the karst massives of the Austrian limestone Alps. Maurin and Zötl (1964) determined by means of lycopodium spores, which were fed into the swallow-holes at Elmsee (Lake in the Toten Gebirge), a connection with 18 springs round about Elmsee. The greatest distance measured 10 km, the shortest 4 km.

In the Dachstein a swallow-hole in Miesboden was fed by Zötl (1957) with red-colored lycopodium spores. They could be identified round about the input in 14 springs at distances between 3.2 and 6.5 km and between 150 and 720 m below the swallow-hole. Two weeks earlier a test was made in a swallow-hole near the cabin Maisenberg which was positive in 11 springs in an angle of approx. 200° and at distances of 5.5-13 km. The height of fall measured 700-1330 m, the v_m was 100 m/h.

Zötl (1960, 1961) came to the conclusion (1961, p. 129) that the results of these investigations gave evidence of "a large connected body of karst water" in the area tested. Zötl rightly doubts that this can be brought into harmony with the hypothesis which has been valid up to now that "karst cavities are independent and have no relation to one another" (O. Lehmann, 1932, p. 15). He stated that according to Stini (1951, p. 230) the number and width of the joints decrease downwards. "This observation reveals the basic fact that within the karst massive there is in general a dense network in the upper regions with smaller open joints reaching much deeper, even below the base level" (Zötl, 1961, p. 130).

The groundwater theories of Davis, Swinnerton, and Thrailkill can be applied analogously to such karst water-bodies, if in Alpine regions the decidedly deeper situation of the piezometric surface and the higher linear flow velocities are taken into consideration. The phreatic zone can rarely be observed directly. In this respect Hölloch (Switzerland), where the contact between the phreatic and the vadose zones can be observed at various points along a 9-km stretch, is an exception for the Alps as is also Lamprechtsofen Cave on the Saalach. In Hölloch the body of karst water is elongated. This

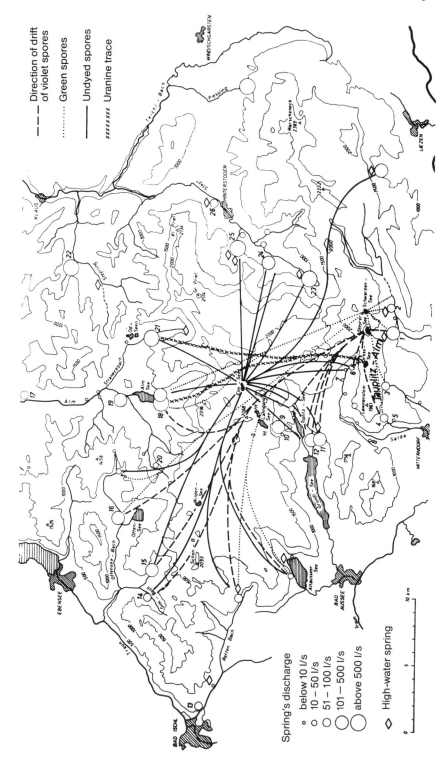

Fig. 6.8. The results of water tracing in the Toten Gebirge (Zötl, 1964, 1974)

Phreatic Zone 109

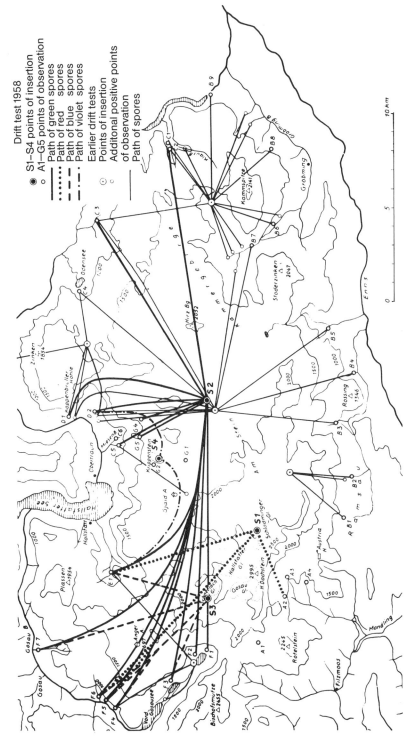

Fig. 6.9. Drift of spores in the Dachstein July/August 1956 (acc. to Zötl, 1957)

gives the impression of a cave river flowing along below the karst water table; it hardly resembles karst groundwater any more. Only the ascending passages (piezometric tubes) give evidence of quiet water. Nevertheless the theory of independent karst water-courses must undergo the correction here too, namely that these show karst-hydrologically active connections. This is not incompatible with the theory, but merely a question of the pressure differences in these connections. If Δp is small or even zero, the corresponding speed of flow is also small or zero.

If a karst water-body is large and deep enough, a shallow phreatic and a deep phreatic part will always be distinguishable, the former with a stronger, the latter with a slighter movement of water. This is only generally valid for there are numerous exceptions. In America 20 m is stated as an average for the thickness of the shallow phreatic zone. It can be very much thicker locally, as shown by the many vauclusian springs which rise up from greater depths. For Silver Spring in Florida, Jordan (1950) mentions the exit of the water into the spring's funnel 60 m below its surface. Certainly eustatic variations play an important role since the present-day shallow phreatic zone was vadose in the cold periods. Bögli (1966) established for Hölloch a former shallow phreatic zone of some 50 m in thickness in which the passages undulate up and down. Ford arrived at similar results for the Mendip Hills in SW England (1965, 1968, 1971).

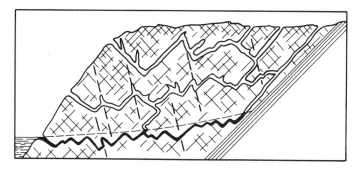

Fig. 6.10. The shallow phreatic water-course according to a scheme by D. Ford (1968)

There hardly seems to be a limit to the depth of the deep phreatic zone, unless it is the lower limit of karstifiable rocks or the transition to the zone of closed joints. In other words, thanks to corrosion by mixing waters, the processes of karstification must still be possible at great depths as long as there are open joints and sufficiently wide interstices. Oil drilling in Florida often strikes large water-filled cavities, among them some as high as 13 m and some situated as deep as 2033 m beneath the piezometric surface (Jordan, 1950, p. 264). Gádoros (1969) observed a similar situation in the Hungarian lowland where, at a depth of 1400-1700 m in the limestone below Budapest, abundant thermal water was drilled; Lang (1969) mentioned such cavities at a depth of 2000 m in the same region. Oil drilling on the peninsula of Hicacos in Cuba struck a karst cave at a depth of 2952 m (Jiménez, 1976).

7 Karst Water – Groundwater

In the previous chapters there has again and again been a noticeable contrast between karst water and groundwater. This is a peculiarity of the German language but is nevertheless of general significance. It is a consequence of Grund's theory of karst groundwater (1903) (see Chap. 6.2.3). Hence an investigation will be made to see whether this is really a question of contrasts. It is characteristic that many karst hydrogeologists in Germany who are concerned with useful water reserves in karst (Weidenbach, 1960; Eissele 1963; Groschopf, 1963) speak of groundwater, but that karst researchers reject this expression. Wagner (1960) writes: "Karst water is a special kind of groundwater" (p. 66).

7.1 Introduction

Keilhack (1917) uses groundwater to embrace all deep water. But already in 1925 Ule writes: ". . . water in loose soils indeed occupies such a special position that we . . . do not wish to apply the term groundwater also to the water in the open joints of rocks" (p. 3). At the congress of the International Association for Scientific Hydrology held in Washington in 1939 groundwater was defined as "water which fills the cavities of the earth continuously and is only subject to gravity and hydraulic pressure. . . . The term groundwater does not depend on whether the concerned parts of the earth's crust are loose or firm, whether they are weathered or unweathered, whether they lie close beneath the earth's surface or at greater depths. Cavities which contain groundwater may be of very different sizes . . . , no upward limit has been determined for the size of a cavity." (Translation from the German edition, Giessler, 1957, p. 30).

In DIN 4049/5.02 groundwater is similarly defined. Pores, joints, and caves count as cavities (DIN 4049/5.31). Underground portions of water-courses on the surface are excepted from the term groundwater. In case of doubt the name cave water or karst water is to be used.

Keller (1962, p. 239) makes the following critical comment on the above: "In limestone and gypsum water behaves quite differently than in other rocks." In Austria there is even the rule: "Water in open joints and joints (joint water) . . . shall not be called groundwater" (Oenorm B2400, quoted by Thurner, 1967, p. 3). This definition shows unmistakably the influence of the karst landscape frequent in Austria, which deeply preoccupies hydrogeologists. It may be mentioned that the water supply of Vienna draws large amounts of karst water out of the Raxalpe, Hochschwab and the Schneealpe (Baur and Zötl, 1972; Zötl, 1974). On the other hand Thurner (1967) adds to the

Austrian definition: "From the hydrological point of view there exists no principal difference, for pores are just as much cavities as are joints, open joints, etc."

Practical and legal requirements are fundamental in the DIN definitions; these do not always satisfy scientific knowledge completely. Some of the conditions for karst groundwater which, compared with the DIN definition, are more restrictive are:

1. The coherent groundwater surface is missing. In its place there are individual water levels at greater distances from one another which cannot be integrated into one geometrically comprehensible surface.

2. The karst water-body does not move forward as a whole, but in individual streams (tube-flow). In spite of their connections they are frequently hydrologically independent of one another.

3. As a result of tube-flow no uniform potential surfaces are to be found in the phreatic zone; there are individual ones.

4. The diameter of karst-hydrologically active cavities can measure from a few millimeters to tens and hundreds of meters. For this reason also the v_m in karst groundwater is higher to the power of ten than groundwater in pores and joints when the pressure gradient is the same.

5. Similar differences in size are also prevalent in networks of cavities.

In conclusion it can be said that the term groundwater means water as defined by the International Association for Scientific Hydrology, or DIN 4049/5.02. It is divided into three different types according to the size of the pores, or cavities, and their interrelationships as well as according to the resultant manner of movement of the water.

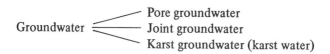

7.2 Underground Water

The main characteristic of a karst landscape is underground drainage in a karstifiable rock. Other types of landscape show underground drainage, also, such as when streams ooze away in valley-fill. Thurner (1967) emphasizes that water in loose rock is not karst water.

Precipitation seeping away crosses the zone of seeping water and reaches the zone of water saturation, the groundwater, from which it again emerges at an appropriate place as a spring. There is a high degree of water accumulation in gravels and sands but a low degree of it in hard rocks, on the other hand, when the reservoir consists of joints; in karst there are great variations.

In the initial phase of karstification karstifiable rocks do not behave very differently from insoluble, hard rocks. They are as such impermeable, and only by means of the joints do they become permeable. Limestone and dolomite are brittle and form tight networks of joints after being exposed to only slight stress. The case with unkarstifiable hard rocks is similar. This network of joints favors the penetration of water and joint groundwater is formed first; this can be equated with the phreatic zone. It extends to very close to the earth's surface. When the interstices are widened, karst hydrological activity sets in and the upper limit of the water-saturated zone sinks. The latter finally comes under the direct influence of the base level, or shallow karst forms with its watercourses on the impermeable underlying bed. The zone of seeping water has reached its greatest thickness. It corresponds to the vadose zone.

The seeping water gathers in underground water-courses which are called *autochthonous*, since their water originates in the karst area itself. Water which comes from outside the karst area is called *allochthonous*. Such is carried into the karst area by surface water systems such as the Reka and the Pivka in Slovenia. Usually they disappear into the underground on reaching the karst or shortly afterwards. Because the loss of water is gradual, this may occur almost unnoticed until the river bed is left dry, e.g., the outflow of the Malham Tarn in Yorkshire (Sweeting, 1972) and the seeping away of the Danube. More rarely allochthonous water enters the underground through a spectacular cave gate, e.g., the Reka entering the Mahorčiceva Jama (Mahorčič Cave), the front cave of the Škocjanske Jame (Slovenia). There can be no doubt that the Reka reaches the phreatic zone and persists in it for the long stretch up to the mouth where it flows out as the Timavo at Duino south of Monfalcone. Nevertheless it keeps its individuality as proved by experiments using various means of water-tracing. One of the best-known examples is the seeping away of the allochthonous Danube between Immendingen and Friedingen. The water disappears in narrow open joints in the Malm limestone and flows underground along various paths to the Aach spring which is 11.7-18.3 km distant, depending on the point of seepage. The speeds of flow measured are between 304 and 422 m/h (Batsche et al., 1970).

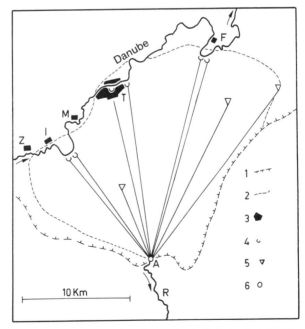

Fig. 7.1. Sketch of the situation of the Danube's seepage points (according to Batsche et al. 1970) *A* Aachtopf; *F* Friedingen; *I* Immendingen; *M* Möhringen; *R* Radolfzell's Aache; *T* Tuttlingen; *Z* Zimmern. *1* karst boundary; *2* limit of catchment area of Aach Spring; *3* towns; *4* swallow-points; *5* dolines; *6* Aachtopf (resurgence)

A significant portion of the underground Danube flows under phreatic conditions; this was corroborated by diving experiments in the Aach Pot. During such an experiment a vadose cave system was reached at a rather great distance from the spring; this was an earlier phase in the development of the Danube's underground course.

7.3 "Karst Barré"

The French term karst barré refers to a small area of karst which is surrounded by impermeable rock. There is no comparable term for this phenomenon in English. The water is dammed up in the karst, which thereby belongs mainly to the phreatic zone. The deep karstification is proof of the validity of Davis' (1930) and Thrailkill's (1968) theories concerning water-courses running through the depths.

Corbel (1957) describes the characteristic example of the quarry of Harehope (County of Durham, North Pennines, England). Limestone is separated from its surroundings by a fissure intrusion (Sp). An allochthonous river used to cross the terrain until recently without seeping through. Work in the quarry removed the barrier. Shortly afterward water broke out 30 m below the surface while at the same time the river dried up. This is only possible if karst-hydrologically active cavities were already in existence at this time. Simultaneously cave passages were discovered which had become dry. The fact that they had not been completely filled in gives evidence of flowing karst water at this depth during the phreatic phase.

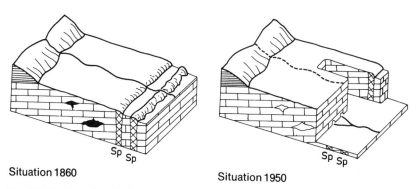

Situation 1860   Situation 1950

Fig. 7.2. The karst barré of Harehope (acc. to Corbel, 1957, p. 285)

In the karst of Montpellier (southern France) faulting created dammed-up karst areas. This karst barré is a typical example and has been dealt with numerous times (Dubois, 1961, 1962, 1964; Avias and Dubois, 1963).

7.4 Blocked Karst

Blocked karst is a common occurrence in periglacial regions with permafrost. In the cold periods of the Pleistocene the water froze in well-developed karst and karst hydrological activity was blocked. In the place of underground drainage there was surface drainage which, aided by frost weathering, formed valleys foreign to karst in karst that was previously water-free; and this in a relatively short time. A good example of this is the dry-valley karst of the Swabian Alb. "Subcutaneous caves" (Ciry, 1959) were formed above the permafrost on terrain with a southern exposure during thawing.

Blocked karst is widespread in central and eastern Siberia. Popov et al. (1972) set the thickness of the permafrost at 20-50 m in the south, at 300-600 m in the north. The maximum depth at which there is thawing is as low as 2.5 m in the south, in the north it is only 1.2 m. The karst springs are small and in the region of Rybinsk they show around 5 l/s, in Angara-Lena Trough up to 100 l/s. This is an indication of the low intensity actual karstifying processes have under permafrost conditions. The development of underground cavities is correspondingly slow and, also only takes place when an actual circulation of water can set in.

Under the permafrost there can occur extensive, slow convection flows. Warm water from the depths rises into the higher regions and is cooled off as it approaches the permafrost, whereby corrosion due to cooling takes place. When the cold water parts sink and become warmer the result should be limestone depositing. Sufficient observations have not yet been made of such processes and their effects.

8 Underground Karst Levels

8.1 Introduction

In Europe underground karst levels were a subject of controversy for a long time. Today their existence is assured although in a few single cases it is still disputed. Swinnerton's conception (1932) of the predominance of underground water-flow through the shallow phreatic zone supplies one of the causes of the formation of karst levels (see Chap. 6.3). It is this which explains the horizontal passages which stretch for kilometers in the Flint Mammoth Cave System in Kentucky (Gardner, 1935). Deike (1967), Miotke and Palmer (1972) have proved the relationship between this cave system and the surface morphology. The chronological identification of the underground karst level with systems of terraces on the surface in the Alps did not meet with approval from all sides. Krieg (1954, 1955) was in support of such levels and used Bock's cave-river theory (1913) as evidence for them. This aroused intense debate in which Arnberger (1955) and Trimmel (1955) rejected both the cave-river theory and the levels as Krieg meant them, as well. Schauberger (1955) compiled the heights of all cave entrances of the greater limestone massives east of the Salzach and thereby confirmed a storeyed structure which was, however, attributed by others to petrographic causes. On the other hand Roglić (1960), who illuminated the relationship of river erosion to the karst process, failed to make any reference to a relationship between the phases of valley formation and underground karst levels. Droppa (1957) has convincingly proved that the storeyed structure is a result of phases in which the valley was deepened in the case of the Demänovské Jaskyne (Demänova Caves) in Slovakia. Bögli (1966) shows the occurrence of cave levels in Hölloch and their relationship with the correlated phases of the Muota Valley's deepening; each deeper phase acted as a base level. Zötl (1958) sums it up as follows: "The question concerning the relationship between levels of erosion and the development of a karst water system is a central problem of karst hydrology" (p. 125).

The underground karst level is directly accessible only in caves, which is the reason why one usually speaks of *cave levels*. This comprises the sum of all karst cavities which have been created by the influence of a given base level and which therefore form a genetic and chronological unity. They can only be identified with certainty in vast caves since smaller ones are only able to provide sections of random heights.

There are two types of cave level:

a) that which is connected to the piezometric surface and oriented towards the base level, called the evolution level by Sawicki (1909), and

b) that which has been created by the underground part of a valley river and is homologous to the systems of terrasses on the earth's surface.

8.2 The Cave Level of the Piezometric-Surface Type — Evolution Level

The first type of underground karst level develops in "deep karst", more precisely in the close proximity of the karst water surface. The latter is controlled primarily by the base level, the overflow spring or, in the underground, by the overflow of water over an impermeable threshold, under consideration of all those factors which cause the height of the karst water table to vary, above all the velocity of flow (or the cross-section of the cavity carrying the water), the arrangement of the narrow passes and the discharge (see Chap. 5.5, 6.3, Fig. 5.11).

Sawicki introduced the term *evolution level* in 1909; it was, however, forgotten during the clash of opinions concerning the cave-river and groundwater theories, although evolution characterizes the essence of an underground karst level strikingly.

Corrosion and erosion form the cavities of the evolution level. At the beginning, when the system of cavities which will produce the cave level is being created, there is only corrosive action. When the width is sufficient first of all the v_m increases greatly, and in addition erosion sets in. At high water erosion may even predominate. If during further development the cross-section of the passage exceeds a certain size — this is determined by local factors and is different for every cave — then the v_m sinks again. Erosion decreases proportionally.

In the shallow phreatic zone there are four possibilities of corrosion. In summer corrosion due to cooling can take place when warmer water flows into the body of karst water; since the difference in temperature is usually slight, there is a correspondingly slight effect. Aggressive water is possible along larger water-courses — larger streams and rivers. At high water unsaturated water can still occur; this works corrosively when in contact with the cave's atmosphere. The form of corrosion which is probably most active is that which results from mixing water. Every surge of high water causes the lime content to sink. In Hölloch (Switzerland) water from melted snow carries only 60 ppm of lime and mixes with the phreatic water with 120 ppm, whereby 3 ppm of limestone are dissolved. That may seem like little. However, joint water with up to 190 ppm is flowing in, indeed in only small amounts, yet constantly. Thus only since the beginning of the Post-glacial Age there have been considerable cavities formed corrosively at the level of the piezometric surface. These figures can be applied to other alpine caves if need be, but never to those with other basic conditions, e.g., to such as are in the "green karst" of the Jura, of southern England or of the midwest of the USA. But valid for all is the fact that at the surface of the phreatic zone, and thus at the evolution level, especially favorable conditions for corrosion dominate.

Because of the pressure flow and local geological conditions the water-courses do not as a rule follow the shortest connection to the karst spring but rather, always following the head, they pursue the shortest hydrological connection. For this reason these courses weave around a median position, both sideways and vertically, and they form an evolution level. This level extends on the one side into the phreatic zone, on the other into the high-water zone; at each high water the vadose conditions are succeeded by phreatic ones and corrosion increases considerably. The number of days with phreatic conditions is the smaller, the closer the place is situated to the upper limit of the high-water zone. In Hölloch the point +180 m above the phreatic zone has been reached only once in the last 25 years, the point +160 m at Bivouac 4 and Bivouac 3 is under water for a few hours every few years, while point +30 m remains under water for months every year. The main passage weaves around this point with differences in height of 30 m when compared to its medium height. Under peneplains, for example in the Flint Mammoth Cave System, the high-water zone is only a few meters thick. There the amplitude of the water-courses in the heights are so small that they are usually situated far below the clearance of the bigger passages and are therefore hardly recognizable. The passages of a level thus normally appear horizontal, except for piles of debris resulting from breakdown or for the formation of sinter.

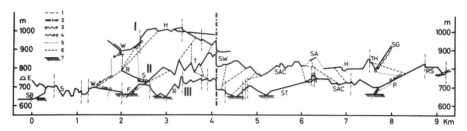

Fig. 8.1. Development of the three main levels in Hölloch (without the high system) according to Bögli (1966). *1* fault; *2* main fault; *3* cave spring with stream; *4* cave stream with sink; *5* connection proved by tracing; *6* passages of a lower order; *7* local karst water-table. *Level I: W* Wasserdom; *H* Himmelsgang. *Level II: R* Riesensaal; *S* Styx; *T* Titanengang; *SW* Seilwand; *SAC* anterior SAC passage; *SA* Sandgang; *H* Hoffnungsgang; *TH* Trughalle; *SG* Schluchtgang. *Level III: E* entrance; *SB* Schleichender Brunnen; *S* Sandhalde; *W* Wegscheide; *F* Fjord; *R* Rabengang; *ST* Schuttunnel; *SAC* rear SAC passage; *P* Pagodengang; *RS* Reinacherstollen

Hölloch shows various levels; in the main section there are three. They were created in connection with the deepening of the Muotatal. Level I with the Himmelsgang region correlates with the preglacial valley system of the Muota. Level II, which can be followed for 6 km, climbs from W to E from 780 to 830 m above sea level. At the west end on the surface at 750 m above sea level lies the remainder of the glacially polished valley bottom of the first interglacial period. In the large interglacial period level III was created. It can be followed for over 9 km (Bögli, 1966, 1968a).

8.3 Cave Levels of the River-Bed Type

When a valley bottom is deepened the water in the geomorphologically hard limestone sometimes finds its way more easily through the rock. Such is the case with the Demänova Caves (Droppa, 1957). These caves are situated in the Demänova Valley on the north side of the Lower Tatra near Liptovsky Mikuláš. At present, part of the river seeps in on the southern boundary of the limestone zone and flows parallel to the course the valley takes, through the Demänova Caves to the spring 4 km further below. The same thing already happened in earlier phases. Droppa counted nine levels. The characteristics of this type are close grouping and the consequently one-sided gradient of the passages which do not weave around a median height as is the case with the evolution type.

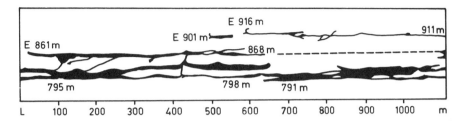

Fig. 8.2. Excerpt from the longitudinal section of the Demänova Caves (acc. to Droppa, 1957). At the lowest level (791 m) the underground valley river flows. *E* entrance; *L* length

Postojnska Jama also shows a storeyed structure caused by the Pivka. In the lower storey the Pivka flows, the higher one is inactive and contains the commercial cave. Here, too, one is struck by the practically horizontal arrangement of the passages (Šerko and Mischler, 1958). Ek (1961) describes a few examples of the riverbed type in the catchment area of the Ourthe, among them the grotto of Remouchamps.

9 Karst Springs

9.1 Introduction

Karst springs are water outlets from karst-hydrologically active cavities in water-soluble rocks, whether they are on the surface or within the earth (cave springs). There are scarcely any other characteristics which apply to them alone. The same lime contents, the same amounts of discharge and the same temperatures can be found in other springs as well. Springs emerging from rocks which are permeable, but practically insoluble and therefore nonkarstifiable (volcanic tuffs, lava, etc.) are exempted from this definition.

In Westermann's *Encyclopedia of Geography* (1970, p. 948) stands: "Cave (karst) springs: springs often with a very strong discharge, in karstified regions." Schmidt (1923) writes: "Karst springs originate when the underground water of karst regions reaches the surface. Karst springs are usually large and can immediately turn millwheels." Karst springs can be large, but there are many small ones. Cave springs do not always emerge on the surface. Therefore these definitions are no longer satisfactory.

O. Lehmann (1932) spoke of the "karst-hydrological contrast" when referring to the contrast between the innumerable seepage points in karst for precipitation and melted snow, and the comparatively few karst springs in the karst area and along its margins. Out of the profusion of interstices originally available, selective corrosion creates a network of water-courses. More and more of these become inactive so that, as a result, there is a concentration of water in only a few of them.

In the early phases of an underground karstified area only marginal parts of the karst water-body are oriented toward a spring; the main part remains undifferentiated as a central karst water-body which passes its water on to the marginal sections (see also Fig. 6.8 and 6.9). With increasing karst-hydrological activity the catchment area of individual springs reaches deeper and deeper and the more efficient tap the others. Thus the smaller karst springs are gradually eliminated. The more advanced underground karstification is, the smaller the number of springs and the larger the average discharge from them.

If the springs are fed from the phreatic zone, the discharge of the lower springs increases at the cost of the higher ones. The faster outflow causes the karst water level to sink and the higher springs are put out of action, one after the other. The catchment area of Hölloch comprises 22 km^2, and when the water level is normal this water emerges only in the Schleichender Brunnen.

Another special aspect of the karst-hydrological contrast is that a stream entering the underground frequently flows out again from very different springs situated a great distance from one another, e.g., in the Dachstein and in the Toten Gebirge (Fig. 6.8 and

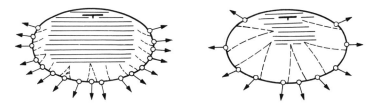

Fig. 9.1. Catchment areas of karst springs in an early and in a late stage of development (*hatched* undifferentiated karst water-body; *dashed lines* underground watershed)

6.9). There is a similar situation with the swallow-holes on the Glattalp (Muotatal, Switzerland; Bögli, 1960b; cf. Fig. 5.2).

9.2 Classification of Karst Springs

Karst springs behave so differently that the general principles of classification for all springs can be applied to them with a few exceptions.

1. Classification according to the outflow.

    a) perennial springs
    b) periodic springs
    c) rhythmically flowing springs, so-called intermittent springs, ebb and flow springs
    d) episodically flowing springs, e.g., Hungerbrunnen.

2. Classification according to geologic and tectonic conditions

    a) *bedding springs*
        aa) contact springs at the contact of an underlying, impermeable bed and a capping, permeable one
        ab) springs on bedding joints inside the permeable rock; usually small
    b) *fracture springs* out of open (widened) joints
    a) and b) are both referred to as *descending springs*
    c) *overflow springs* drain a phreatic zone with an impermeable base which sinks in toward the mountain
    d) *ascending springs;* when the outflow is heavy these are also called vauclusian springs after the original of the type: Fontaine de Vaucluse east of Avignon
        da) ascending rock springs which drain the phreatic zone along open joints or ascending strata
        db) alluvial karst springs drain a phreatic zone in the karst across a valley-fill, e.g., in the Alps in the glacially over-deepened valleys (Fig. 9.7)

Fig. 9.2. Contact spring

Fig. 9.3. Spring on bedding joint

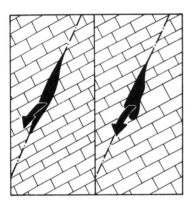

Fig. 9.4. Fracture springs

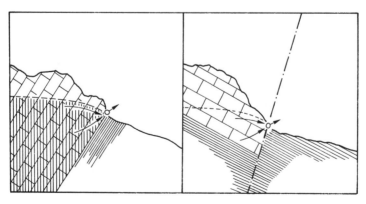

Fig. 9.5. Overflow springs

3. Classification according to the origin of the water
   (common in French, occasionally used in German and English)

   a) *emergence:* larger karst springs without further evidence of the origin of the water
   b) *resurgence:* the re-emergence of a swallet stream at the surface
   c) *exsurgence:* autochthonous seepage water flowing out

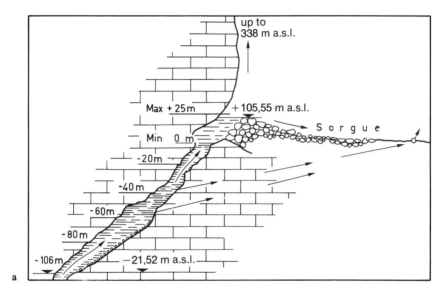

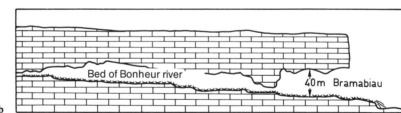

Fig. 9.6. a The Vaucluse Spring east of Avignon according to Flandrin and Paloc (1969); **b** Grotte de Bramabiau according to Guide Michelin (1951), longitudinal section

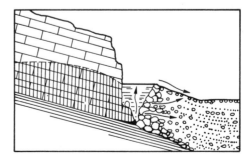

Fig. 9.7. "False" vauclusian spring

To these types of spring there can still be added individual types which do not fit so easily into a general scheme: *subaqueous karst springs* (sublacustrine and submarine); karst springs can also be found in the underground in every form: *cave springs.* The question whether water which does not emerge on the earth's surface can be termed a spring can be answered by the fact that the visual impression in a cave is just as unmistakable as on the earth's surface. Karst springs which have been covered by boulders or by gravel betray their presence by the considerable flow of water. Small karst springs under loose material can, however, not be distinguished from talus springs. In karst areas springs which flow out of a valley-fill may also be effusions of groundwater — decisions as to classification must be made on the basis of chemistry and temperature. *Estavelles* function alternately as sinks and as karst springs, thus they are also temporary springs. O. Lehmann (1932) interprets them as relieving an underground stream during high water, Grund sees them as emerging groundwater. Depending on the conditions one or other of the two opposing opinions is correct.

9.3 Vauclusian Springs and Other Large Karst Springs

Vauclusian springs are karst springs which have an outflow the size of a river. In earlier days their water was used to power mills, today they sometimes power smaller electricity plants, e.g., at the Areuse Spring in Neuenburg's Jura (Switzerland), at the Brenztopf and the Blautopf (Fig. 5.12) in the Swabian Jura and at the Ombla Spring close to Dubrovnik. Cave rivers which emerge at the surface through a gate in the rock form large springs such as the Pivka, which flows out of the Malograjska Jama (Kleinhäusler Grotto) into the Polje of Planina as resurgence, the Bournillon out of the cave of the same name below the northern rim of the Vercors in the Vallée de la Bourne and the resurgence of the Bonheur River out of the Grotto of Bramabiau on the eastern edge of the Causse noir (Fig. 9.6 b).

Trombe (1952) defines the vauclusian spring as an outflow from the ascending branch of a siphon. In a genuine vauclusian spring the water course runs upwards through the rock (Fontaine de Vaucluse, Blautopf), in a "false" one the water is forced to rise by gravel, moraine or scree (Schleichender Brunnen in the Muotatal). Genetically this distinction is completely justified, yet in many cases no decision can be reached without costly investigations.

Thurner (1967, p. 94) erroneously equates vauclusian springs with intermittent springs because the water does not always flow out over the final boulder wall (see Chap. 9.4.2 and Fig. 9.6).

When the local base level is laid deeper a new type of spring develops which only flows during high water. The water is drained to the lower outflow and the former vauclusian spring becomes dry. Now the ascending branch acts as a piezometric tube. The larger the supply of water from the inside of the mountain, the higher the pressure necessary to achieve the faster rate of outflow required — then the water rises in the piezometric tube. If the water level reaches the earlier rim of the spring, the water will flow out again: *high-water relief spring.* Under the water pressure which now remains constant, the outflow of the lower spring remains constant, too. Whether this high-water relief can be called episodic or periodic depends on the duration, the frequency, and the regularity of the outflow. Almost every larger karst spring shows higher spring openings which are partly or completely out of use. High-water relief takes place in the case of Hölloch (Switzerland) through the present-day entrance, 96 m above the present base level, the Schleichenden Brunnen.

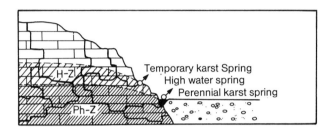

Fig. 9.8. Cross-section of a high-water relief spring. H-Z high-water zone; Ph-Z phreatic zone

Table 9.1. Outflow of several vauclusian springs (partly acc. to Bakalowicz, 1973, p. 38)

	Extreme values (m³/s)	$Q_{min}:Q_{max}$	Average (m³/s)
Aachquelle (Germany's largest spring)	1.3 - 24.1	1 : 19	8.8
Blautopf (Germ.)	0.35- 26.2	1 : 75	2.16
Brenztopf (Germ.)	0.15- 17	1 : 115	1.2
Rhumequelle in the Harz (Germ.)	1.3 - 4.7	1 : 3.6	–
Areusequelle (Switz.)	0.18-100	1 : 500	4.0
Various Yugoslavian ones		1 : 2 to 1 : 20	
Bournillon (Fr.)	0 - 40	–	–
Fontaine de Vaucluse (Fr.)	4.5 -200	1 : 44	29
Silver Spring (Florida)	Scarcely any	1 : 1.1	22.9
Rainbow Spring (Florida)	variation	1 : 1.1	21
Peschiera (Italy)	–	–	18
Timavo (Italy)	9 -138	1 : 15.3	17.3
Big Springs (Ozarks, USA)	–	–	12.3
La Touvre (Fr.)	1.4 - 4.5	1 : 32	12
La Loue (Fr.)	1.2 - 90	1 : 75	10.8

9.4 Periodic Springs – Ebb and Flow Springs (Intermittent Springs)

9.4.1 Periodic Springs

As can be seen in Table 9.1, variations in the outflow of a spring are normal occurrences. The variations are a result of precipitation, more precisely a result of the portion of precipitation which seeps in (A), of the difference between liquid precipitation + melted snow N and evapotranspiration E. The underground water stores have a balancing effect.

$$A = N - E$$

Extensive parts of Central Europe show a balanced distribution of precipitation; yet evapotranspiration, which is dependent on temperature, creates an excess of water in the winter and a deficit in summer and thereby effects a periodic increase and decrease in the amount of water carried in underground courses. This causes the surface of the karst water-body to rise and fall, too. In alpine regions the contrast between winter's dryness and early summer's high water (melting snow) is striking. In the Mediterranean climate, tropical steppe, savannah, and dry forest, or in regions with cold winters, the curve of discharge shows a strongly periodic, annual rhythm. Superimposed on these periodic variations are aperiodic ones resulting from single precipitations. Springs which show in their average discharge a pronounced periodicity that is dependent on climate and is therefore lengthy, are called *periodic springs*.

9.4.2 Intermittent Springs, Ebb and Flow Springs

There is some confusion concerning the names of springs in the group with a strongly varying outflow; this is probably because such intermittent springs are rare. Intermittent implies a rhythmical process and does not mean a suspension of flow, but rather a temporary, short-periodic interruption of the constant situation. Intermittent springs show a rhythmical periodicity of a few minutes or hours with acute maxima and/or minima. It is common that the outflow increases abruptly, remains at the high-water mark usually for only a short time, and falls quickly back down to the old value. In an extreme case the amount of water carried can sink to zero (Fig. 9.9).

The cause of the rhythm is not dependent on climate, but solely on hydrological physics. Katzer (1909) was the first to recognize this and he explains the phenomenon by a siphon system. The water gathers inside the rock in a basin B from which a tube with an elbow in it first rises in the direction of the spring and then falls again. If the water table in the collection basin rises above the culmination K in the tube, the water flows over and fills the descending branch. If the drain down to the spring is narrow enough below, a column of water builds up. It can finally suck the water out of the basin until the basin's water-table has sunk below the level A of the point of suction. Air follows after and the flow of water to the spring is abruptly interrupted (Fig. 9.10).

If the siphon is to function correctly, the descending branch must be so narrow that when the water runs over at K, it fills the whole tube. There must be a siphon or an especially narrow place inserted below, moreover, so that no air enters the tube from below. In the area of the elbow tube above the level of the outflow A there cannot be any open joints which carry air, for otherwise the column of water will break off at this point. These additional conditions make it understandable that intermittent springs are rare.

In 1924 Bridge published a treatise concerning the *Ebb and Flow Springs* in the area of the Ozarks (MO, USA) in which he made an essential contribution to the problem. He also collected in it the number of ebb and flow springs which were known at that time:

USA:	Virginia	2
	Missouri	4 plus one doubtful one
	Nevada	1
	New Mexico	1
Europe		7 plus one doubtful one
		15 plus two doubtful ones

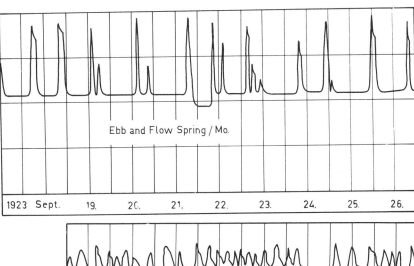

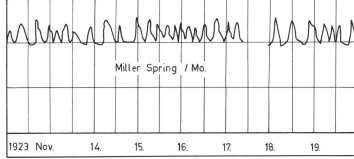

Fig. 9.9. Diagrams of drainage according to Bridge (1924)

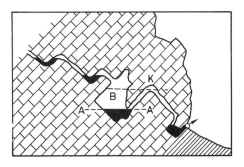

Fig. 9.10. Cross-section of an intermittent spring

The siphon phenomenon is not compelled to be in the direct vicinity of a spring. If the siphon is close, the spring reacts with a sharp change from a surge of water to be the suspension of it. If it is further away, the rhythm remains intact but a suspension of flow no longer takes place. Frequently the intermittent stream joins with another one with a steady flow of water. The rhythm is superimposed on the constant flow and the outbreaks out of the siphon stand out as peaks. Bridge describes a spring in the Ozarks which has complete suspension, two others with 150 l/s as their basic outflow with peaks of up to five times as much. Moreover, in one other there are variations in the length of the period between 1 and 6 h with single, longer periods of 13 and 20 h. The drainage of Lilburn Cave in Sequoja National Park belongs to the same type; its basic discharge is superimposed by a rhythmical swelling and subsiding over an interval of several hours (personal observation).

Ule (1925) describes the Idzuk Spring in the Bihar Mountains (western Romania), which suspends its flow every 10-15 min in rainy periods, in dry periods on the other hand, every 20-30 min.

Many intermittent springs do not show the siphon effect at all possible water levels:

a) the intermittent action continues the whole time. The length of the periods changes with the water supply, since this determines the duration until the basin is full

b) at high water the siphon becomes flooded and does not function because of the lack of air: intermittent at low and mean water

c) the basin is not watertight. When the influx of water is slight it cannot be filled over the level K (Fig. 9.10). With an increasing supply of water the siphon action sets in: intermittent at mean and high water

d) a combination of (b) and (c) results in an intermittent spring only at mean water

According to Verdeil (1962) the springs of Fontestorbes (Ariège), the Fontaine Ronde (Jura) and others, belong to type (b); to type (c) the Gouffre (shaft) de Poudak (Htes Pyrénées), rising 15 min, high-level stage 3 min, descent 4 min and low-level stage 36 min. For the springs of Fontestorbes he registered for an outflow of 560 l/s an increase in the amount of water during 15 min 4 s, a halt of 6 min, and a decrease and a continuation at the old amount during 35 min 36 s, thus a cycle of 56 min 40 s; in the Fontaine Ronde the water swells for 4 min, subsides for 2-2 min 30 s, and rests for 1 min 30 s, corresponding to a cycle of 8 min (p. 65). He mentions six such springs for France, nine for the rest of Europe.

9.4.3 Episodic Springs

Episodic springs flow extremely irregularly, sometimes they do not flow for years. There is no periodicity. They flow when there is an extremely high water level in the karst water-body, which is only the case in wet years. In the Swabian Alb wet years have, however, always been years of bad harvests, years of hunger; this is the homeland of the so-called Hungerbrunnen, the hunger springs.

Binder (1957) recorded the following for the Hungerbrunnen southwest of Heidenheim (eastern Swabian Alb):

Table 9.2. Flow periods of the Hungerbrunnen between 1867 and 1957

1867	3.5 months	1939-1941	Continuously from
1896	? short		April 1939 to Jan. 1942
1897	? short	1942	6.5 months
1906	11 months	1944/45	Nov.-April, 5.5 months
1914/15	12 months	1947	A few days
1916/17	11 months	1955	7 months
1925	6 months	1956	3 months
1927/28	9 months	1957	2.5 months
1937	8 months		

Eissele (1963) describes the hunger spring in Liebental east of Sigmaringen. It is a case of piezometric tubes above an underground water course; only when there is heavy precipitation does the water reach the surface through the tubes.

9.5 Subaqueous Springs

Subaqueous springs can be divided into:
   a) sublacustrine springs in lakes
   b) submarine springs or vruljas

9.5.1 Sublacustrine Springs

Water which emerges under water is found in many karst lakes and very often also in limestone-alpine valley lakes that cut into karstified limestones, e.g., in the Lac d'Annecy (80 m deep), in Thunersee west of Interlaken at the Bätterich, in Hallstättersee the Köhbrunnen (Morton, 1963). Water emerging subaqueously can be recognized by the bubbling up of the water surface Wallerquellen, Boiling Springs according to Stini (1933), by the difference in temperature to that of the rest of the lake, by the different lime content, at high water also by the color of the water, by its cloudiness or also by its bacteria content — karst water is not filtered.

There are three possibilities of its origin:

1. Creation of the spring after the formation of the lake. This is possible when the water pressure in the rock surpasses the hydrostatic pressure of the lake water. No research has been done on this subject.

2. Division of an already existing underground water course by erosion, e.g., when alpine valleys were glacially overdeepened in the Pleistocene Period, or when river erosion has taken place and the valley is later dammed by slides, alluvial cones, or moraine ramparts.

3. Later submergence of a karst spring at the margin of a valley bottom by other morphological and/or tectonic processes.

These processes are also valid for subaqueous springs in the groundwater of an alluvial valley bottom.

9.5.2 Submarine Springs – Vruljas

A submarine spring is called a vrulja in Yugoslavia; the word has been taken into German terminology as Vrulje. Submarine springs emerge partly from open joints in rocks (Petrik, 1958), partly from submarine karst caves (Port Miou; Gallocher, 1954; Roques, 1956; Martin, 1968); however, they emerge most frequently from funnel-shaped openings in the floor of the sea (Gulf of Kaštela, Alfirević, 1958; Etang de Thau, Dubois et al., 1963).

On the Yugoslavian coast 30 m is taken to be the lower limit for the frequent occurrence of vrulje (Petrik, 1958). This slight depth is an indication that the creation of the springs was under mainland conditions which were a result of the eustatic fall of the sea's surface during the cold periods of the Pleistocene. Evidence of this is given also by the submarine spring off Port Miou, which issues from a cave at a depth of 12 m under the water. Divers have found stalactites in the cave, which is proof that it was formed under vadose conditions (Martin, 1968).

In Yugoslavia the water of the vrulje originates in the hinterland of the coast; from there part of it flows 40-50 km under the bordering mountain chain through to the sea. Vrulje are – as an exception – freshwater springs, yet they generally carry brackish water which eliminates the possibility of using them as drinking or irrigation water.

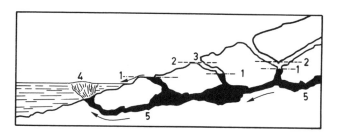

Fig. 9.11. Freshwater vrulja with high-water relief, diagrammatic. *1* piezometric surface at normal water level; *2* p.s. at high-water; *3* temporary karst spring (high-water relief); *4* vrulja; *5* freshwater stream (*black*)

According to the Bernoulli equations lower pressures are created in narrow passes because of the higher speed of flow ($\Delta h_v = \frac{v^2}{2g}$). Open joints which take such a narrow pass as their point of departure have a lower pressure than the sea's level and suck in sea water. In this way the freshwater becomes brackish (Fig. 9.12).

Sea water can also be sucked in when the spring is not submarine, but is situated somewhat above sea level: brackish-water spring (Fig. 9.13).

If the suction joint ends immediately at the surface of the sea, the piezometric surface lies somewhat deeper and sea water flows into the rock: salt-water swallow-hole (Fig. 9.14).

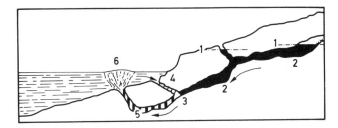

Fig. 9.12. Brackish water vrulja. *1* piezometric surface; *2* freshwater; *3* narrow pass; *4* sucking tube for sea water; *5* brackish water and brackish-water spring; *6* vrulja; *7* salt-water swallow-hole

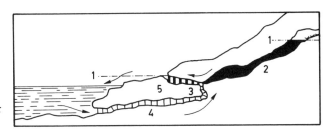

Fig. 9.13. Brackish-water spring; legend see Fig. 9.12

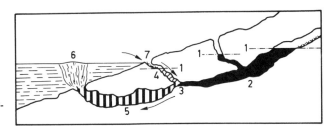

Fig. 9.14. Salt-water swallow-hole; legend see Fig. 9.12

9.5.3 The Sea Mills of Argostoli

The sea mills on Kephallenia (Ionic Islands) are the best-known example of sea water swallow-holes. The sea water, which flows land-inward and disappears there into open joints, used to power mills; today there is one lonely water wheel as a tourist attraction. Depending on the season the water table lies 75-135 cm under sea level. Wiebel (1874) explained this phenomenon in the sense of Fig. 9.14. Moreover he observed that it is not the numerous, closest springs on the other side of the bay which come into question for resurgences — they are freshwater springs with a smaller outflow than should correspond to the sea water streaming in; rather he assumed that the brackish water springs on the other side of the island in the Gulf of Sami were the points of outflow. O. Lehmann (1932, p. 81 ff.) violently attacks Wiebel für this assumption without being able to give a better solution.

The hydrogeological investigation done by Maurin and Zötl (1960, 1963), who made use of uranine for tracing, confirms Wiebel's hypothesis.

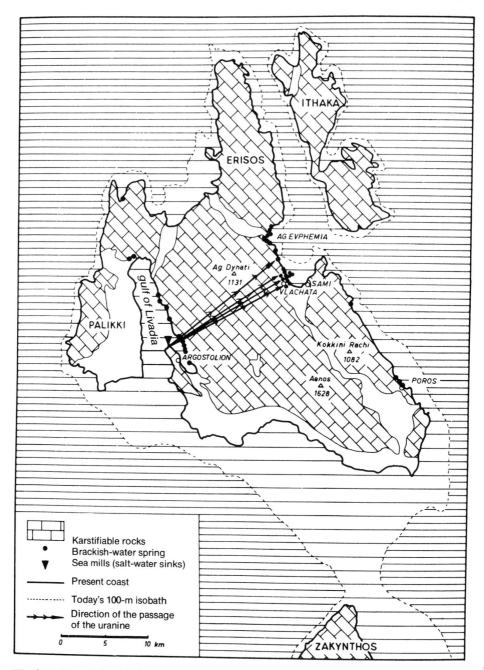

Fig. 9.15. The relationship between the sea mills and the Gulf of Sami (acc. to Maurin and Zötl, 1963)

Glanz (1965) explains sea mills by the ejector effect rather than by the suction principle, and corroborates his opinion with a physical model. In the period of the eustatically deep water level a tectonically controlled system of karst water-courses came into being which leads from W to E. Today it is submerged beneath the sea's surface. A stalactite from a depth of 3 m under the sea's surface is 16,000 years old; another from a depth of 26 m is even 20,000 years old (^{14}C-dating by Münnich, 2nd Phys. Inst. Univ. Heidelberg); these corroborate Glanz's theory. Precipitation and melted snow sink down into the depths from the more than 1000-m-high island mountain of cretaceous limestone (Aenos 1628 m above sea level). Flowing from W to E, it joins the water in the cave system, accelerating it. This can happen any arbitrary number of times and the effect is cumulative. Although the theory is very comprehensible in this case, its general applicability to all vrulja which carry brackish water is to be doubted.

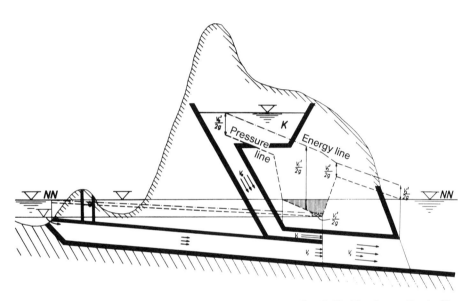

Fig. 9.16. The ejector process in the karst water system Argostoli — Gulf of Sami according to Glanz (1965, p. 124). Strongly exaggerated height

The distance h_v from energy line and pressure line is always $\frac{v^2}{2g}$ (see Chap. 5.5.5). v_a is the speed in the drainage tube in the mountain, v_1 the speed in the narrow tube sucking in sea water and v_3 that in the part of the cavity system crossing the island which carries brackish water. $h_v = \frac{v_2^2}{2g}$ is the difference in height between sea level and the piezometric level in the sea mill. Since v_2 is the smaller, the less freshwater flows in variations of h_v result in rhythm with precipitation or with the seasons respectively.

The phenomenon of the sea mills has been observed in other places as well but has not met with such great scientific interest as has Argostoli. Von Knebel (1906) names the Teufelsbrunnen (Devil's Springs) south of Opatija: "At three neighboring points there disappear here far more than 100 liters per second. ... It has been recognized that sea mills are completely limited to karst regions alone" (p. 107 f.). Reclus (1881, p. 262 f.) portrays a sea-water swallow-hole north of Cette (Sète) in the Lagoon of Thau (France), the Gouffre d'Enversac (inversae aquae!). In winter's rainy season freshwater streams out of it, from the end of April to the beginning of winter, however, sea water flows, swirling into the depths.

9.6 Physico-chemical Properties of the Water of a Karst Spring

By means of random sampling the momentary physico-chemical state of a spring and of the spring's water can be measured. Measurements during a full year for any single parameter are able to tell much more but they demand an input which can rarely be afforded.

9.6.1 Discharge

The discharge of a spring is the amount of water emerging in a second in l/s or m³/s (Table 9.1). Small amounts of water up to 0.5 l/s can best be measured with a container; the time it takes to fill it is measured with a stop-watch.

$$Q = \frac{V}{t} \ [l/s]$$

Q: Amount of water in l/s
V: Volume of container in l
t: Time to fill container

For greater amounts of water the brim-board can be used or some other instrument. Estimates are difficult and demand a lot of experience.

The discharge of a karst spring lies between a few l/min and 200 m³/s (Table 9.1). Extreme values are always the result of extraordinary circumstances. Of economic importance (water supplies, electricity plants) is the mean discharge, calculated over weeks or months. There is a close relationship between outflow A and the amount of liquid precipitation N_R, of snow-melt water S, of evapotranspiration E and of storage-formation (−V) or storage-tapping (+V).

$$A = (N_R + S) +/- V - E$$

Water storage is included in the evapotranspiration curve for the Blautopf according to Binder (1960). The relatively high values in autumn and winter he explains by the snow covering which is a form of storage. The negative value in March is explained by the activation of the snow reserves.

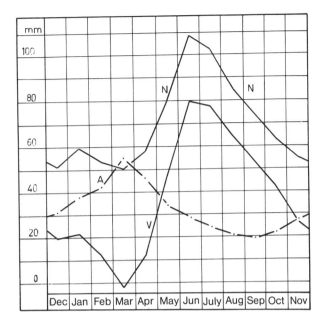

Fig. 9.17. Month's mean discharge (A), precipitation (N), and evapotranspiration (V) in the Blautopf area and its hinterland (1925-59) according to Binder (1960)

Prolonged periods of low precipitation result in characteristic curves of discharge. Their analysis is of especial interest to researchers in the French language (Salvayre, 1969; Nicod, 1970; Tripet, 1972, etc). They see the spring's discharge as the result of the combination of

1. water out of karst-hydrologically-active cavities
2. water seeping out of fissures
3. water from superficial streams sinking into the underground.

At high-water the cavities fill with type 1, and their piezometric surface rises quickly (in Hölloch as an extreme case 100 m in 12 h). The narrow fissures follow along only slowly, whereby a hydraulic gradient is created toward them. Water is pressed into them. When the high-water peak is past, the water surface in (1) begins to sink, while the water surface in (2) continues to rise until it is on the same level as in (1). If the water surface in (1) continues to fall, the hydraulic gradient turns round and the fissures begin to supply water to the cavities. If the water storage is exhausted in the cavities, the spring will be fed essentially only by the water from the fissures and will slowly continue to decrease its outflow (Fig. 9.18).

If in Fig. 9.19 the surface of the inflow rises, the hydrostatic pressure increases. The pressure wave rushes ahead of the penetrating water, making the water level of the spring rise long before the afflux has reached it: high-water situation. The Noiraigue (Jura Neuchâtelois, Switzerland) is a karst spring which carries light brown water. Burger (1959) observed that there is no change in the color when high water sets in. Only after a rather long time does it become almost abruptly darker brown, when the water which created the pressure wave reaches the spring. By then the high-water peak has strongly subsided, however. Tintant (1958) measured the rate of flow of the water in the Grotte de Bèze in the French Jura at 170 m/h, but states a velocity of 1100 m/h for high water. That is evidently the velocity of the pressure wave of the high water.

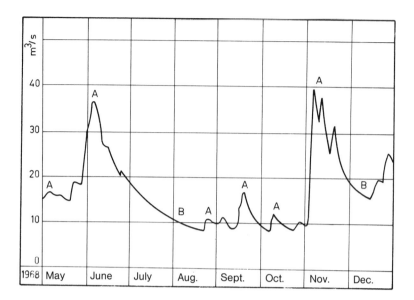

Fig. 9.18. Section from the curve of discharge of the Vaucluse Spring according to Flandrin and Paloc (p. Hy 13/5). *A* outflow mainly from karst-hydrologically active cavities; *B* outflow originating in the fissures

Fig. 9.19. The origin of the water of a karst spring. *a* narrow open joints, fissures; *b* karst-hydrologically active cavities; *S, H* vadose zone with percolation zone *S* (seepage zone) and high-water zone *H*; *ps, pt* shallow, deep phreatic zone; *K* karst water surface; *HF* high-water surface; *Hq* high-water spring

Bourgin's figures (1946) for the Vernaison, in which the pressure wave was measured at 1500 m/h correspond to the above.

9.6.2 Variations in Temperature

In general karst water shows a very balanced temperature, the more balanced, the deeper the water-courses reach down into the phreatic zone. They are usually somewhat above the annual average air temperature in the main catchment area of the water. In Hölloch (Switzerland) at an average depth of 600 m (extreme 900 m) the temperatures in the cave sink from the beginning of the snow-melt at the end of March until toward the end of July by about 1°C from 6° to 5°C, and then they gradually rise again to the higher value. That is 2°-3°C more than the mean value on the surface. Flandrin and Paloc (1969) quote temperatures between 11.9° and 13°C for the Fontaine de Vaucluse. Three days after the high-water peak in the spring the temperature sinks; this is an indication of the fact that so much time lapses between the arrival of the pressure wave and the water.

9.6.3 Chemistry of Spring Water

When the water remains in the underground for a longer time, a physicochemical equilibrium with the surrounding rock occurs (Holland et al., 1964). Where physical processes of dissolution take place, e.g., when sulphates (gypsum) are dissolved, the concentrations approach saturation more and more closely. On the other hand, wherever reversible chemical processes take place (corrosion), an equilibrium is reached; it is, however, disturbed by any change in the concentration of one component or by a change in temperature and must readjust.

Water flows slowly through the karst-hydrologically active cavities in the phreatic zone at low water; at high water the flow is correspondingly faster, up to ten times and more. The contact of this water with the rock is reduced by the same factor so that when it reaches the spring it clearly contains less lime. The longer the way through the phreatic zone, the later the less-concentrated water arrives after the high-water peak.

Other parameters, such as the content of tritium, electrical conductivity, and the pH are of importance. They all show a dependence on the amount of water and on the rate of flow indicated by it. For treatment of this subject turn to specialized literature on it, especially to Zötl (1974) where there is also a rich fund of information on other literature.

10 Tracers

It may be of scientific as well as of economic interest to know the origin of a springs's water, or where losses of water in the underground flow to. Problems arising from such questions can often not be solved directly, even though, with the help of geology, an underground catchment area is roughly comprehensible. It is not unusual that watersheds at the surface are not identical with underground watersheds. The showpiece is the Danube's loss of water, in which case the water runs under the European watershed to flow into the Untersee near Constance and into the Rhine.

Occasionally natural phenomena help expose relationships. Local downpours make distant springs rise; seeping, dirty water unexpectedly clouds emerging water, sawdust is swept away by high water into the ground and appears in a karst spring. When a factory which makes aniseed liqueur (Pernod) burned near Pontarlier, the barrels burst: some time later the Loue Spring smelled of aniseed.

10.1 Tracers

Experiments with artificial marking materials or tracers were preceded by natural aids. Thus the origin of the Timavo close to Triest was determined by means of marked eels which were put into the Reka at the Grotto of St. Kanzian (Škocjanske Jame). They took 55 days to cover the distance, thus they swam about a kilometer a day (Wagner, 1954). Sawdust and salt were used for investigations. For a long time marking materials were used only in karst areas; today they are being used to help discover other losses of water, e.g., in testing the water-tightness of a dam or in the search for the cause of impurities in groundwater. Zötl (1974), who published the most complete compilation on the subject of tracers up to now, emphasizes: "that the attempt at marking represents the last link in the chain of karst hydrogeological investigations of an area" (p. 89).

Main groups of tracers
a) *Salts:* kitchen salt, NaCl, is the most-used marking salt; it is simple and safe to use. However, considerable amounts are required, at least 300 kg for 1 million cubic meters of water. Thus NaCl is about 20 times more expensive and 15,000 times heavier than uranine (Käss, 1967, p. 131). Qualitative evidence results from a 5% solution of silver nitrate (Lukas, 1959b). An elegant, quick, and precise method is the measurement of the electrical resistance. With the aid of tables the concentrations can be determined (Mitter, 1959).

During the investigation of the Danube's sinks by Batsche et al. (1970) potassium chloride was also used and the K was measured with a spectral photometer. According to Käss the expenses were, however, 30 times higher than the use of uranine would have been (1967, p. 131).

Karst researchers in Triest marked with lithium chloride, LiCl, as early as 1907/1908. In 1911 markings were done in the Reka. The Li^+ appeared 211 h later at Duino in the Timavo. The proof resulted spectroscopically (Wagner, 1954). This salt was recently used for a marking project in the Muota Valley (central Switzerland).

b) *Test dyes:* their number is large. The most useful are fluoresceine and its sodium salt, uranine. Since this disintegrates in acid water, red-colored sulforhodamine is used in such cases today. Dissolved in water it can be traced at $1:10^{12}$ (1 g to 1 million t). Uranine can be recognized as a green color by the bare eye in a solution of $1:10^7$, with the analytical quartz lamp at $1:10^{10}$ and with activated carbon even at $1:10^{12}$ (Lallemand and Palloc, 1964). Such a high degree of sensitivity demands the utmost care when the test is taken and analyzed; any contact with the coloring material must be avoided in order to prevent the contamination of other water. For details see Lukas (1959).

According to Thurner (1967) the following formula is valid for the amount of coloring material to be used:

$M = \overline{EK} \cdot Q$

M: amount of dye in kg
$\overline{EK}$: distance place of entrance – karst spring in km
Q: spring's discharge in m^3/s

c) *Radio isotopes:* originally only radioactive marking materials were called tracers but the word is used for the others as well today. The use of radioactive indicators demands care since any damage done by radioactivity is noticeable only later. The radio isotopes used are short-lived and have half-life periods of 12 h up to 3 months; tritium is in a different class with its 12.3 years. The maximum dose permitted should be as large as possible, yet one should absolutely avoid exceeding the limit in the belief, for instance, that the amount will be very diluted in the underground. ^{24}Na, ^{82}Br and ^{131}J have been used most frequently up to now. Today a new complex has been added, ^{51}Cr in combination with ^3T, which is being used more and more (Knutson et al., 1963; Knutson, 1967; Batsche et al., 1970). The results are "substantially more precise than for example with coloration" and more versatile since one can read from them "not only the facts concerning the connection between swallow-hole and spring, but also exact details regarding the distribution of the water to various springs, regarding water flowing in and out and regarding the average rate of flow" (Mitter, 1959, p. 124; Table 10.1).

Naturally also nonradioactive isotopes can frequently make a contribution today, since they give evidence of the age and/or the temperature of the water at which condensation takes place (climate).

d) *Tracers which can be activated:* the analysis works with substances that do not give off radioactivity but are activated by means of neutrons. After they have been activated they allow an analysis which is just as exact as that done with radio isotopes. The disadvantage is that not only the tracers but also many other elements are activated. The activated tracer can either be established by its own rays, or it is separated chemically first and then further examined. During the investigations of the Danube's sink use was made of NH$_4$Br with Br as the activated substance and La(NO$_3$)$_2$ · 6H$_2$O with Lanthan which can be activated (Batsche et al. 1970).

Table 10.1. The most important radiotracers according to Mitter (1959), complemented by ^{3}T and ^{51}Cr

Isotope	Half-life period	Max. dose permitted (μC/cm^3)	Ray energy (MeV)	Substance
^{24}Na	15.10 h	0.8 · 10^{-2}	1.39	NaCl
^{32}P	14.07 d	2 · 10^{-4}	1.69	Orthophosphate
^{35}S	88.00 d	5 · 10^{-3}	0.167	Na$_2$SO$_4$ + HCl
^{42}K	12.44 h	10^{-2}	3.5/1.9	K$_2$CO$_3$ + KCl 56 : 52
^{82}Br	35.87 h		0.465/0.54-1.45	NH$_4$Br
^{86}Rb	19.5 d	3 · 10^{-3}	1.80/0.724	RbCl
^{131}J	8.14 d	6 · 10^{-5}	0.606/0.306	NaJ
^{3}T	12.3 a	–	0.018	HTO
^{51}Cr	18 d	–	–	EDTA-complex

e) *Aromatic substances and detergents:* although little use has been made of them, aromatic substances and detergents can occasionally be quite useful. For the investigation of the Danube's loss of water two aromatic substances were employed: dipenten (which smells of lemon although in the Aach Spring it smells of turpentine), and isobornyl acetate (spruce aroma). The threshold of perceptibility is 1:10^7 for dipenten, 1:5 · 10^7 for isobornyl acetate and for isoamyl salicylate (orchid aroma), which is also usable, 1:10^8 (Schnitzer and Wagner, 1967). Such tracers have already been used with success but not often enough that they can be judged conclusively.

Detergents have proved to be cheap tracers, easy-to-use but imprecise. They are alcylbenzol sulfonates which create a lot of foam and are biologically decomposable. By means of the addition of substances to stabilize it, the foam can be made to last for days. Detergents can be detected up to a dilution of 1:2 · 10^6 (Schnitzer, 1965, 1967).

The detergent content can, however, also be measured colorimetrically (Batsche et al., 1970 p. 112 ff.), which is sufficiently precise. Water which is organically heavily contaminated decomposes "soft" detergents biologically so that losses will occur when the water remains a longer time in the ground before emerging at the spring.

f) *Drifting materials:* insoluble bodies which are sufficiently small can be used in karst as drifting material. A beginning was made with sawdust which was then followed by synthetic material in powder form, but it was not until spores and bacteria were used that this method won popularity. In 1953 Mayr first employed lycopodium spores, which are light and also cheap; their diameter is 30 μm. Zötl (1953) improved this method by means of coloration. When differently colored spores are used they can be put into the water at different points at one time. With plankton nets they are filtered out of the spring's water. The various colors can be counted out in one operation.

The spores are mixed in water to a paste with a substance to moisten them, e.g., a detergent. When the flow of the water is at least 1 l/s, the paste (of spores) can be introduced into the stream directly. Otherwise it is made to flow into the sink through a hose which is well flushed afterwards. For the investigations which covered an extensive area of the Dachstein region Zötl needed 12 kg, in the Buchkogel region near Graz, 3 kg. Further details of this procedure can be found in Maurin and Zötl (1959).

Nonpathogenic bacteria which are strongly colored also belong to the useful drifting materials. Those serving as test bacteria are: chromobacterium violaceum with violet colonies, which do indeed have a slow growth and are in danger of being overgrown; serratia marcescens (bacterium prodigiosum) with scarlet colonies, which grow faster (Lukas, 1959, D2).

Batsche et al. (1970) describe the use of serratia marcescens in the Danube's sinks. "The facts that (these bacteria) are harmless to humans, have a striking red color on a dry culture medium and are not aquatic allow them to be used . . . as a tracer" and: "The separated, bright red cultures were made into a paste in 1 l of the river's water and introduced into the largest ponor at the end of the Danube's course in one gush. In this paste there were about 30 trillion bacteria" (p. 129).

10.2 The Tracer-Diagram

The curve of concentration of the tracer in the spring's discharge has a number of uses. It includes, besides the concentrations, also the times the tracers were introduced, the times they reached the spring and the maximum concentration, which are all facts of importance. In karst water there is usually a rapid rise to the maximum and a following, slower decline. If only one maximum occurs, the underground course is simple, without any great complications.

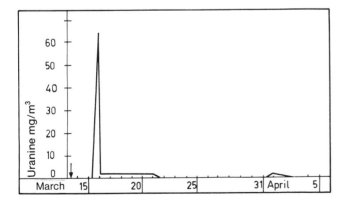

Fig. 10.1. Simple curve of concentration in the Aach Spring after introduction of a dye at Neuhausen o.E. on 14.3.1962 according to Käss (1965, p. 45). The second wave of dye (from 1.4.62) is the result of an increase in the amount of water to three times as much after three days of precipitation (30.3.62-1.4.62)

This allows the medium linear velocity of flow v_m to be calculated:

$$v_m = \frac{a}{t}$$

a: distance $\overline{SK}$ (swallow-point – karst spring)
t: time of passage from introduction to appearance of tracers

When the factor or prolongation is known $f = \frac{L}{a}$ the result is the medium real velocity v'_m:

$$v'_m = f\frac{a}{t} = \frac{L}{t} \text{ (see also Chaps. 5.1 and 5.5.7).}$$

If the discharge is known, the volume V of the moistened cavity can be calculated:

$$V = Q \cdot t_m$$

According to Prandtl (1969) $v_{max} = 1.3\ v_m$ (see Chap. 5.5.5). The earliest arrival of the tracer after the passage of time t_1 is to be put in as v_{max}, thus a/t_1. Since a is constant, the average velocity v_m is given by the time of passage t_m. And

$$t_m = 1.3\ t_1.$$

From this it follows that

$$V = 1.3 \cdot Q \cdot t_1.$$

The average moistened cross-sectional surface is then:

$$A_m = \frac{Q}{v_m} = 1.3\ \frac{Q}{v_{max}}.$$

This is, however, only an approximate formula for V and A_m refer to a tube. In nature there are often bulges or even in- and outflows which are situated close to one another in a lake so that no surrounding areas of any great extent can immediately be included in the occurrence. Experience shows that every case is situated differently, but that one can draw conclusions concerning the inaccessible phreatic zone from the situation in the accessible vadose zone. This makes an improvement in the results possible.

The tracer-diagram often shows additional maxima. An unpublished study of the Centre d'Hydrologie in Neuchâtel (Switzerland) concerning a spring on Thunersee (Switzerland) shows such a diagram with three maximums (Fig. 10.2).

The introduction of a dye was made in the Septemberschacht at 1740 m above sea level, the detection in the Gelben Brunnen at 560 m above sea level, at a horizontal distance of 12 km and with a gradient of 9.5%. Thirty-three h after introduction uranine appeared in the Gelben Brunnen but that was already the declining branch of a small maximum (point A), followed by a larger maximum (B) after 46 h, upon which came a small third (C). After 9 d the investigation was terminated at $5 \cdot 10^{-10}$ g/cm³ uranine. The rates of flow calculated for the horizontal distance were 364 m/h and 261 m/h respectively (Simeoni, 1973).

The Tracer-Diagram

Fig. 10.2. Tracer-diagram for uranine in the Gelben Brunnen on Thunersee (Switzerland) according to Simeoni (1973). Explanation in text

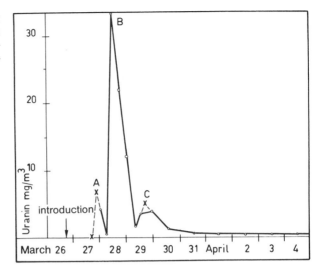

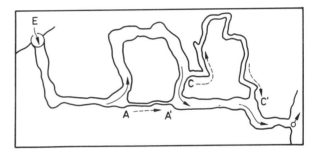

Fig. 10.3. Model of a network of passages. Explanation in text

An interpretation of three maximums is problematic. In Fig. 10.3 a model is shown, the original of which is also to be found in Hölloch (Switzerland), and which could be worked out in calculations. But here it is only a question of a qualitative statement.

The model shows phreatic conditions and the flow is so slow that losses through friction do not have a negative effect. The main passage is twisted and in the loop at AA' it is closed for a short stretch by a narrow connection, while between C and C' a similar narrow detour exists. Therefore the main amount of water flows through the main passage. A dye-front arriving with the water at A lets a small portion pass through the short connection. This portion reaches A' before the dye-front in the main passage and mixes with the water which has not been colored: in the spring this color appears as a first, small maximum. Meanwhile the dye-front in the main passage has moved on and reaches C, where similarly a portion again branches off into the side passage. When the dye-wave in the main passage reaches the spring it creates the large maximum there. When the concentration decreases the third, small dye-wave at C' reaches the main passage, creating a new peak of concentration. The concentration then diminishes and fades away.

11 Incasion, Breakdown

Incasion, breakdown, comprises all processes which cause the walls and ceilings of an underground cavity to break down naturally, and also as an exception the passage floor (bumps). In French the terms effondrement, éboulement, affaissement, décollement are used, in German Deckenbruch, Versturz, Einsturz, Höhlenverfall, Ablösung, Bergschlag, etc. The term incasion is derived from the Latin incadere, which means to crash into or to break into; the prefix in points out that the crashing rock falls into a cavity. Incasion is equal to the two other cave forming factors: corrosion and erosion.

In the development of caves incasion, or breakdown, is a symptom of age in the sense of a succession of forms and not as a stage in a speleomorphological cycle in the sense of W.M. Davis. It happens when underground cavities have become so extended by corrosion and/or erosion that the strength of the rock is overtaxed. An exception to this rule is breakdown due to tensions in the rock, spontaneous rock detachment and bumps, which can take place very early. Incasion increases parallelly to the age of development and finally leads to degeneration of the cave (see Chap. 14). The beginning of breakdown depends on lithological factors and on the tectonic stress on the rock. When the latter is heavy, breakdown can already begin in a cavity the width of 1 m; on the other hand it does not yet occur in a span more than 10 m in width when the rock is thickly layered and poorly jointed. For research it is of the utmost importance that breakdown destroys old structures or conceals them with boulders (see Chap. 14.3).

Incasion has its own laws of form-giving. Ceilings and walls can only break down when the pressure of the mountain or the weight of rock portions in the ceiling overtaxes the local strength of the rock. The break is directed along already-present interstices and cleavage, yet without any regard to whether these are in any way of karst-hydrological importance or not. Other criteria are valid for the potential karst-hydrological activity of an interstice than for the detachment of a block. The result of this is the possibility that interstices — whether bedding or joint interstices is of no matter — are exposed without their being of the slightest importance for karstification. It is obvious that such a visually dominant side of an interstice is looked upon as the cause of the formation of the passage although it was and is of no importance for the underground karstification. The wrong decisions made on this account are serious apropos of speleogenesis, for a bedding-plane passage of phreatic origin would then be classed as a joint passage of vadose formation. Such wrong interpretations are frequent especially in alpine caves. This may be a reason why joint passages were spoken of for so long while bedding-plane passages were not mentioned.

The innumerable joints and interstices in karstifiable rocks are the result of former or still continuing inner tensions. Deep inside the rock there is generally a balance of pressure, comparable to hydrostatic pressure. As the surface is approached, disturbances of morphological origin make themselves noticeable: limestone and dolomite on a plastic foundation (marl, slate) begin to slide; where a valley is cut into limestone the rock presses outward into the evacuated space. In mountains where considerable erosion occurs — it can be many hundreds or thousands of meters of rock — the relief causes interstices to form parallel to the surface (according Renault, 1967, p. 229).

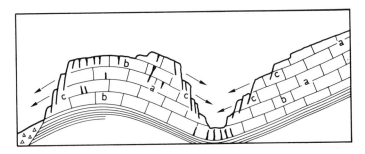

Fig. 11.1. Interstices and open joints in a limestone massive. *a* bedding planes; *b* jointing due to tractive force; *c* tension-cracks of morphological origin

In a mountain which has reached an equilibrium there is an opposing pressure for every pressure — thus no further movements take place. The main pressure is produced by the weight of the superimposed masses of rock. It is proportional to the specific gravity and to the height of the pillar of rock:

$$p = \frac{\rho \cdot g \cdot h \cdot A}{A} = \gamma h$$

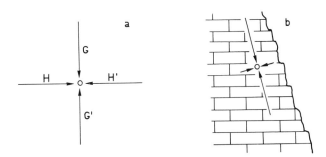

Fig. 11.2. a, b. Equilibrium of the forces in a rock. **a** in the inside of the mountain according to Davies (1951); **b** near the flank of the valley according to Goguel (1953). H or H′ respectively: *a* horizontal components when there is a hydrostatic distribution of pressure; *b* as long as the rock's breaking strength is not overtaxed the "horizontal" components are the same. G or G′ respectively: pressure on, or counter-pressure due to the rock's mass

In the rock the pressure lines generally run in the direction of gravitational acceleration. However, they diverge from it as they approach the sides of the valley, for that is the side from which the hydrostatic pressure is disturbed (Goguel, 1953).

Every cavity, whether it is a mere open joint 5 mm in width or a cave passage 5 m across, is a disturbance in the structure of the stress lines. Since in the cavity the pressure is zero, i.e., the same as that of the atmosphere, the stress lines turn aside and thereby create a pressure-free space which extends above the cavity in the pressure dome.

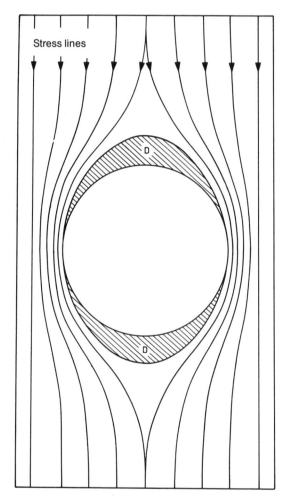

Fig. 11.3. Stress lines around a cavity with a round cross-section. *D* pressure dome

By means of artificial cavities, e.g., tunnels, serious disturbances in the pressure structure are created in no time at all and they can only gradually adjust themselves. Pressure tensions are thereby created parallel to the surface of the cavity; these can lead to the splitting-off of thin rock sheets. The process is called detachment, when the piece is hurled away it is called a bump. Bumps occur especially on the side walls, less often on the roof, still more rarely on the floor. Irregularities in the rock lead to tensions even in the zone of the theoretically pressure-free pressure dome. According to a description by Stini (1950), the side walls "hurl small to very large slabs which grow thin toward their edges, into the cavity without any warning. An intense cracking noise and earth tremor accompany the burst. . . .Sheets fly off usually only a few hours or days after the rock's surface has been laid bare. Detachment takes place parallel to the wall of the cave without regard to the cleavage of stratification or jointing of the rock" (p. 274 f). Thus the form of the cavity survives to a certain extent, and the rock's surface is relatively smooth.

Karst cavities have a long period of formation behind them. In general tensions have adjusted but for a few remaining ones. This is the reason why, in contrast to artificial tunnelling, bumps are extremely rare. In an earlier stage there must have been more of them, especially detachments, too. An analysis of cross-sectional forms shows this. Indeed it is difficult to distinguish an ellipse — the usual name for a lens-shaped cross-section — derived from a flat ellipse by means of detachments from one that was formed by corrosion, if the rock's surfaces show corrosion forms created under phreatic conditions (Plate 7.1).

Bedding-plane-controlled passages created by corrosion first show narrow, lens-shaped, flatly elliptical cross-sections. When the width is sufficient rock sheets begin to crack off. The elliptical form survives but becomes asymmetrical because the upper arch is strengthened. Moreover, tension interstices form which in time develop corrosively into rock scales, thrusting out of the rock's surface at a very sharp angle (Plate 6.4). Such indicators were usually not recognized or given no consideration and therefore ignored. Detachments can also often be recognized from the debris. When slab-like debris is created out of thickly layered limestone it is an indication that pressure tensions have created it. This is certain if the surface of such debris is slightly convex. In the commercial part of Hölloch there lies on the side of the path a strongly bent, thin pressure-sheet which diminishes toward its edges (Plate 6.3). At the Polyp branching of the Passage of the Titans behind Bivouac 1 pressure-sheets formed in the floor, which is 10 m wide there.

Wherever the sides of the ceiling approach the floor of the passage, detachments usually break away and rock stumps remain which are a further proof of the part pressure tensions play in the formation of the cross-section of a passage (Plate 6.4).

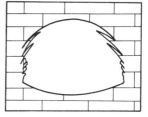

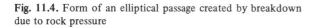
Fig. 11.4. Form of an elliptical passage created by breakdown due to rock pressure

Theoretically the creation of rock detachments can be expected in the high-water zone. The rapid change from the water-filled state to that of dried-up gives rise to fluctuations in pressure. When the cavity is filled with water up to the surface there is at every point a relationship of hydrostatic pressure p_H to the pressure in the rock p_G of 1:2.75 or rather

$$p_H = 0.36\, p_G$$

When the high-water zone is 10 m deep and 100 m below the surface the fluctuations of pressure are only 3 1/2% in reference to the surface bordering on the phreatic zone. In the back section of Hölloch, however, the maximum depth of the high-water zone is 180 m with a basis of 800 m below the surface. Here there results a pressure difference of 8.2% to the low-water level. Such high water occurs several times every year. However, there have not been any noticeable results of its influence in the last 30 years, neither in the deepest nor in the higher parts of the high-water zone. Attention must be drawn to the fact that in comparison to other karst regions a high-water zone of the depth of 180 m is extraordinarily large.

Due to the fact that the pressure lines turn aside to go around a cavity, there is a pressure-free area above and below the cavity, the so-called tension dome. As a form it has the most stable cross-section (Fig. 11.3). If the rock breaks down easily, as is the case with thinly layered limestone, this form is approximately attained. In the upper tension dome — this one alone has consequences for the formation of the cavity — the rocks are burdened only by their own weight, but they are not under the pressure of the total column of rock (Plate 6.1). If the strength of the rock is overtaxed, parts of it fall down. The surface form, as opposed to detachments, is of no importance; important are stratification, jointing, and cleavage. The newly created surface is composed therefore of facets of joints and bedding planes mixed with surfaces of fractures so that the surface as a whole is angular and irregular. This kind of breakdown destroys the original form completely (Plate 7.2).

In most of the caves in the USA the limestone shows horizontal stratification. In the ceiling the strata are situated in the pressure-free area of the tension dome and form rock beams. Under the influence of their own weight they move down, causing tensile pressure and shear stresses and corresponding cracks. As long as these cracks do not penetrate the sagging beam of rock, it works as a support. If the beam breaks, it forms two cantilevers opposite one another. According to Davies (1951) there are two formulas which can be used to calculate the required minimum thickness of the rock beam d for a given span b, so that it will not break under its own weight and fall down.

Beam: $\quad d_T = 0.75 \dfrac{\gamma b_T^2}{f}$ $\qquad$ f: bending strength
$\qquad\qquad\qquad\qquad\qquad\qquad\quad$ γ: specific weight

Cantilever: $\quad d_K = 3 \dfrac{\gamma b_K^2}{f}$ $\qquad$ b_T: length of the span of the beam
$\qquad\qquad\qquad\qquad\qquad\qquad\quad$ b_K: length of the span of the cantilever

When the beam is divided through its center then $b_K = 1/2\, b_T$. The necessary minimum thickness of the rock beam which thus became cantilevers is unchanged. However, if the break is at the full distance between the supports (break at the opposite abutment) then the thickness of the cantilever must be at least 4 times the mimimum thickness of the beam, or else incasion occurs.

When the limestone layers are horizontally stratified and at slight depth (up to 100 m) the cross-sections of passages often become rectangular. When an ellipse is formed along a bedding plane and reaches the next higher one, a rock triangle comes into being, which as a cantilever easily breaks off along fissures. The beam above it begins to bend, and tension cracks develop. These are often interpreted as tectonic joints, joints therefore which are supposed to control the formation of the passage. But that is by no means what these cracks are. Because of the breakdown of these beams a rectangular cross-section results which can be created under both phreatic and vadose conditions. When water flows through, whether it is a high-water stream or a phreatic current, the debris is dissolved and washed away; on the walls and ceiling forms of corrosion emerge.

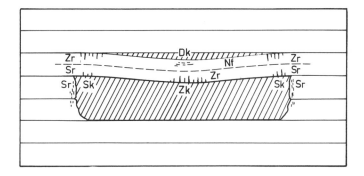

Fig. 11.5. Forces on the ceiling of a cavity with a rectangular cross-section, according to Davies (1951). *Dk* pressure forces; *Sk* shear forces; *Zk* tensile forces; *Sr* shear cracks; *Zr* stress cracks; *Nf* neutral plane

In caves with a flat ceiling the "junction effect" can be observed. Because of the sudden increase in the span at a junction breakdown is more likely there. The angle between diverging passages acts as a pillar which supports the ceiling. Between the supports circular incasion occurs, each additional layer with a steadily smaller diameter so that a flat dome develops.

Most knowledge of the fundamentals of breakdown is gained from the physics of mine and tunnel construction where these problems are of vital importance. However, it may be applied only in a limited way to natural underground cavities, as important conditions differ fundamentally, even if the principles remain unchanged. Artificial cavities are made quickly by the use of force which essentially loosens the rock (blasting). Added to this there will be adjustments to the changes in pressure. Therefore the walls of an artificial tunnel remain in motion for quite some time so that incasion happens as a rule in new tunnels without reinforcement. In the case of karstification, on the other hand, the extension of underground cavities occurs slowly and gently. During the extended period of time since its generation, a cave's tensions have mostly been long balanced by the time one can enter it. Only rarely have visitors experienced a spontaneous breakdown. During 30 years of research in Hölloch with close to 200,000 h of research and 100,000 visiting hours not a single spontaneous case of breakdown could be observed, not even any traces of it during the periods in between. As for changes which could not be directly observed, a single one was reported. It was caused by the interference of man who had effected a change in the direction and the velocity of flow of the high water by the construction of a wall across a passage. The degree of stability of natural cavities is typical. But human intervention may lead to incasion even here, e.g., when one brushes against loose rocks in passing, blasts, or clears a passage.

Incasion does not primarily increase the volume of subterranean cavities; it rather separates the existing space into many small portions. It reaches the effect of increasing the volume only in conjunction with the other cavity-forming forces: corrosion and erosion. Thus blocks which fall into the water are dissolved or eroded. Cavities may thereby be extended upward, taking on the forms of domes or shafts. Incasion is of equal importance for underground karstification as are corrosion and erosion.

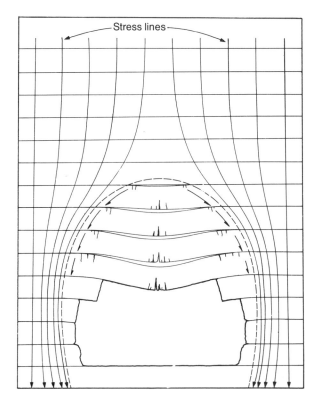

Fig. 11.6. Behavior of rock layers in a tension dome according to Davies (1951)

12 Speleomorphology, the World of Forms Created by the Subterranean Removal of Matter

12.1 Introduction

Speleomorphology is cave-oriented geomorphology, the description and interpretation of the forms in the underground which are due to corrosion, erosion, and breakdown. The wealth of forms which are the result of sedimentation (see Chap. 13) are not included in it, in contrast to surface morphology which also embraces all forms due to sedimentation.

This world of forms is classified into large forms, which embrace the smallest tube up to the largest giant cavities and the most elaborately branched cave systems, and small forms, which originate on the surface of the large forms. Caves and cave systems are dealt with in Chap. 17.

12.2 Large Forms

Table 12.1. The more important large forms: 1, ph: phreatic, v: vadose; ph/v: combined; 2: normal corrosion (exclusively vadose); 3: mixing corrosion; 4: erosion; 5: incasion; XXX: predominant or exclusive; XX: dominant; X: minor; (X) exceptional

		1	2	3	4	5
12.2.1	Passages	ph/v				
	Double passage/labyrinth, maze	ph	–	XXX	X	X
	Garland passage	ph	–	XXX	–	X
	Blind passage, pocket passage	ph	–	XXX	–	X
	Cross-sections: ellipse	ph	–	XX	–	(X)
	Asymmetrical ellipse	ph/(v)	–	XX	–	XX
	Rectangular	ph/v	X	XX	–	XX
	Keyhole profile	ph + v	X	X	X	–
	Canyon	v/ph + v	X	X	XX	X
	Joint passage (tectonic)	ph/v	X	X	X	X
	Isolated cavity	ph	–	XXX	–	X
12.2.2	Ceiling half-tube passage	ph	X	X	X	–
12.2.3	Hall, chamber	ph/v	XX	XX	(X)	XX
	Dome	v/ph + v	X	X	(X)	XXX
	Bell-shaped dome, collapse type	v	X	X	–	XXX
	pothole type (see also dome)	v	(X)	(X)	XXX	X
12.2.4	Shaft	ph/v	XX	XX	X	X
	Chimney	ph/v	XX	XX	(X)	X

12.2.1 Passages – Passage Cross-Sections

Passages are "nearly horizontal or moderately inclined cave sections" (Fink, 1973). This definition of passages is essentially valid for inaccessible (small) cavities and tubes, also.

Double passages and mazes lie in principle on the same plane whether this is a joint or bedding plane. They are phreatic forms and therefore have originated as a rule by corrosion due to mixing waters, rarely by cooling corrosion or aggressive water.

Elliptical (lenticular) passages (Bögli, 1956a; Plates 1.3, 7.3) are primarily of phreatic origin. Symmetrical elliptical cross-sections may be taken as key forms of corrosion due to mixing waters and of phreatic origin (Bögli, 1964a-c). A key form with more than one cause appears at first to be contradictory. As a decision must be reached in classifying alternatives: vadose – phreatic, and normal corrosion – mixing corrosion, it is nevertheless unambiguous.

Asymmetric elliptical cross-sections (Plate 7.2) with distinctly pronounced vaults are key forms of tension effects in the rock (see Chap. 11, Incasion). If the stronger bend is in the floor of the passage, it indicates an erosive effect; sand and gravel were carried away, mechanically deepening the floor. If the rocks on the two sides of the bedding plane behave differently, asymmetric elliptical passages can likewise be created from them.

Garland passages are forms which are derived from elliptical passages and are therefore also key forms of mixing corrosion (Bögli, 1971a). Under phreatic conditions water flows out of the generative interstice toward the side of a joint or bedding plane-controlled passage and mixes with the water-fill; this leads to mixing corrosion. The ellipse of the passage is thereby extended by curved, flat niches. In the horizontal plan – in joint passages in the profile view – the garland-formed side is very prominent while the other has no concavities, since, as it lies deeper, no water flows out into the passage from it.

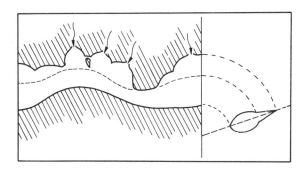

Fig. 12.1. Plan of a garland passage of Hölloch (Innominata Passage) according to Bögli (1971a). *arrows* flow of water out of bedding interstices

Blind passages (pocket passages) (Bögli, 1964a-c) begin abruptly with a semi-circular, apse-like cavity which usually continues on as an elliptical passage in the direction of the water flowing away. In the apse there are two or more small inflows of water approaching, a few centimeters in size, more rarely a few decimeters. These passages, which are usually only short, may often originate from the mixing of two or several types of water of different concentrations; these have continued to flow on one interstice, widening it to a passage.

Isolated cavities, appearing to be closed, are closely related to the blind passage; they also can only be explained by mixing corrosion. They are occasionally found in limestone quarries and tunnels (Bögli, 1964a, 1965; Fink, 1968). They have a diameter ranging from a few decimeters to many meters; they are often isometric, can, however, also be stretched in length along a joint or they can diminish in the direction of the out-flowing water. Cave karren and speleothems which occur in them betray the continuation of development in a later vadose phase.

Rectangular passages occur mainly at slight depths and when the bedding approaches the horizontal. They are the dominant form of passage in the giant caves between the Appalachians and the Rocky Mountains. They can have their origin under phreatic as well as under vadose conditions but their formation is always combined with breakdown (see Fig. 11.7).

Canyons are usually the result of a deepening under vadose conditions. If the disposition is vadose, the canyon then has one phase (see Chap. 14.4), and was created out of an open joint. There are, however, also canyons with one phase which were phreatically created by upward corrosion, called inverse corrosion by Maucci (1973). These seldom occur. If the floor of a passage becomes covered with sediments, corrosion only takes place on the ceiling. If fine, loose material is continuously swept in, the cavity will grow upward. If the sediments are carried away (erosion), a canyon-like passage will remain; it will usually have a high rectangular cross-section. The floor of the passage often runs up and down, which is a clear sign of its phreatic origin. The Paläotraun in Dachstein Mammoth Cave is a typical example (personal information from F. Bauer, 1961).

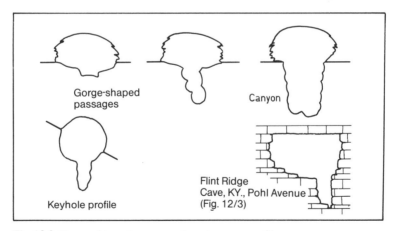

Fig. 12.2. Types of two-phase gorge-shaped passage profiles

Gorge-shaped passages and canyons can, however, also have two phases. First an elliptical or rectangular passage is formed on a bedding plane or on a joint. Under vadose conditions such a passage can continue to carry water. The freely flowing stream cuts in and forms a canyon (Plate 8.1). If the phreatic form is well-preserved, one speaks of a *keyhole profile,* whereas if the state of preservation is poor or also if the canyon strongly predominates, the term gorge passage is to be preferred (Bögli, 1956a).

In a rectangular passage the stream often meanders on the passage floor, thereby sometimes undercutting the side walls or even leaving the area of the passage. In Pohl Avenue in Flint Ridge Cave (Mammoth Cave National Park) Columbian Avenue begins as a floor canyon which becomes deeper and deeper in the direction of the outflow. It leaves Pohl Avenue then, sinking somewhat further down and becomes an elliptical passage (Columbian Avenue), an indication that here the vadose water-course dipped into the phreatic zone. After the karst water-surface sank further, however, the cave stream went on flowing for a while and thereby left behind a floor channel in the elliptical profile.

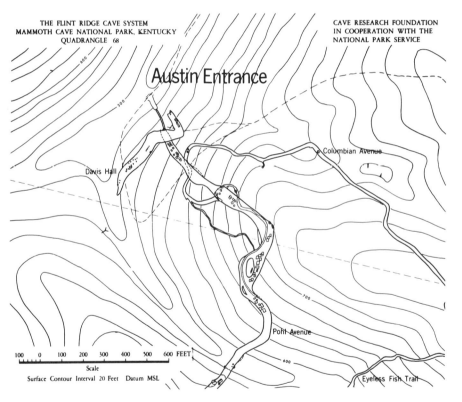

Fig. 12.3. Columbian Avenue is a canyon in Pohl Avenue, however it changes shortly afterwards into an elliptical passage (plan according to CRF Surveyance 1964)

Joint passages are, according to form and direction, always determined by joints and they are always higher than they are wide. Under phreatic conditions steeply inclined elliptical passages are created which often merge into an open joint at the top

(Plate 7.3). Under vadose conditions passages on open joints are created with heights that can be 10-20 times their width (single-phased gorge passages; Plate 7.4). Even though such passages are said to have developed from open joints, the term open-joint cave must not be used for them as it is reserved, according to Trimmel (1965), for caves which are "open toward the top" – "in this sense the big joints created by sliding rock are considered to be open-joint caves" (p. 90).

12.2.2 Ceiling Half-Tube Passage (Plate 8.2)

In the ceiling of descending elliptical passages there is frequently a channel with a cross-section in the form of a semi-circular arch to be found; that is the ceiling half-tube passage (Wirbelkanalgang, Bögli, 1956a). Such passages begin in the zone of a former piezometric surface, where the gravitational flow becomes pressure flow. When the passage dips gently the rapidly flowing water can take up considerable amounts of air and carry it along into the depths (see Chap. 15.1.4). When the passages are steeper an eddy forms on the water's surface through which air is sucked in. The water must, however, flow fast enough to be able to compensate for the upward movement of the air. The steeper the passage, the higher the speed necessary.

The air is carried along under the ceiling in the thread of maximum velocity of the water (Fig. 12.4). The pressure thereby increases by the amount of water pressure. At a depth of 10 m there is an additional pressure of 1 atm, at 20 m, 2 atm. Since in the enclosed air $p \cdot V$ is constant, the p_{CO_2} increases to double or triple its value – it must be emphasized that this is only true of air bubbles, not for the CO_2 which has already been dissolved in the water. The result is a surplus of CO_2 in the air so that the rock is corroded by the mixture of air and water; a half-tube passage forms on the ceiling. The whirling mixture of water and air probably also attacks the rock by erosion. As soon as the half-tube passage begins to form, the zone of mixing of water and air is concentrated on this area so that it stands out more and more sharply from its surroundings. This theory of the creation of such a ceiling-channel type has not yet been tested experimentally.

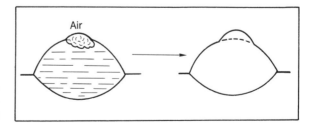

Fig. 12.4. The creation of a ceiling half-tube in a passage (Bögli, 1956a). For explanation see text

Wherever the passage continues straight ahead or has only gentle curves, the ceiling channel (half-tube) follows the highest area of the arch; if the passage makes a sharp turn in direction, the ceiling channel diverges from the direction of the passage's axis towards the outside, following the inertia of the water and the stream's thread of maximum velocity; however it balances out again at the highest point in the elliptical cross-section.

12.2.3 Dome – Bell-Shaped Dome – Chamber

Domes are large spaces with high, arched ceilings, e.g., bell-shaped domes; chambers, on the other hand, have low flat ceilings. These forms are the present result of adaptations made to the conditions of tension in the rock. Flat ceilings form in thick layers which are under little tectonic stress, domes and bell-shaped domes in thinly layered rocks which may or may not be under heavier tectonic stress. Here breakdown often reaches into nonkarstifiable strata. These forms frequently show considerable amounts of debris which can prevent access from below to the cavity created above them. Before the beginning of breakdown the original empty space must have been larger, since part of it is claimed by the many gaps between the boulders.

Domes and bell-shaped domes show all transitions to shafts; they also often show a continuation as a shaft at the top. These shafts may have been the primary form out of which an optically independent part was created by under-cutting and breakdown.

12.2.4 Shafts

Shafts are generally counted among caves with water movement in a downward direction. Today this is certainly correct but it was not necessarily the case during the period of their formation, for there are also rising springs. They can be phreatic as well as vadose in their origin, and all three form-giving factors can play a role in the creation of shafts. Tectonics always have a part in it, whether it is a matter of open joints or steep bedding planes. Because of the many overlapping parameters there is also no generally valid scheme of classification.

Morphographically the *purely vertical* cave parts are vertical to very steep (over 65°). Such are shafts in the mining sense of the term. Then there are *stepped shafts* where flatter, although short stretches of joint passages or canyons link the vertical sections.

Not only the forms which lead to the surface are called shafts but also vertical to steep connections between two cave levels. Seen from above shafts are pits or open pits, seen from below, domes or chimneys. These expressions only state the line of sight.

Gèze (1953) distinguishes morphogenetically:

a) *Tectonic shafts;* those which are created by the extension of joints and steeply positioned bedding planes. Gèze obviously understands thereby shafts in which tectonic conditions dominate physiognomically.

b) *Collapse shafts* are created by the collapse of underground cavities as long as the ceiling is not too thick to prevent a break-through to the surface. This restriction according to Gèze can be disregarded because such shafts by no means always reach through to the surface. Such a cavity in Hölloch, the Schwarzer Dom, with a volume of 400,000 m^3 and a height of 75 m is only the part in a debris-filled shaft 225 m in depth which has remained free; moreover it is 400 m below the earth's surface.

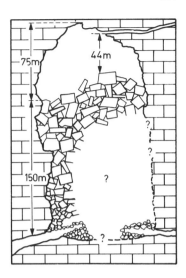

Fig. 12.5. Schwarzer Dom in Hölloch

c) *Swallow shafts* are created by water flowing or seeping in. Genuine ponors which are still active today belong to these.

d) *Spring shafts* (Fr: gouffres émissifs) have been formed by water welling up. The spring shaft of the Vaucluse Spring was explored down to a depth of more than 100 m, yet it reaches even further down (see Fig. 9.6a).

e) *Piezometric shafts* are created by the rising and falling of the water table (pressure table) in the high-water zone. Above all they are shafts between two cave levels which were created during the development of the lower system of passages. However, there are also some that reach as far as to the surface of the earth, e.g., the gouffre de Poudak (see Chap. 9.4.2). Estavelles (alternate karst holes) present a mixed form of piezometric shafts, swallow and spring shafts.

Types (c), (d) and (e) can be classed together as *water-formed shafts*. They occur frequently in poljes, either singly or in combinations. If they become inactive, they can hardly be distinguished from one another. Scallops (see Chap. 12.3.3) can help with clarification.

Gortani (1953) added to these five types:

f) *Shafts of hydrothermal origin*, e.g., the shafts on Monte San Calògero at Sciacca on the southern coast of Sicily. Also the thermal open joints in the underground beneath Budapest belong to this type.

A special type of shaft occurs in table lands, e.g., in Kentucky; Pohl (1955) names it simply and aptly a *vertical shaft*. This type is found fully developed in Mammoth Cave National Park in places where carbon limestones are covered by quartz sandstone. Vertical shafts are absolutely vertical; in their cross-section they are almost the same size from top to bottom and their cross-section is round, or more often elongated toward the joint. The surface of the walls is smooth, interrupted only for a few centimeters at the bedding planes, and straight corrosion grooves run from top to bottom. The unrelated proximity of horizontal passages and vertical shafts is striking. The shafts were unmistakably formed later. They cut through the passages purely fortuitously, whether in one broad part, or just barely, so that only peep-holes betray their presence.

Besides morphological criteria, the absence of *troglobionts* (cave animals) in the pools at the bottom of the shafts speaks for their later creation since troglobionts are otherwise frequently found in Kentucky.

Creation: water low in lime seeps through the quartz sandstone ceiling, therefore remaining low in lime. Along a joint it then crosses the limestone under phreatic conditions — therefore corrosion does not take place. If the interstice cuts through an air-filled cavity, normal corrosion results. The water seeping out dissolves the limestone and a cupola is formed. This moves upward toward the water flow until the quartz sandstone is reached. The walls of these shafts are wet even in caves which are otherwise dry. In the pools on the bottom lie boulders of quartz sandstone and chert nodules which have been loosened, but there is no limestone debris. This is a sign that breakdown in the limestone did not play a part in the formation of the shafts. Bögli (1969c) assumes that the water, which is low in lime content, seeping into the depths (lacking air) creates a primary cavity when it comes into contact with the groundwater, which has a high lime content, by means of mixing corrosion. As soon as the karst water table sinks sufficiently, air enters this cavity. Then corrosion in Pohl's sense begins and also growth upward.

12.3 Small Forms

Small forms are created with few exceptions by only one single form-giving factor. Corrosion causes the greatest wealth of forms. These predominate on the ceiling and on the walls. Erosion is limited to the lower parts of the wall and to the passage floor, e.g., potholes and, with the assistance of corrosion, scallops. Breakdown exposes bedding and jointing planes and creates angular surfaces of fractures.

12.3.1 Cave Karren

Cave karren are solution features according to the definition of karren. Bretz described various ceiling karren in 1942 and in his work *Caves of Missouri* (1956); Bögli (1963a) presented a first collection of the cave karren which are most frequent in the Alps.

Anastomoses (Bretz, 1956) are intricately twisted cavities which are connected to one another; they range from a few centimeters to up to 20 cm in diameter and are found on outcropping bedding planes in caves. Probably slowly seeping interstice water mixes with the water of the passage already within the interstice so that mixing corrosion results. A corrosive mixture of water seeping into interstices produces the same forms. Anastomoses are not frequently observed in Europe, indeed they are usually overlooked. Occasionally they are found in Hölloch, e.g., in the Titanengang. In the USA they are frequent and are given a great deal of notice as a result of Bretz's influence. The holes which appear in outcropping bedding planes in Mammoth Cave are called pigeon holes.

Pendants according to Bretz (1956), described by H. Lehmann (1956) as *Deckenzapfen (ceiling cones of rock)* are 10-100 cm long, rarely more (Plate 8.4). Neighboring ceiling cones end at the same height, a proof of the fact that these cones are remainders

Table 12.2. The more important small forms (see Table 12.1 for the symbols)

	1	2	3	4	5
12.3.1 Cave karren					
Ceiling karren					
Anastomoses	ph	–	XXX	–	–
Pendents	ph	–	XXX	–	–
Spongework	ph	–	XXX	–	–
Ceiling dimples	v	XXX	–	–	–
Cave rills	v	XXX	–	–	–
Cave grooves	v	XXX	–	–	–
12.3.2 Potholes					
Potholes due to mixing corrosion (inverse solution pockets)	ph	–	XXX	–	–
Ceiling pockets, wall pockets	ph	–	XXX	–	–
Vadose ceiling pockets (rare)	v	XXX	–	–	–
Floor potholes, erosion pockets	ph/v	(X)	(X)	XXX	–
Floor potholes with central cone	ph	(X)	(X)	XXX	–
12.3.3 Scallops and ceiling dents					
Scallops	ph/v	XX	XX	XX	–
Ceiling dents	ph	–	XXX	–	–
Ceiling half-tube	ph	XX	–	X	–
Exposed interstice surfaces	ph/v	–	–	–	XXX

of a limestone stratum in which dissolution had created winding cavities in the form of channels, moving from the bottom upward; ordinarily these forms must be counted among the anastomoses. Bretz includes them among the vadose forms, yet the formation of the cavity demands that the passage be filled with water, phreatic conditions then – whether in the vadose zone (high-water zone) or in the phreatic zone is irrelevant. Warwick (1962) assumes that the channels were created on the ceiling when sediments filled the cavity and water had to force its way between them and the ceiling. If it is a matter of individual channels only, *inverse ceiling channels* are formed, e.g., in Endless Caverns (VA, USA). In Beatushöhle (Switzerland) they are developed to the perfection of a model; they are 10 cm wide, up to 30 cm deep, and have a length of many meters. Wherever they connect in a network pattern single cones are created, pendants.

Drops of condensation water create *ceiling dimples*. Their form resembles a fingerprint. The border is generally flat. This type is frequently observed, e.g., it covers large areas of the ceilings of the elliptical passages in Hölloch. When condensation is rapid the drops may dissolve deep pits with sharp ridges and peaks; these generally occur on the underside of practically horizontal slabs. They are found to perfection in Dome Home Cave (KY, USA, outside Mammoth Cave National Park).

If the condensation water flows away *cave rills* form on steep limestone surfaces; shallow and up to 3 cm wide, they run side by side without interruption. It is assumed that they also form out of wash stripes in the clay covering on cave walls (Bögli, 1963a), when they are covered under phreatic conditions by corrosive water.

On slightly inclined surfaces the drops of water join forming *cave grooves*; these are the equivalent of exokarst's meandering karren. If unsaturated water emerges from bedding or joint planes, it immediately takes up CO_2 and corrodes. In this manner large cave grooves are formed which begin directly at the point where the water emerges. In ice caves it is not unusual to find twisted vein-like rills 1-5 mm wide and even less deep, on steep to overhanging walls: *ice-water grooves.* They can be short, only a few centimeters in length, they may, however, occasionally attain several meters in length, e.g. in Dobschau Ice Cave (Dobšinská L'adová Jaskyňa, ČSSR). They are created by the water which, held by capillary forces, seeps away between the rock wall and the ice.

When snow is blown into shafts and the rock is moistened over the whole surface *cave pinnacles* may be created, which can be razor-edged (Bögli, 1963a). Similar forms are generated in vertical shafts where water which is low in lime emerges from the open joints of the quartz sandstone and reaches the limestone, e.g., in Dome Home Cave (KY, USA).

12.3.2 Potholes, Inverse Solution Pockets

Inverse solution pockets (Plate 1.4) are located on the ceiling and on the side walls where water emerges from interstices or from isolated points: ceiling and wall pockets. A water whirlpool is not necessary for the generation of the form but it may play a part in it.

Inverse pockets form only under phreatic conditions, besides in the phreatic zone, also in the vadose where the passage is permanently filled with water (siphons) or during high water (temporary phreatic conditions).

Originally such inverse pockets were explained by erosion (evorsion), by strong, local water whirling against the ceiling (Trimmel, 1965). Bretz (1956) explains them by corrosion, which acts differentially, therefore selectively in whirling water. The usually elliptical form of the cross-section speaks against this interpretation; it is normally extended in the direction of the joint carrying the water, even when the joint crosses the passage and thereby the direction of flow. This indicates a low velocity of flow.

The fact that the pocket is localized around the point of emergence of the interstice betrays that it was created by mixing corrosion — interstice water plus passage water: *pockets due to corrosion by mixing waters.* In Hölloch during the snow melt the high water with 60 ppm of lime or less meets interstice water with 120-200 ppm in the passages. When the high water recedes and the passage becomes dry, the water with the high lime content continues to seep out of the interstices. Calcareous sinter (speleothems) is deposited. For this reason one often finds sinter formations in the inverse pockets due to mixing corrosion, especially in covered karst with its water rich in lime, e.g., in the Grotto of Castellana south of Bari. Formerly these sinters were interpreted as the effect of a change in climate.

The depth and the diameter of inverse pockets vary greatly (Plate 1.4). They range from a few centimeters up to many meters. Chimneys up to 20 m in height which are more or less rounded at the top or which end in a peak where they show small points of emerging water are to be interpreted as large inverse pockets.

As an exception there are also ceiling pockets which form in an air-filled cavity. Water low in lime emerging from the ceiling begins to corrode there and creates these pockets. Franke (1963a) attempted to clarify the laws of these formations. Generally

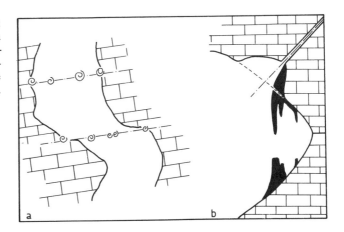

Fig. 12.6. Inverse pocket due to mixing corrosion. **a** plan of passages (*spirals:* inverse ceiling pockets); **b** section of a side of a passage with an inverse pocket. *Black* speleothems

they show additional solution grooves; in this way they differ from the pockets of mixing corrosion. Wherever such grooves are lacking it is a pocket due to mixing corrosion, if its surroundings have been formed aqueously. The most extreme example of a vadose solution pocket might well be Pohl's "vertical shafts" (1955; see Chap. 12.2.4).

Corrosion pockets are rare on the floor of the passage as seeping water usually does not rise. However, dripping water which is low in lime can attack the floor; such pockets (potholes) give evidence of a considerable proportion of erosion in their creation; and also they form only under vadose conditions. They usually reveal themselves by additional runoff channels.

Potholes due to erosion (Plate 9.1) are created by stationary water whirlpools which are laden with sand and gravel. These attack the rock base and deepen it. Centrifugal forces drive the solid particles over the floor of the pothole onto the periphery where they smooth the walls. Screw-shaped (helicoidal) holes thereby develop, reaching into the rock. These occur most frequently where water with a high velocity flows into water with a slower one, i.e., behind narrow passes. If the passage widens directly, the water flowing through it is flanked by a reverse flow so that between the two, stationary whirlpools can form. As a result of centrifugal forces the pothole's center loses its sand and gravel and erosion there is but slight. Under phreatic conditions a cone of rock is created in that place. In Hölloch (commercial part) the central cones are 5-20 cm high, but as an exception they even reach almost 1 m. Potholes due to erosion which have been created under vadose conditions never show such a cone because the air sucked into the center of the whirlpool effects additional erosion in the bottom of the pothole.

12.3.3 Scallops and Ceiling Dents

The separating surface between two media which move at different speeds is wavy. Water, flowing away over limestone, therefore creates forms of flow, so-called scallops by means of corrosion or erosion (Fr. vagues d'érosion, Germ. Fließfazetten).

There are three types of scallop, shell-shaped ones on the passage floor and on the foot of walls which are not strongly sloped (Plate 9.3), wave-like ones on the lower parts of walls, called flutes, and large flat ceiling dents.

The scallops of the passage floor are shell-shaped, following on one another without interruption. They are 5-25 cm wide and usually somewhat longer. As an exception the relationship can become w:1 = 1:5. Every scallop has a steep side and an adjoining, more gently sloping surface.

The steeper inclination indicates the direction the water flows away. This fact is of great importance for karst hydrology. Since scallops require a certain time for their formation, they are the expression of the behavior of flow during the last longer phase of water-covering.

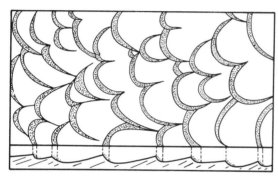

Fig. 12.7. View and longitudinal section of scallops (acc. to a photograph taken in Hölloch). *Dotted* clay deposits under 1 mm in thickness on the steep side

Bock (1913a) specified scallops as forms of erosion and set up a formula to determine the velocity of flow which is necessary to cause a certain size of scallop. Bretz (1942), on the other hand, believes their origin is due to corrosion. Maxson (1940) observed such forms on insoluble rocks; this is in agreement with the author's observations in the bed of the Aare on the granite of Handegg-Grimsel (Switzerland).

The effect of erosion can be recognized from such observations. Renault (1961a, 1961b) is of the opinion that scallops are related to insoluble gravels lying on the limestone. According to him the water must flow around the rounded stones whereby the specific scallops arise. Opposing this theory is the situation in Hölloch where there is to be sure no hard, insoluble gravel, but uninterrupted fields of scallops which are hundreds of square meters in area. The manner of flow of the water is decisive in the creation and further development of scallops.

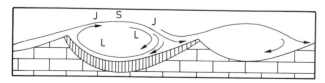

Fig. 12.8. Movement of water over scallops according to Blumberg (1970). *S* shearing surface; *L* lee-whirl; *J* jet stream; *hatched* corrosion or erosion

Fig. 12.9. Stream lines on the surface of scallops according to Allen (1972, p. 10)

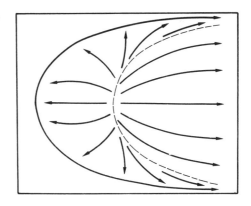

Curl (1966, 1974, 1975), Goodchild (1971) and Wigley (1972) attempted to solve the problem of the dependency of size experimentally and mathematically, assuming a constant velocity of flow v_m and a sufficiently long, straight section for the water flow. Since such conditions are not often given in nature, the results are only approximately correct; still they are of great assistance in the interpretation of karst-hydrological relationships. When the v_m is variable, the formation of scallops takes place fastest when the speeds are high. For literature see Curl (1966, 1974) and Allen (1972).

To deduct the formulas Curl uses especially flutes, which show a greater regularity than scallops. The form is delimited by edges which run from above downward. The distance L from edge to edge is indirectly proportional to the v_m. The basic equation according to Curl (1966, p. 127 f.) is:

$$\frac{\rho}{\mu} v_m \cdot L = Re_L$$

Re_L: Reynold figure for flutes
μ: coefficient of viscosity
ρ: density

$\frac{\mu}{\rho}$ is the kinematic viscosity ν of the water.

Table 12.3. Kinematic viscosity ν for water

	0°C	5°C	10°C	15°C	20°C	25°C	30°C
ν:	0.0179	0.0152	0.0131	0.0114	0.0100	0.0090	0.0080

Re_L can be determined with the aid of the diagram according to Curl (1974, p. 4); in it Re_L is set opposite to the ratio of the passage's diameter D to the average length L of the flutes. The Reynold figure is somewhat higher for a rectangular cross-section than for a circular one.

Since D and L are directly measurable, Re_L is easily determined.

Example: D of a rectangular passage measures 300 cm, the average length L of the flutes when the smallest are excluded is 15 cm, the water temperature 5°C.

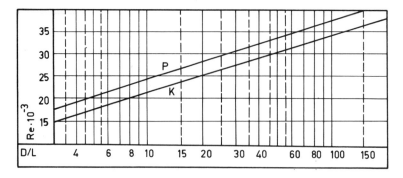

Fig. 12.10. Relationship between Re_L and D/L according to Curl (1974). P passage with rectangular cross-section; K passage with circular cross-section

D/L = 20 from which Re_L can be determined to be 28,000.
$v_m \cdot L = \nu \cdot Re_L = 0.0152 \cdot 28,000 = 425.6$ cm²/s,
$v_m = 28.4$ cm/s or 1021.4 m/h

A passage with a circular cross-section results in Re_L of 25,000 when D and L are otherwise equal and v_m is therefore 25.3 cm/s or 912 m/h, which is about 11% lower than in the case of the rectangular passage.

Halliday (1962) determined in Boyden Cave (CA) an L of 6.5 cm and calculated the velocity at about 50 cm/s at 8°C ($\nu = 0.0138$ cm²/s); this concurs very well with the actual velocities.

Curl based his investigations completely on corrosion. There are, however, many indications that the actual scallops are created mainly by mechanical erosion, e.g., by water containing sand (corrasion). If the forms were created corrosively, surely the roof of the cave would also have to show scallops similar to the others in size and shape: this is, however, only so in exceptional cases and only in narrow passages where solid particles (sand) carried by the water are swept upwards to the roof. The distribution of the forms in the passage cross-section indicates that gravitational forces play an important part here. Corrosion takes place without the vital participation of any such forces. Allen (1971), one of the best authorities on the forms created on rock surfaces by removal, who classes flutes and scallops as erosion forms, nevertheless writes: "These marks, now recognized to be of solutional (mass transfer) origin..." (p. 177).

Thus the manner of flow determines not only the amount of corrosion, but also of erosion, so that the latter must generate similar forms. It must be taken into consideration that ceiling dents, which are formed by corrosion only, are several times longer than the flutes or scallops in the same cross-section. A general equation for v_m which would also take erosion into account would not be of any great use because of the difficulty of determining the parameters.

13 Cave Sediments

The term cave sediments embraces all clastic, organic, and chemical deposits occurring in subterranean cavities and cave entrances. They are the same three main groups as on the earth's surface. The clastic deposits hardly differ from those on the surface. Organic and chemical sediments differ less chemically than morphographically than the corresponding ones on the surface. For cave ice refer to Chapter 16.

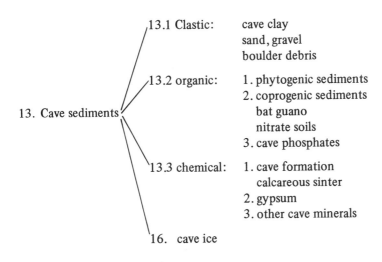

13.1 Clastic Sediments

Clastic sediments form from deposits of fragments of pre-existing rocks which were mechanically destroyed. Not the origin of the components, but only the place where they have been deposited is decisive in their classification as cave sediments. Allochthonous here means an origin outside the cave.

The kind of clastic cave sediment is often determined by the climate on the earth's surface; this is especially true of those in cave entrances (Schmid, 1958; Riedl, 1961). Frost loosens the rock and increased breakdown takes place. Many cave entrances therefore have a wide opening although at the back of the entrance hall only small passages lead into the mountain, if any. Congelifracts — angular rock fragments produced by frost action — are to be found in many alpine caves. They form under periglacial conditions in caves which are situated close to the surface, in so-called subcutaneous caves (Ciry, 1959 and chap. 7.4), however they occur in ice caves, also, especially in dynamic ones (see Chap. 16). In Eisriesenwelt (Austria) the floor behind today's ice section is covered with congelifracts for hundreds of meters, which are proof of the much more extensive frost zone of earlier days. In caves where today the temperature never sinks below 0°C congelifracts are an indication of earlier periods of cold. Debris from breakdown and frost is always autochthonous.

13.1.1 Coarse Clastic Cave Sediments

Gravel from karst rocks can be allochthonous as well as autochthonous. The proportion of autochthonous components increases rapidly with granulation and with the distance from the point of entrance of the cave water and, with the exception of quartz pebbles, e.g., Augenstein, reaches 100%. The causes are the special conditions of erosion in cave waters. According to Brinkmann (1967, p. 67) limestone debris is rounded after 1-5 km, and after 10-50 km ground down to half its size. However, in caves the distance required to reduce the debris to half its size is extremely short: only a few 100 m when there are no ascents. If the water's course includes ascents, coarse rock fragments remain lying at their foot, narrowing the cross-section of the passage. The higher velocity of flow caused by this results in a portion of the coarse components being carried along up the ascent behind the narrow place. In the following wider section, where the rate of flow diminishes, the gravel rolls back again and is ground down until it is small enough to be carried away, i.e., as pebbles and coarse-grained sand. It must be pointed out that the rate of flow varies greatly at the narrow passes; from low water to high water from 1:5 up to 1:50, rarely higher under phreatic conditions. Thus the distance necessary to halve the diameter of limestone pebbles is frequently reduced to under 100 m.

Pebbles are broken when the velocity of flow is high and when they fall over steps. Therefore well-rounded components are rare in many caves but on the other hand subangular ones are frequent.

Of the nonkarstic rocks especially quartzites and magmatic rocks are resistant and are only slightly eroded, e.g., the Tertiary quartzite pebbles of many Austrian caves, called Augensteine. Chert nodules are resistant and usually autochthonous since they occur in many limestones, as in the Mississippian in the USA or in the chalk strata of Western Europe. Debris from breakdown is also included in coarse, clastic sediments; its components range between the size of a grain of sand and angular boulders of several cubic meters. Davies (1949) classified breakdown debris into four groups: blocks, plates, slabs and chips.

13.1.2 Fine Clastic Sediments

In warm, humid climates chemical weathering predominates. Thus in these regions a lot of fine material is carried in; in cold, humid climates mainly coarse clastic material. Fine clastic sediments — as opposed to the coarse clastic ones — have a relatively large surface and the rate at which they sink in water is correspondingly slower so that they can be carried far in the cave without becoming noticeably smaller. Only a small portion of them stems from residue of the limestone dissolved in the cave. Limestones which are very karstifiable are pure, usually containing less than 5% clay impurities,

Fine Clastic Sediments

occasionally less than 1%. Yet the volume of clay deposited in a cave often exceeds 5% of the volume of the underground cavities. Moreover there are also great losses of fine material due to its being carried away out of the cave. For these reasons the main portion of cave clay must consist of allochthonous components; this conclusion is supported by its mineral composition and its content of organic substances.

Table 13.1. Sizes of grains of clastic sediments according to DIN 4022

Loose sediment	Diameter of grain	
Clay	Less than 0.002 mm	Pelite (mudstone)
Silt	0.002-0.06 mm	(siltstone)
Fine sand	0.06-0.2 mm	Psammite (sandstone)
Medium sand	0.2-0.6 mm	
Coarse sand	0.6-2 mm	
Fine pebbles	2- 6 mm	Psephite (conglomerate, fine breccia)
Medium pebbles	6-20 mm	
Coarse pebbles	20-60 mm	
Stones	Over 60 mm $\emptyset$	(Coarse conglomerate, coarse breccia)
Breakdown debris	Up to several m $\emptyset$	

When serial investigations are undertaken the distribution of the organic material in a series of sediments allows some conclusions to be drawn concerning the situation of the place of sedimentation in reference to the surface of the water. Organic materials, e.g., particles of humus, have a density of a little over 1 g/cm^3, fresh components of plants, for example needles, seeds, have a little under 1 g/cm^3. Therefore the latter float and are occasionally deposited on the edge of the water's surface as suggestions of small littoral banks. The former, however, are incorporated into the sediments, those deposited near the surface of the water contain approximately three times as much organic material as those in deeper regions of the water.

Table 13.2. Composition of three cave sediments according to Groner (1977); recent: Hölloch (Styx section)

Situation and behavior	Org. Mat.	< 0.002 mm	0.002-0.063 mm	0.063-2 mm
Sample 19. Rutschbahn; almost calm water; occasionally small influx	2.2%	12%	65.5%	22.5%
Sample 21. Rutschbahn; 15 m under water's surface; almost calm water	1.7%	10%	75%	15%
Sample 23. Sylvia Lake, below Rutschbahn; near outflow, water somewhat agitated	0.7%	9%	73%	18%

13.1.3 Conditions of Sedimentation

The approximate velocity of flow of the water at the moment of sedimentation can be determined from granulometry. The smaller the diameter of the components, the longer the time necessary to deposit them. The velocity with which the particles are transported is always less than the velocity which is necessary to tear them away from an agglomeration of loose sediments; it is, however, greater than the velocity at which the particles are deposited. In the case of coarse pebbles the velocity of the water is 50% greater for corrasion than it is for deposition; in the case of coarse sand, however, it is approximately three times as great. The forces of cohesion increase parallel to the degree of fineness of the grains. In the case of fine sand of 0.1 mm ∅ the limits of velocity during accumulation and erosion are 1:20; in the case of silt of 0.01 mm ∅ they are 1:1000 (according to Hjulström's Curve, 1935).

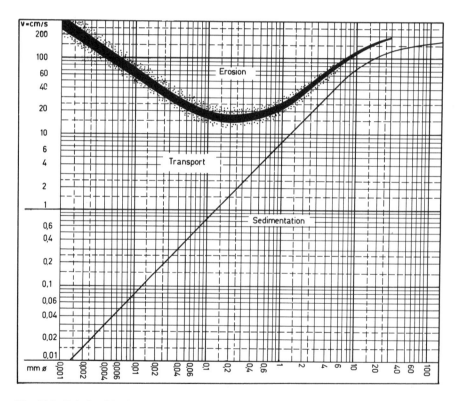

Fig. 13.1. Relationships between velocities of flow, removal and sedimentation according to Hjulström (1935). Mean values

Conditions of sedimentation are different in the tubes in the phreatic zone and in the high-water zone, and they are essentially different again in piezometric tubes.

In the *phreatic zone* the velocity increases in the tubes during the rising of high water (H), remains at the maximum for a certain time and decreases to the former value.

Conditions of Sedimentation

The diameter of the components is therefore at first fine (F) — clay to silt — becoming coarser (G), coarse silt to fine sand or even to coarse sand, depending on the situation, and reverses again when the high water subsides.

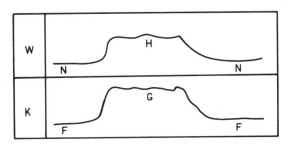

Fig. 13.2. Water level (W) (or velocity of flow) and size of the grains (K) in the sediments deposited during a high-water phase (H) in the tubes of flow in the phreatic zone (N low water) (see text, also)

Every high-water phase results in two layers consisting of finely and coarsely grained material, respectively. The appearance of such sediments recalls varved clay (Plate 10.1). The term varved clay should be avoided as this is not a case of fluvioglacial or seasonal variations in the size of the grain. Bull comes to similar conclusions (1977).

In narrower sections of passages the velocity of flow so increases during high water that old, fine sediments are also swept away. If, for example, v is 60 cm/s, silt of 0.01 mm ∅ is eroded, but at the same time gravel of 8 mm ∅ is deposited. The sedimentation of coarse sand on silty clays can frequently be observed without any signs of erosion, (see Fig. 13.9 and accompanying text).

In the *high-water zone* the water-course lies dry during the period of normal water level. When the water begins to rise the fine sediments in the feeder tubes are washed away if the gradient is one-sided and only the coarse components are left there. When the water level rises the passage is filled and conditions become phreatic. If the cave stream reaches the karst water surface and thereby the karst water-body, the velocity is immediately reduced. The debris that has been dragged along is dropped. During the phreatic phase not only fine materials are deposited by the cloudy water, but also organic substances and clay remain on the ceilings and walls through cohesion — sedimentation contrary to gravitation.

When vadose conditions begin, the feeder washes the fine material out of the debris so that the stones lying in the splash zone become clean; those beyond, however, remain covered with clay.

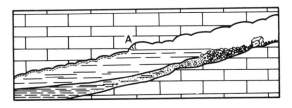

Fig. 13.3. Sedimentation and hollowing (A) where the stream passes into the karst water surface

Sedimentation proceeds differently in passages with varying directions of incline where as a result depressions occur which become siphons during high water. In the descending branch of the first depression any coarse material that has been carried along remains behind. Since the depression (siphon) fills with water, the rate of flow usually diminishes to such an extent that fine material, too, is deposited, especially in the ascending branch. If the water now flows over into the next depression, the descending branch is washed clean, while in the ascending branch fine material is again deposited. When the high water decreases the water in the depressions seeps into any available open joints or flows away downward through side passages (see Chap. 5.4 and Fig. 5.7).

The depressions can, however, also fill up from below. The overflowing water — the direction is unessential — will cause a flushing phase in the neighboring depressions. An analysis of the distribution of sediments therefore provides information concerning the behavior of karst water during high water.

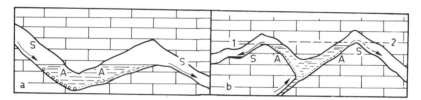

Fig. 13.4. Accumulation and flushing in passages with varying directions of incline. *A* accumulation, coarse: *dotted*; fine: *hatched*. *S* flushing (rock clean). **a** by feeder (*circles* gravel) **b** by water bubbling up. *1* first phase of filling; *2* second phase of filling

In every high-water course there are interpolated narrow passes, sections of higher velocities of flow. The coarse material which is swept through such a narrow place remains lying in the following wider section. If the narrow places are ascending and if the wider section coincides with a considerable diminishing of the slope, high gravel banks form there with their axis in the direction of the current. The water, which is now forced to go around the gravel bank, flows on the side along the wall.

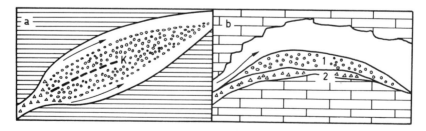

Fig. 13.5. **a, b.** Gravel bank, Kiesburgen in Hölloch, schematicly. **a** ground-plan; *K* crest of the gravel bank. **b** longitudinal section; *1* young pebbles; *2* old deposits, stoney-rocky

Conditions of Sedimentation

Piezometric tubes are ascending passages which do not have water flowing through but are filled from below up to the piezometric surface. If they begin in the phreatic zone, the deepest part always lies under water and shows correspondingly large accumulations of sediments which are usually clayey-silty. Piezometric tubes which begin in the vadose zone fill up only after the water begins to rise. There is only a current in piezometric tubes when water moves through them to fill newly attained cavities. As opposed to passages which have a flow of water through them, the water in them comes to rest during high water and remains in them as long as the high water level does not essentially change. The finest material is deposited. When the high water recedes again a channel is washed out of these sediments. Therefore in the phreatic zone every high-water phase effects a double sequence of fine and coarse deposits, one while it is rising and one while it is subsiding.

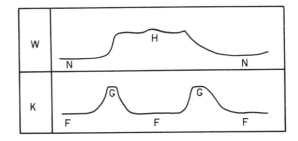

Fig. 13.6. Water-level (*W*) and size of the grains (*K*) of loose sediments in a piezometric tube ascending out of the phreatic zone (see Fig. 13.2)

The distribution and the sizes of the grains of fluvial sediments provide information concerning former velocities of flow, the situation of the high-water zone and the main connections. For example, the Titanengang in Hölloch is clean as it once was a high-water course; the piezometric tubes which depart from it are, however, generally coated or filled with clay. A granulometric analysis of the sediments almost always brings rewards. The results are usually shown in tabular form and/or in gradated curves of the percentile proportions of the grain fractions.

The central value (median Md) indicates that position in the gradation curve where the total weight is halved and divided into areas with finer and coarser components. The median is closely related to the velocity of flow (Sindowski, 1961). For sample 13 of the sediments of Mammoth Cave (KY, USA) this point lies at a grain diameter of 0.0034 mm, for sample 27 at 0.1 mm and for sample 33 at 0.17 mm. With the aid of the Hjulström Curve (Fig. 13.1) one could from this arrive at velocities of flow of 0.025 cm/s, 0.75 and 1.4 cm/s respectively. These are values which do not differ very greatly from those reached by other means (Fig. 13.8). Thus for a first orientation they are quite useful.

By division into four parts Q_1 (25%) and Q_3 (75%) join the median; they give information concerning the degree of grading (grading coefficient S) and concerning their asymmetry, their so-called skewness (Sk).

$$S = \sqrt{\frac{Q_3}{Q_1}}.$$

The grading coefficient results in values above 1 since $Q_3 \geqslant Q_1$. A simple division classes $S < 2.5$ as graded, such > 3.5 as ungraded. Sindowski (1961) makes further differentiations:

$S < 1.20$	very well graded
$S\ 1.20-150$	well graded
$S\ 1.50-2.50$	moderately graded
$S > 2.50$	badly graded

The skewness Sk shows the situation of the curve's center of gravity. At $Sk > 1$ the fine portion is better graded, at $Sk < 1$ the coarse portion.

$$Sk = \frac{Q_1 \cdot Q_3}{Md^2} \quad \text{(acc. to Pettijohn, 1957).}$$

Table 13.3. Evaluation of the gradation curve of cave sediments from Mammoth Cave (Fig. 13.7), Md, S and Sk

Samples:	13	27	33
Md	0.0034 mm	0.1 mm	0.17 mm
Q_1	0.0014 mm	0.023 mm	0.11 mm
Q_3	0.011 mm	0.24 mm	0.23 mm
S	2.83	3.22	1.41
	Badly graded	Ungraded	Well graded
Sk	1.3	0.6	0.9
	Fine fraction better graded	Coarse fraction better graded	

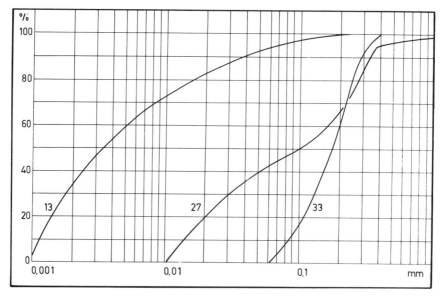

Fig. 13.7. Gradation curves of sediments from Mammoth Cave (KY, USA) according to Davies and Chao (1959). *13* silty clay, Boone Avenue; *27* silty sand, Pensacola Ave.; *33* sand fill, Sims Pit Ave

Interpretation of a Sediment Profile

Granulometric gradation curves, capable as they may be in giving information, are not suitable for the analysis of velocities of flow, as the peaks are not clearly recognizable. However, in the percentage histogram the peaks are clearly in evidence.

Fig. 13.8. Percentage histograms to Fig. 13.7

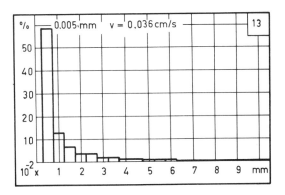
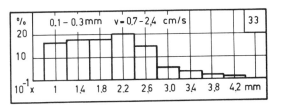
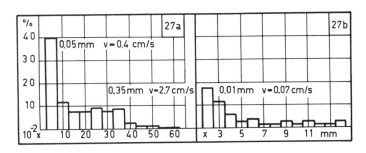

The velocities of flow during the sedimentations according to the Hjulström Curve for samples 13 and 33 are around 0.036 cm/s, and between 0.7 and 2.4 cm/s respectively. The more complex sample 27 shows a main peak which is to be associated with velocities between 0.07-0.4 cm/s, and a subordinate range of velocities between 0.4 and 2.7 cm/s.

13.1.4 Interpretation of a Sediment Profile

Clastic sediments reflect a part of the history of development of a cave from the beginning of sedimentation. The chronology remains relative, however, as long as there are no criteria to fix the age absolutely by means of correlations. In the following example the passage correlates to the preglacial valley bottom which, in the role of local base level, determined the altitude of the cavity. The sediments are therefore at the most of preglacial age. Mainly there is an absence in these cave sediments of any key fossils. Moreover the age of the sediments is much too great for ^{14}C-dating to be considered.

Example: the Rote Gang (Red Passage) of the upper system in Hölloch corresponds to the late Pliocene terrasses in Muota Valley (which are termed preglacial by Swiss geomorphologists). In it there is a series of deposits which betrays interesting relationships.

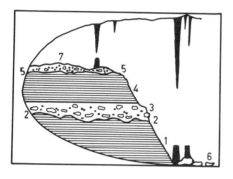

Fig. 13.9. Sediments from the Roten Gang, Hölloch (for details, see text)

Above the rock there lies a reddish-brown, ferruginous clay with 63% granulometric clay fraction; it is practically free of organic substances (1 in Fig. 13.9). Its upper limit is determined by the surface of erosion (2-2), covered by gravels up to 6 cm ∅ and coarse sand in a clayey matrix (3). With some certainty 2 and 3 are syngenetic, for the granulation of the clay deposited indicates a rate of flow of at least 170 cm/s during erosion, whereby coarse gravel of approx. 6 cm ∅ is deposited. Cold periods, seen as a whole, have been recognized as being periods of increased erosion of fine material and of sedimentation of coarse clastic sediments in many caves. Since the clay had to be classed as preglacial, this layer corresponds to the first cold period. Above it the first interglacial period left a yellowish-brown, silty clay (4), such as one finds as the oldest deposit in several places in the upper cave system. It shows a clay fraction of 36.8% and a low humus content. After this phase there followed a new period of thoroughgoing erosion (5-5) which removed large portions of the sediments previously deposited right down to the passage floor. Some pebbles have remained (6). The age of this clearance cannot be determined. Some time later well-washed, grey, medium sand was deposited with some gravel (7). These are lying in thin layers mainly on the older sediments, less often on the passage floor, where they were more easily washed away. The stalactites are the most recent formation. Tests to determine their age with the ^{14}C-method have resulted in a maximum age of 12,000 years (unpublished investigations of Fantidis); they are postglacial.

13.1.5 Clay Minerals — Heavy Minerals

Research concerning *clay minerals* has not been conducted to any great extent. During his investigations in Douglas Cave (N.S. Wales, Australia), Frank (1969) found that, from the top downward in a profile, kaolinite decreased, chlorite increased, while the illite content remained quite constant. He interpreted this as the result of climatic changes during the last 30,000 years. In a humid climate that means an increase in warmth. In Hölloch illite and mixed layers of illite-montmorillonite were detected. Up to then only illite, which indicates a humid climate, had been found in Pleistocene cave clays, whereas the mixed layers, which have only been found in "preglacial" clays (layer 1 in Fig. 13.9) give evidence of a somewhat drier climate.

Nor have many analyses been made of *heavy minerals.* Davies and Chao (1959) analyzed the sediments of Mammoth Cave. They agreed approximately with those of the surface sediments. There were found zirkon, tourmaline, rutile, goethite, leucoxene, and chromite. In the medium and higher cave levels these were joined by authigenous barite and celestite, which indicate that these strata are older. In the ČSSR the analysis of heavy minerals is integrated in paleogeographical and stratigraphical research (Misik, 1956).

13.1.6 Small Forms of Fine Clastic Sediments

When formed under water, the surface of cave clay either runs parallel to the surface of the rock under it, or is horizontal. Lamination shows the same arrangement, also.

Mud stalagmites (Plate 10.3) are created by dripping water which splashes fine material to the side, thereby building up a wall around the splash-cup. In the inside, a crater a few centimeters in diameter is formed. Turbid high water deposits new mud in it, which is again flung toward the embankment. Mallot and Shrock (1933) investigated mud stalagmites up to 45 cm high and 15 cm in diameter in Lost River Cave (IN, USA). Often the dripping water deposits some lime which makes the mud stalagmite more resistant. In silty-sandy cave clays the water seeps through. If it is rich in lime, a portion of the CO_2 volatizes and sinter is deposited. This strengthens the sediment under the mud stalagmite and forms a hard, inverted cone which is called a *negative stalagmite* or a *conulite* (Amer.) (Fig. 13.10).

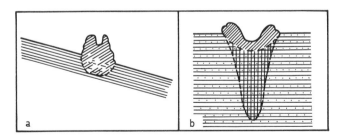

Fig. 13.10. a Mud stalagmite; **b** conulite

During periods in which the water is at rest a portion of the material in suspension is deposited, creating a thin covering layer of clay. When this dries, it shrinks, cracking open so that scales of clay are formed which curve upward at the edges, measuring from a few cm up to 20 cm in diameter (Fig. 13.11).

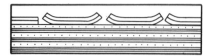

Fig. 13.11. Clay scales on banded cave clay

During sedimentation of the finer granulometric components, especially of the silt fraction, there is an increase in the density in the water on the steeper surfaces of the cave passage (water + not yet deposited sediment); this results in a downward movement of the water. Therefore the degree of sedimentation increases downward. In addition, grooves are formed in the direction of the slope (Plates 10.2, 10.3). They are not created so much by erosion as by the varying amount of sedimentation. On a gradient of 20° they become more and more distinct with an increasing angle of inclination. Ordinarily they are in rows, one after the other without a gap, and they have a V-shaped cross-section. If they are wider than 5-10 cm, small, secondary furrows occur on their flanks; these also cover the surface without any gaps and run parallel toward the median line. Such furrows are created in the phreatic as well as in the vadose zone. Occasionally they are defined on their upper end by a horizontal line, the border of the former water table. Such grooves are found in many caves (Bögli, 1976a) but they are frequently overlooked. The sediment contains not only silt but also clay in changing portions. Groner (1977) found in sample 34 that the sediment from the wall of the Osiris passage (Hölloch) contained 78% silt and 19% clay. The clay is carried along by means of cohesive agglomeration on the silt particles. The particles are caught on vertical walls also by means of cohesion. The appearance of the surface changes on steep walls — cave clay forms striped, drop-like bulges. Bull (1977) investigated the clay furrows, which he calls surge marks, of Ogof Ffynnon Ddu and in Agen Allwedd in southern Wales; he established the dependence of the forms on the degree of inclination of the surface. Under 40° they are lacking, between 40° and 60° they are frequently dentritically branched (which corresponds to the above-mentioned clay furrows) and on even steeper inclinations they run exactly along the slope, straight and parallel.

Vermiculations occasionally occur on walls and ceilings; they are crooked, sinuous or hieroglyphic stripes and networks of clay and organic material. The diameter of the individual meshes lies between a few millimeters and 10 cm, their thickness is under 1 mm. The range of color extends from light yellowish brown to black; light vermiculations are usually overlooked. The nature of these has not been sufficiently investigated. Renault (1953) interprets them as thin coatings of residual clay on the rock's surface, through which the rock's moisture seeps. Yet this does not seem very probable. Besides Renault (1953, 1963), Choppy (1955), Pommier and Garnier (1955), and others in France have been engaged in studying these forms. The author has found vermiculations of black and greyish yellow colors in many places in Hölloch on impermeable limestone (Schrattenkalk, Lower Cretaceous). Here they have formed from particles as fine as dust under the influence of the surface tension of water drops. On the Gargano Peninsula black vermiculations decorate the ceiling of a Marian grotto; they were created by the soot of burning candles, which gathered around drops of condensation water. Renault (1963) mentions such in the catacombs of Paris and in the station for experimental speleobiology in Naples. In a small cave near the Grotta del Vento (Prov. of Ancona, Italy) they occur as a brownish-black network and in Pivka Jama (Postojna, Slovenia) they are large, showing the most diverse forms, even radiating points. In all these cases the material obviously has its origin in the atmosphere; its dust particles have come to rest on the damp surface dotted with drops of condensation water. In Pivka Jama the material might have originated in the spray of the turbid water of the wildly frothing Pivka. The drops of condensation water trickle down the steep surfaces so that vermiculations are elongated.

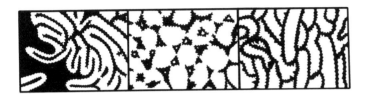

Fig. 13.12. Main types of vermiculation. *Left* Hölloch; *middle:* Pivka Jama; *right* lower Grotta del Vento/Ancona

13.2 Organic Sediments

Organic sediments in caves are those that either originate directly from organic substances, e.g., guano, or those that have obtained their characteristics from organic admixtures and their resulting products, e.g., bone breccia and phosphate soils.

13.2.1 Phytogenic Sediments

Phytogenic sediments are created from the remains of plants. They are usually found close to the surface: leaves that have been blown in, tree trunks that have fallen into shafts. In silvan karst tree trunks mixed with boulders occasionally form the bottom of the shaft and are covered by a layer of litter which is often several meters thick. In cool, humid climates decomposition often progresses even more slowly than the increase in new material. In warmer regions nests of animals form accumulations of plant remains, thus in the USA there are nests of cave rats to be found, e.g., in Kentucky. Beyond that, however, it rarely happens that layers of phytogenic sediments are developed.

Organic substances, mostly humus, are normal components of Pleistocene and younger cave sediments in Central Europe. The proportion of them in Hölloch varies between 0.2% and 5% (Bögli, 1961a; Groner, 1977).

Fragments of plants decay when air has access to them or, in the case of the exclusion of air, they remain preserved with increasing carbonification. The first is the case with small parts of plants washed in, e.g., leaves, needles, seeds, which moreover provide food for cave animals; the second results in permanently wet cave clay; wood embedded in it becomes black and brittle.

13.2.2 Coprogenic Sediments

Coprogenic sediments have their origin in excrements. Guano, in the original sense, namely the excrement of sea birds, is found occasionally in shelter-caves along coastal cliffs protected from the rain. Cave guano is, however, as a rule converted bat droppings which collects in meter-high strata beneath the places where these animals hang. Abel (1922) suggested the name chiropterite for it (Chiroptera = bat) but it did not become adopted. Cave guano is removed in wartime as valuable fertilizer, e.g., in New Cave (NM, USA; Hill, 1976). Rarely dung balls of cave rats are found in larger amounts in Kentucky and Indiana.

Through the accumulation of excrements and urine the soil is enriched with compounds of nitrogen which become oxidized and transformed into nitrates. Because of their solubility they are only retained in dry caves. Kraus (1894) mentions finds in Brazil and in Luray Cave in the USA. However, the nitrate soils of many caves in the Midwest were of greater importance. Such soils from Mammoth Cave were already used by the early Indians and later by Civil War troops for the manufacture of gunpowder, (Faust, 1967; Watson, 1969, 1974).

Coprolites which are little decayed are very suitable for the ^{14}C-dating method. Long and Martin (1974) determined the age of the dung piles of the extinct mammal *Nothrotheriops shastense* Sincl. in order to establish the extinction of this species. The aridity of the localities of the finds (AZ, USA, among others Rampart Cave in Grand Canyon) preserved the excrements from decay by fungi and bacteria. The most recent remains show a mean age of 11,070 years, while some from a depth of 99 cm show 36,200 ± 6000 years and the oldest, those lying at a depth of 132 cm, more than 40,000 years; thus they are older than the limits of detection.

13.2.3 Cave Phosphates

Cave phosphates are almost without exception of organic origin. They also originate in animal remains. In primary deposits they are found in situ, thus in guano, in cadavers and bones, in bone breccia, and bone soils. Secondarily they are due to the reaction of seeping water containing phosphates with cave sediments: phosphate weathering, *phosphate soils* develop. *Bone soils* are transitions between bones and phosphate soils. Common cave phosphates are usually Ca compounds. Where still other available cations occur in sediments, especially in regions with ores, compounds form with Al^{3+}, Fe^{3+}, Mg^{2+}, Zn^{2+}, K^+, and NH_4^+ out of guano. As additional anions F^-, OH^-, CO_3^{2-}, and SO_4^{2-} are found occasionally.

the biphosphates, are more frequent, above all brushite, in addition struvite (Hill, 1976). In the Drachenhöhle (Dragon Cave) of Mixnitz 60%-70% biphosphates occur (Kyrle, 1923).

In central Europe cave phosphates were removed from many sites during and after the First World War, mostly in the eastern Alps, less often in the Carpathians. These deposits were created above all by cave bears whose bones are to be found in each of them. The most famous is the Drachenhöhle of Mixnitz in Styria, where the thickness of the deposits in part reached 9 m. It contained 25,000 m^3 of phosphate soils. In addition there are many smaller caves in Austria. The mean P_2O_5 content is 10%. From 1919-

Table 13.4. Cave phosphates (see Mineral List Chap. 13.3.3)

Ca-phosphates

Apatite:	fluorapatite	$3Ca_3(PO_4)_2 \cdot CaF_2$
	hydroxylapatite	$3Ca_3(PO_4)_2 \cdot Ca(OH)_2$
	carbonate-apatite	$3Ca_3(PO_4)_2 \cdot CaCO_3 \cdot H_2O$
Martinite		$Ca_3(PO_4)_2 \cdot CaCO_3 \cdot Ca(OH)_2$
Whitlockite		$Ca_3(PO_4)_2$
Ardealite		$CaHPO_4 \cdot CaSO_4 \cdot H_2O$
Brushite		$CaHPO_4 \cdot 2H_2O$
Monetite		$CaHPO_4$

Phosphates containing Al

Crandallite	$CaAl_3(PO_4)_2(OH)_5 \cdot H_2O$
Taranakite	$6AlPO_4 \cdot 2KOH \cdot 18H_2O$
Variscite	$AlPO_4 \cdot 2H_2O$

Phosphates containing Fe

Diadochite	$4FePO_4 \cdot Fe_2(SO_4)_3 \cdot 2Fe(OH)_3 \cdot 24H_2O$
Tinticite	$2FePO_4 \cdot Fe(OH)_3 \cdot 3H_2O$

Phosphates containing Zn

Hibbenite	$2Zn_3(PO_4)_2 \cdot Zn(OH)_2 \cdot 7H_2O$
Hopeite	$Zn_3(PO_4)_2 \cdot 4H_2O$
Parahopeite	$Zn_3(PO_4)_2 \cdot H_2O$
Scholzite	$Ca_3(PO_4)_2 \cdot Zn(OH)_2 \cdot H_2O$
Spencerite	$Zn_3(PO_4)_2 \cdot Zn(OH)_2 \cdot 3H_2O$
Tarbuttite	$Zn_3(PO_4)_2 \cdot Zn(OH)_2$

Phosphates containing Mg and NH_4

Biphosphammite	$NH_4H_2PO_4$
Newberyite	$MgHPO_4 \cdot 3H_2O$
Struvite	$Mg(NH_4)(PO_4) \cdot 6H_2O$

1925 alone, 24,000 tons of cave fertilizer were removed (Götzinger, 1928; Saar, 1931). The oldest information concerning cave phosphates and their exploitations is given by Buckland (1823) and refers to the caves of the Franconian Alb.

In the Carpathians a supply of 15,000 t of phosphate soil with 8% P_2O_5 was identified in 1923 in the cave of Vypustek (ČSSR, Mähren), (Frodl). According to Götzinger (1919) almost 30,000 t of cave fertilizer had been removed from the Cave of Csoklovina (Siebenbürgen) up to 1919.

13.3 Chemical Sediments

13.3.1 Limestone Deposits

When $CaCO_3$ is precipitated from freshwater calcareous mud (Germ. Seekreide), calcareous tufa, calcareous sinter and moonmilk are formed. No calcareous mud is precipitated underground, no moonmilk forms above ground.

Carbonates are deposited in a reaction at the interface liquid/solid in the system carbonate-CO_2-water. This process has its place in a system of numerous equilibria. Every change in the temperature or in the concentration of any component in the system, from the CO_2 in the atmosphere to the interface liquid/solid, is the cause of a readjustment of all equilibria. In an equilibrated solution every loss of CO_2 leads to an oversaturation with Ca^{2+} and finally to the deposit of carbonate.

Roques' experiments (1964) have shown that a metastable state exists, which explains the occurrence of oversaturated solutions in nature (Thrailkill, 1968). The smaller the excess amount of Ca^{2+} and CO_3^{2-} is, the longer this state continues. His laboratory experiments showed — supposing that experimental conditions remain constant — that solutions with a $CaCO_3$-concentration between 150 and 200 ppm need 20 d to reach equilibrium with the CO_2 in the atmosphere, but after 45 d there was still no visible formation of calcite-nuclei in spite of oversaturation. With 250 ppm of $CaCO_3$ the CO_2 equilibrium was reached similarly in 20 d, but after further 10 d crystal nuclei became visible. At 350 ppm the formation of nuclei was reached after 9 d, at 400 ppm after 1 1/2 d.

In the practical situation when the temperature remains constant, a change in the p_{CO_2} is the most important reason for the disturbance of the equilibrium, and therefore for precipitation, or corrosion, respectively.

$$Ca^{2+} + 2HCO_3^- \rightarrow CO_2\uparrow + H_2O + CaCO_3\downarrow.$$

CO_2 losses are only possible under vadose conditions, and then only when the CO_2 in the water has a higher concentration than corresponds to the p_{CO_2} in the atmosphere. Then there follows an oversaturation of Ca^{2+} and CO_3^{2-} and, after termination of the metastable phase, $CaCO_3$ precipitation. However, assimilating plants may also take their CO_2 from the water rather than from the atmosphere. This, too, leads to the precipitation of lime. It is deposited rapidly which is why a porous, friable sediment forms, which includes numerous plant particles: calcareous tufa.

But changes in temperature also disturb the equilibrium. An increase in temperature leads to oversaturation. Where cold water flows out of springs and cave entrances which warms up in summer the formation of calcareous tufa results. In endokarst the temperature rarely increases as the water carries away the earth's warmth, cooling the rock.

Evaporation also causes calcareous deposits by constricting the volume of the solution and thus creating oversaturation. Rapid evaporation of water flowing away in a film results in the rapid precipitation of $CaCO_3$ as a deposit similar to calcareous tufa. In

Central Europe adequate conditions exist only on the surface and in the entrances of caves, thus in places where assimilating plants also grow. Evaporation also takes place in dry areas inside the cave which is why friable, tufa-like coatings are occasionally found there.

13.3.1.1 Polymorphy of $CaCO_3$

$CaCO_3$ is tetramorphous and occurs as calcite, aragonite, vaterite, and calcium carbonate jelly.

Table 13.5. Polymorphy of $CaCO_3$

		Hardness	Density (g/cm^3)
Calcite	Rhombohedral	3	2.75
Aragonite	Orthorhombic	3 1/2-4	2.95
Vaterite	Hexagonal	?	2.5
$CaCO_3$-jelly	Amorphous	?	2.25-2.45

In nature $CaCO_3$ occurs practically only as calcite and aragonite. In caves calcite usually forms steep scalenohedrals which are placed together in calcite clusters. The crystals can form small points which are almost unrecognizable, or they can have well-developed forms up to 15 cm in length. Rhombohedrons are quite rare. In moonmilk calcite occurs in microscopically small needles, which are called lublinite. Aragonite, on the other hand, often appears in clusters of silky, gleaming, long needles (anthodates) and in branched aggregates of short needles. The two polymorphs occur as coarse-crystalline sinter and as helictites. Aragonite helictites are called Eisenblüte in German (iron blossoms) since they are to be found in the iron-ore deposits in Erzberg (Styria), where Fe^{3+} seems to be crucial for their creation.

If the two minerals cannot be distinguished by the form of their crystals, nor by their density or hardness, the Meigen Test can help: samples of the material are pulverized and boiled in a watery solution of $Co(NO_3)_2$. Minerals of the calcite group remain uncolored, those of the aragonite and vaterite become lilac-colored. However, since vaterite is not found in rocks, this procedure can be used specifically for aragonite in karst. For more precise detection the X-ray analysis must be applied (Graf and Lamar, 1955).

Calcite is stable, aragonite is metastable and therefore 16% more soluble. Vaterite is unstable and is converted into secondary aragonite. $CaCO_3$ jelly is still more unstable and is rapidly converted into vaterite:

$CaCO_3$-jelly → Vaterite → secondary aragonite.

For the formation of primary aragonite from a solution an oversaturation of $CaCO_3$ is required in both calcite and aragonite. This oversaturation can be attained by means of CO_2 losses, evaporation, or even an increase in temperature. It is probable that warmth not only causes a change in the $CO_2/CaCO_3$ equilibrium but also influences the formation of aragonite directly (Curl, 1962). Certain foreign ions have similar effects (Murray, 1954; see Chap. 13.3.1.1.c).

a) Influence of the Loss of CO_2 and of Evaporation

Roques (1964) in his investigation of the fields of existence of the four polymorphs under laboratory conditions found that in a solution of 400 ppm $CaCO_3$ (equilibrium-p_{CO_2} : $9.5 \cdot 10^{-2}$ atm) which was exposed to the open atmosphere (p_{CO_2} : $5 \cdot 10^{-4}$ atm) a metastable oversaturation occurred at first through losses of CO_2; this was followed by a spontaneous deposit of calcite and then a deposit of vaterite. In this process no aragonite was formed primarily, only later, secondarily from vaterite.

In the first place evaporation effects an increasing concentration and oversaturation of Ca^{2+} and dissolved CO_2 which induces CO_2 to be given off. When the conditions are otherwise constant — temperature and humidity of the atmosphere unchanged with no movement of air except the convection current — the ratio of the evaporating surface A to the volume V of the solution determines the total time of evaporation t_{max} and the rise in the concentration of Ca^{2+}.

$$V_t = V_o - K_E \cdot t. \qquad \begin{aligned} &V_o\text{: volume at the beginning} \\ &V_t\text{: volume at the time t} \\ &K_E\text{: constant of evaporation when A is constant} \end{aligned} \qquad (1)$$

At t_{max} all the water is evaporated and $V_t = 0$.
Then:

$$V_o = K_E \cdot t_{max},$$
$$t_{max} = \frac{V_o}{K_E}, \qquad K_E = K'_E \cdot A. \qquad (2)$$

In nature the evaporating surface frequently occurs on a thin film of water. A/V is then especially high whereby K_E also becomes high. Moreover as long as there is no precipitation the following is valid:

$$V_o [Ca^{2+}]_o = V_t [Ca^{2+}]_t, \qquad (3a)$$

$$\frac{[Ca^{2+}]_o}{[Ca^{2+}]_t} = \frac{V_t}{V_o}. \qquad (3b)$$

If Eq. (1) is integrated in Eq. (3a) then:

$$[Ca^{2+}]_t = \frac{V_o}{V_o - K_E \cdot t} [Ca^{2+}]_o. \qquad (4)$$

The curve of concentration for $[Ca^{2+}]$ shows a hyperbolic approach to t_{max} during evaporation, as long as no precipitation occurs.

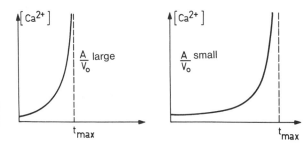

Fig. 13.13. The concentration of $[Ca^{2+}]$ as function of the time of evaporation t

If, however, the precipitation of $CaCO_3$ sets in, the concentration sinks and the curve of concentration deviates from the asymptotic one. But since further evaporation causes the constriction of the solution to increase more and more rapidly, precipitation soon cannot keep pace and a renewed increase of $[Ca^{2+}]$ results. Wherever the concentration of Ca^{2+} exceeds the limits of metastability the amorphous $CaCO_3$ again forms which is converted into vaterite; this is in turn converted into secondary aragonite (Roques, 1964).

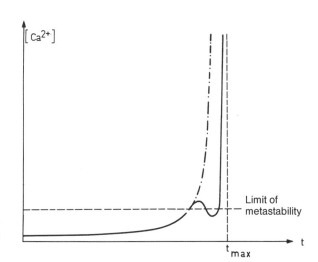

Fig. 13.14. Curve of concentration for $[Ca^{2+}]$ according to Roques (1964)

b) Influence of Temperature

As the temperature rises the tendency to the formation of aragonite increases. According to Moore (1956) there are in the SW of the USA numerous caves with aragonites which are coated with calcite. South of the 16°C-isotherm the precipitation of calcite does not take place, north of the 8°C-isotherm the precipitation of aragonite does not. He concludes from this that the formation of aragonite requires a temperature higher than 16°C and that during the warm period 6000 years ago it must have been approx. 8°C warmer than today. The calcite-aragonite method for the measuring of paleotemperatures is, however, controversial, since rightly the importance of other factors in the formation of aragonite must be evaluated much higher than temperature.

According to Gèze (1965) in reference to calcite and aragonite an oversaturation of the solution causes the formation of aragonite, e.g., when a draught of air evokes the rapid evaporation of water slowly seeping in; he attributes the aragonite in southern France to this. The same explanation can be applied to the caves of the SW of the USA. Then Moore's observation could be interpreted as the result of a change in climate with an increase in the humidity of the atmosphere and therefore a decrease in evaporation toward the north so that calcite was deposited on aragonite. Pobequin (1954) and Roques (1964) obtained aragonite already at 10°C by means of rapid evaporation.

c) Influence of Other Ions in the Solution

The formation of aragonite is clearly dependent on certain other ions in the solution. Murray already showed in 1954 that aragonite forms much more easily in the presence of Mg^{2+}, Sr^{2+} and Pb^{4+}. Roques' experiments in vitro (1964) confirm these results. To a solution of 740 ppm of $CaCO_3$ in a beaker there were added 12 ppm of Mg^{2+} (as $MgCl_2$) which produced a residue of 20% aragonite in 15 days, with the addition of 40 ppm of Mg^{2+} the aragonite content reached 75%, with more than 150 ppm even 100%.

13.3.1.2 Calcareous Tufa (Calc-Tufa)

According to Pia (1953) calc-tufa is a soft, porous rock to the formation of which plants contribute; calcareous sinter is, on the other hand, nonporous, crystalline, and hard, occurring only where there are no assimilating plants. Tufa is derived from the Latin tofus referring to porous, crumbling, freshwater limestone and to volcanic tuff (Stirn, 1964). Both are friable, which explains how two genetically so foreign rocks could be united under one name. A distinction between limestone tufa and calcareous sinter is commonly made by karst morphologists and speleologists in the French and German languages (Gèze, 1973), whereas English and American geologists (Am. Geol. Inst., 1962; Monroe, 1970) frequently use calcareous sinter as the generic name.

In German *calc-tufa* is to be defined as friable, porous, freshwater limestone which forms, as a rule, with contributions from assimilating plants, mostly mosses and algae.

Calc-tufa consists, apart from impurities, of $CaCO_3$ for the most part, even in dolomite regions. Stirn (1964) published for the Swabian-Franconian Alb a proportion of less than 0.7% of $MgCO_3$ in five analyses, a proportion of over 1% (up to 1.8%) in three further ones. Impurities and organic substances occur in strongly varying proportions.

In Central Europe there are considerable, localized tufa deposits as valley-fill, but they are not continuing to form today. With the aid of the ^{14}C-dating and pollen analyses they have been placed in Central Europe in the Atlantic Warm Period (5000-2500 B.C.) and partly in the earliest Subatlantic Period (Groschopf, 1952, 1961). Such intensive limestone deposition is only possible in warm, humid periods, when great biological activity coincides with an increased rate of reaction in the $CaCO_3$-CO_2 system.

In summer spring- and cave water, which is relatively cold, warms up greatly at the surface especially in cascades; calc-tufa, called spring-tufa in German, is deposited. Cavities become enclosed by it. One of the largest tufa caves is situated near Baar (Switzerland, Ct. Zug), Höllgrotte (see Chap. 17.2), a commercial cave, not to be confused with the karst cave Hölloch.

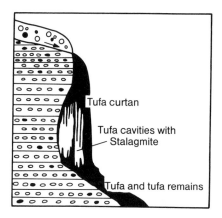

Fig. 13.15. Schematic cross-section through Höllgrotte

As soon as the cavities were closed by the growing tufa the crystalline, hard but porous travertine (polishable, usually porous calcareous sinter, cave onyx) was formed in place of the soft tufa. This change took place because of the high humidity of the air (little evaporation), the constant temperature, and the higher CO_2 content of the cave air.

The most famous tufa formations in Europe are the tufa barriers of Yugoslavia on the Krka (Krka Falls) and on the Korana (Lakes of Plitwitz).

13.3.1.3 Morphology of Calcareous Sinter – Speleothems

Under calc-sinter one understands hard, coarsely crystalline deposits of $CaCO_3$ by water; their surface is not formed by crystal-faces. Calc-sinter can be massive, i.e., free of larger pores, or else it can be porous; this depends on local circumstances during sedimentation. Occasionally connected cavities occur – a result of the process of growth – through which the water seeps. The term *travertine* – Lat. lapis tiburtinus – has not been clearly defined and subsequently it is used with quite different meanings. Formerly the term designated a deposit similar to tufa, which can be processed easily in the primary humid state, but which becomes hard when it dries. Tivoli east of Rome is the locus typicus for travertine. In English travertine is often used with the narrow meaning of a deposit in thermal springs (Stamp, 1963), while for American geologists it is synonymous with calc-sinter (Am. Geol. Inst., 1962, p. 511), they sometimes even broaden it to mean calcareous tufa. Only banded calc-sinter, massive to porous, should be termed travertine.

Calcareous sinter – cave formations – is very rich in forms. One must distinguish between flowstone, dripstone, rimstone, and pool deposits. Flowstone is stratified sinter deposited by flowing water; dripstone is created by dripping water; rimstone forms bars on the cave floor behind which pools are dammed up (called gours in the U.K.). The pool deposits are precipitations in such pools, mainly as calcite crystals.

The following description of forms applies a classification which relies on the English one: flowstone, rimstone, pool deposits, dripstone, cave formations due to capillarity, cave pearls. These groups are of various sizes — dripstones comprise more forms than all the other groups together.

A. Flowstone

Water rich in $CaCO_3$ flowing over the rock in a thin film has a large surface and gives off CO_2 forming *flowstone*. If the water pursues a channel, this gradually fills with flowstone until the same level is reached as in the neighboring rock; then the water spreads out in a thin film over the whole surface.

Depending upon the amount of impurities carried in the water either in dissolved form or in suspension, variously colored strata form: organic substances create yellow to reddish and dark brown and black layers, iron hydrate creates bright ocher to brown, mangan hydrate black, copper compounds blue to green (rare) — pure water creates white speleothems. Cu-ions create light blue flowstone in a marble lens inside granite and gneiss in Lilburn Cave (CA, USA). Nicely banded, polishable flowstone is called cave marble and, if it is translucent, also cave onyx (in comparison common onyx is a cryptocrystalline SiO_2 variety and much harder).

The bed of flowstone can be the naked rock, cave clay, gravel or sand. It can be removed by corrosion or mechanically. If the flowstone is thick, it survives and freely projects out into the cavity or forms a bridge in it. However, such a bridge usually breaks off; the remainders of it can occur as flowstone ledges or survive on cave walls as crusts of flowstone of slight thickness. They differ from ledges of calcareous sinter formed in other ways on the break by their striation which is vertical to the surface, for in calcareous sinter the calcite crystal's main axis is vertical to the surface.

B. Rimstone and Rimstone Pool Deposits

One speaks of *rimstone* when calc-sinter bars the water causing a pool or coats such a pool with a thick layer. It is created where water which is rich in lime trickles down over rock in a thin film. When it flows over the rim there is an increase in the loss of CO_2 and calc-sinter is deposited. Where more water flows over, more sinter is formed. The rim where the water overflows gradually becomes horizontal thereby and the water is distributed over its whole breadth.

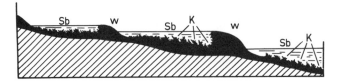

Fig. 13.16. Longitudinal section of a row of rimstone pools. *Sb* rimstone pools; *W* rimstone bar; *K* calcite crystals

Rimstone pools (French and Eng. gours) have a diameter of from 10 cm to many meters; the rimstone bar which delimits them is a few centimeters up to 1 m high, rarely higher. There is a row of large rimstone pools, one above the other, in Brunnen Grotte in the Škocianske jame (Slovenia).

The same mechanism creates on flowstone the small *microgours* (German: Sinterschalen), a few millimeters up to a few decimeters in width, and up to a few centimeters in depth. They occur frequently but are usually overlooked because of their small size. There are some especially large and beautiful ones in Katerloch near Weiz (Styria).

Pool deposits comprise calcite ledges projecting freely into the water, attached calcite crystals as well as loose components. Clusters of calcite crystals are called calcite roses (Plate 11.1). In Hölloch the individual crystals in a cluster are up to 5 cm long, in Eisriesenwelt (Austria) up to 15 cm. The prerequisite of their formation is quiet water. The more the water moves, the more numerous the crystal nuclei are, and the closer the calcite crystals stand. Sometimes one can distinguish the individual crystals only with a magnifying glass. If limestone is rapidly deposited in moving water, rounded surfaces like cauliflower are created.

The CO_2 diffusing into the air from the water's surface results in a thin layer of high Ca^{2+} concentration and thus in calcite nuclei. Microscopic, flat, skeletal crystals grow out of these; they are kept swimming by the surface tension: *floating calcite*. They sink to the bottom when the water is in motion (ρ: 2.75 g/cm^3). There they become the point of departure for subaqueous *calcite deposits*. If the floating calcite strikes against the rim of the pool or against boulders which project out of the pool, the water is somewhat raised by capillarity; as a result additional calcite is deposited which cements the former floating calcite, forming ledges. These *calcite ledges* continue to grow, becoming thicker and broader. They are an indisputable indication of actual or former permanent water level.

C. Dripstone

Dripstone comprises "all standing or hanging forms which are attributable to dripping water" (Fink, 1973). The forms which are created by this water running off do not belong to the same group but to flowstone.

C.1 Hanging Dripstone

Floury-looking, *white sinter stripes* on cave ceilings make visible the rock's network of interstices. Here water held capillarily in these interstices evaporates. These stripes are up to 2 cm wide and only fractions of millimeters thick. They are rare in the caves of Central Europe. They belong to the dripstones only conditionally since it rarely comes to the development of drops, but they are a preliminary stage of them.

The prototype of the hanging dripstone is the stalactite (Gr. stalaktos = dripping; Plate 12.4). It forms wherever water drips freely, whether on ceilings or on overhanging parts of walls. If the place from which the water drips is not on an interstice, the stalactite is massive in its cross-section: first type. As a rule this type (Fig. 13.17; Plate 12.1) is in the company of other forms, e.g., of draperies, sinter flags, medusa dripstones (Plates 12.2, 12.3). Stalactites which grow on an interstice have a central canal inside, 2-3 mm in width: second type (Fig. 13.17; Plates 11.2, 11.3). Often the central canal is empty, but it can also be filled with calcite crystals formed subsequently or with fine material that was washed in. The starting form is a small sinter tube, a "straw stalactite."

Sinter tubes, "soda straws", "maccaroni stalactites" are the form which has till now been best investigated (Plate 11.3). Their diameter is between 4 (wall thickness 0.3 mm according to Krieg, 1975) and 6 mm (up to 1 mm wall thickness). The cross-section hardly changes along the whole length, whether they measure only 1 cm or 450 cm as in the Grotte de la Clamouse (S. France) — these may be the longest in the world. Occasionally there are thousands of them hanging on a cave ceiling, in such a case they determine the physiognomy of the space.

Origin: water emerging from an interstice in drops deposits a circle of calcite or aragonite around the rim of the drop. When the deposit is steadily made a sinter tube is formed. In the tube itself the water cannot give off any CO_2, therefore no $CaCO_3$ is deposited and the tube remains open. However, on the hanging drop at the other end oversaturations of Ca^{2+} and CO_3^{2-} result as the water begins to give off CO_2 (see Figs. 2.19 and 2.21); these are accumulated on the lower rim of the tube, controlled by the crystal lattice of the tube wall. It may be polycrystalline or monocrystalline (Gèze, 1965). The losses of CO_2 from the emerging drop were described in more detail in Chap. 13.2.2.2. According to it the forming drop loses 10% of the CO_2 gradient between the emerging water and the cave's atmosphere when the drop hangs for 1-10 s; when it hangs for 100 s it is about 25% and for 1 h 62% (see Fig. 2.20). The amount of the greatest possible lime deposit can be determined with the aid of the curve of equilibrium of the system CO_2-$CaCO_3$ (Fig. 2.23). The faster the cadency of the drip, the larger the proportion is of undeposited, excess ions, since the time for their movement to the centers of crystallization on the rim of the tube is then too short.

If the water flows over the outside of a "maccaroni", calcite (aragonite) is deposited, which has as a rule the same crystallographic orientation as the tube. Basset and Basset (1962) describe monocrystalline stalactites with a hexagonal cross-section in Rushmore Cave, S. Dakota. With increasing thickness, however, the number of disturbances increases also, but the axes of the crystals are always arranged vertically to the surface.

The difference between the CO_2 of the solution and the CO_2 of the cave's atmosphere, and thus the $CaCO_3$ deposit also, decreases exponentially as the water flows away. The result of this is always a cone becoming more slender toward the bottom. As long as water flows through the sinter tube which has now become the central canal, the stalactite develops as type 2; otherwise the sinter grows over the tube's exit and the continuation develops as type 1.

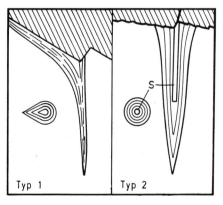

Fig. 13.17. Stalactites of *type 1* and *type 2*, longitudinal and cross-sections. *S* primary sinter tube

In the tropics and subtropics there are also *exterior stalactites* which hang down from the rock walls. They are analogous to the tufa draperies of cool-temperate climates, yet there are not usually any green plants involved in their formation. In Europe good examples of these are found in Italy but there are especially typical ones near Banyalbufar on Mallorca. The causes are high temperatures, a high lime content in the water, and a high degree of evaporation. The greater warmth increases the rate of diffusion on the one hand, and biological activity on the other hand, thereby also the $CaCO_3$ content of the seeping water. Under the conditions of Central and Northern Europe (cool-humid to cold-humid) stalactites form only in caves.

Sinter flags and draperies (curtains) are formed on slanted ceilings and overhanging walls by drops which run down them. These primarily follow the incline of the surface but they avoid all uneven spots so that their course is winding. This at first creates a sinter band which is called so up to a width of 5 cm (Trimmel, 1965). When it is wide enough it continues independently to develop into sinter flags and curtains (Plate 12.1). These can become up to 1 m wide and several meters long, whereas their thickness is between 0.25 and 1 cm. Therefore draperies are translucent. As the drops run down they deposit sinter along the edge, smaller or greater amounts depending on the lime content; this gives rise to stripes visible on the surface. Impurities carried in the water effect colorful stripes. Structures which run vertically to these bands at a distance of about 1 cm from one another are often found in sinter flags. They are connected with the indentations on the edge of the sinter flags; large drops frequently hang in these indentations. When the drops exceed a certain size they empty into the next indentation; a toothed edge forms: *saw-tooth sinter.*

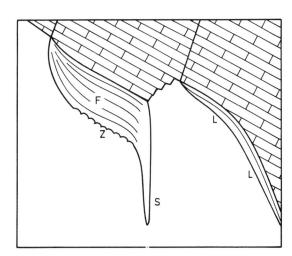

Fig. 13.18. Sinter flag (*F*) with saw-tooth sinter (*Z*) and stalactite (*S*) of the 1st type as well as sinter band (*L*)

Sinter flags end either in the wall itself (wall sinter flags) or on a projection in which case the lower end continues as a stalactite of the 1st type.

Flowstone frequently forms on projecting parts of walls. When an overhanging portion follows, first sinter flags form which, however, draw to a head in stalactites already after 10-20 cm and on the next projection there is another transition into flowstone.

This can be endlessly repeated. A form is thereby created which is called *medusa dripstone* (Plates 12.2, 12.3). It is the most frequent combination of flowstone and dripstone in warmer climates and is also found frequently in Central Europe.

C.2 Standing Dripstone, Stalagmites

The main type of standing dripstone is the stalagmite (Gr. stalagmos = dripped off), the counterpart of the stalactite (Plates 11.2-11.3, 12.4). In spite of the multiplicity of their forms the upward striving column occurs again and again in the foreground. Curl (1972, 1973) and Franke (1961a, 1961b, 1963b, 1965b, 1968, 1975) especially have investigated the laws which control the formation of stalagmites. Franke was the first to apply physicochemical calculations to the formation of dripstone. He begins with the simple case of a convex stalagmite tip when the drops have little height of fall and finds a linear dependence of the cross-section on the amount of water carried to it, and a linear dependence of the staglamite's height on the Ca^{2+}-concentration. The applicability of this formula to more complicated cases is controversial. Curl (1973) pursued the problem further and determined the minimal diameter of a stalagmite to be about 3 cm; this agrees well with the actual facts. Franke (1975) did indeed find so-called subminimal stalagmites with diameters of 1.5 cm when the distance the drops had to fall was a matter of millimeters.

Stalagmites are built of cap-shaped layers on top of one another. When the height of the fall is slight the tip is convex, becoming flat as it increases, and is finally transformed into a splash cup.

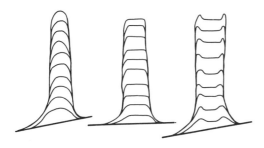

Fig. 13.19. Stalagmites' longitudinal sections; *from left* to *right* with increasing distance of fall of the water drops

Single-crystal stalagmites occur rarely, but more often than monocrystalline stalactites; nevertheless they also show a convex arrangement; this proves that it is not the crystal structure which causes this arrangement, but color differences in the impurities, e.g., Grotta di Castellana in Apulia south of Bari. The base of stalagmites consists of flowstone which sometimes extends only a few centimeters but which is also traceable far in the direction of the inclination.

When the height of fall is great, *splash stalagmites* (Fr: pile d'assiettes), form, e.g., in the huge bell-shaped caves of France, Aven Armand and Aven d'Orgnac, where they become about 30 m high. The falling drops are sprayed in every direction when they splash down and thus the stalagmite grows toward the outside along the rim of the splash-cup, forming a "plate". If the sprayed water is caught because of the increasing diameter of the splash-cup so that it finds its way to the middle, the further development of the plate is arrested. Limestone is now deposited in the splash-cup so that a new plate can form when the base has been sufficiently raised. This explains the French term pile d'assiettes = pile of plates.

In the case of *palm-stem stalagmites* (Fr. tronc de palmier), which occur under the same conditions and in the same places, the "leaves" become much larger than the "plates" of the other form because here water rising capillarily moves to the rim where it evaporates (Gèze, 1965).

If the waterdrops, rich in lime, fall on a permeable surface of silt or of fine sand, the water seeps in and deposits the limestone in this substratum: *negative stalagmite or conulite* (Chap. 13.1.5, Fig. 13.10). Often a normal stalagmite builds up on it.

Stalactites and stalagmites grow toward one another and can unite in a *dripstone column*. These belong to the largest dripstone formations, thus in New Cave (NM, USA) and in the Grotte des Demoiselles in southern France.

C.3 Growth and Age of Dripstones

There are no rules which reveal a relationship between age and length. There are periods of rapid growth and others of slow growth, even periods of inactivity. Growth can be measured in exceptional cases. In a cave in the city of Buxton (South Pennines, GB) there is a stalagmite of a height of 10 cm sitting on a gas pipe which was installed 100 years ago. In the Grotte des Demoiselles (southern France) a stalactite reached the same length in 20 years.

Table 13.6. Mean yearly growth of dripstone in mm according to the direct method

Cave	Author	Sinter tube	Stalactites	Stalagmites
Tunnel near Stuttgart FRG	Bauer, 1971	40	–	–
Aven Armand F	Martel, 1948	–	–	0.1
Grotte de Han BE	Trimmel, 1968	2	–	–
Grotte C. Doria I	Polli, 1962	–	0.1-0.55	–
Ingleborough Cave GB	Dawkins, 1874	–	0.17	6.2
New Cave IRE	Coleman, 1945	–	2.12	–
Grand Caverns VA USA	Johnston, 1916	–	–	1.02
Domica Cave ČSSR	Petranek, 1951	–	–	0.02-0.005 [a]

[a] periods of inactivity included

An incontestable method of determining their age is also possible with the aid of radiocarbon ^{14}C. Franke pointed this out already in 1951. He stated that CO_2 which originated in the air was required for the limestone solution. Therefore when limestone is deposited an average of 50% of the atmospheric CO_2 is transferred into the crystal lattice and thereby also ^{14}C which is in the atmosphere in a relationship of $1:10^{12}$. In the case of dissolved limestone there is still an exchange of CO_2 with the cave's atmosphere to reckon with, whereby the ^{14}C content of the water increase somewhat. The recent standard of calcareous sinter lies at 85%. ^{14}C was accredited a half-life period of 5570 years. This value is still retained today even though more recent investigations quote 5730 ± 40 years (Libby, 1969, p. 8).

In Central Europe there is evidence of an older generation of dripstones; they can be traced back to about 40,000 B.P., lasting until about 20,000 B.P. Still older finds cannot be dated because they have too small a concentration of ^{14}C. There is a gap between 20,000 and 12,000 B.P. which corresponds to the Würm Ice Period. Measurements of samples from the Mediterranean region show that there was a remarkable decrease in the formation of speleothems. The younger dripstone generation continues immediately after this; its first phase lasted from 12,600-9500 and a second phase from 8000-3500 B.P. (Franke, 1966; Franke and Geyh, 1971).

With the help of the radiocarbon method the rates of growth can also be determined.

Table 13.7. Mean growth of dripstones in mm/y according to Broeker et al. (1959) and Franke (1968) using the ^{14}C method

	Growth/y	Period of Growth (y)
Moaning Cave (CA, USA)	0.064	1400
Katerloch (Styria, Austria)	0.42	885
Bärenhöhle (S. Germany)	0.087	4490
Prinzenhöhle (S. Germany)	0.26	1970

The values in the above table are smaller than in Table 13.6 since here, in the long periods of time, interruptions occasionally occur in the deposition of calcareous sinter.

Up to now the uranium method has been applied much less often for age detection. The prerequisites are sufficient uranium content, the possibility of determining the original concentrations, and the absence of subsequent disturbances (Thurber et al., 1965). The following details are based mainly on Harmon et al. (1975). According to them the ^{230}Th/^{234}U-method is the best. It covers the period of time from 2000 B.P. back to 350,000 B.P. Uranium occurs in cave water as a $[UO_2(CO_3)_3]^{4-}$ complex; it decomposes when calcareous sinter is deposited as a result of the loss of CO_2 and UO_2^{2+} is formed. Uranium is free of thorium and is deposited together with the sinter. U concentrations are between 10 ppb and 100 ppm. ^{234}U in the uranium in solution is enriched and therefore it is also enriched in the sinter. It decomposes into ^{230}Th, among other things, which is used for age detection.

The oldest speleothems investigated up to now are from the caves of the South Nahanna Region (Canada, N.W.T.); they are over 350,000 years old (Ford and Schwarcz, 1976). A stalagmite from Norman Bone Cave (WV, USA) has in its base sinter which is 199,000 ± 8800 years old, in its tip sinter which is 163,000 ± 6800 years old. A stalagmite from Sotano de Tinaja (Mex.) began to grow 48,900 ± 800 years ago and its tip (+ 96 cm) is 8200 ± 400 years old, corresponding to a mean growth of 2.41 cm/10^3y.

D. Speleothems Due to Capillarity

Capillarity is a widespread, important factor in the formation of speleothems. The main groups are the eccentrics, the coral formations and the shields. Finally, the leaves of the palm-stem stalagmites belong to this group.

D.1 Eccentric Formations, Helictites

Eccentric formations belong to the speleothems most appreciated by speleologists (Plates 11.3, 11.4). Their name refers to their eccentric appearance which seems to contradict the force of gravity; their eccentricity is not just a matter of appearance, it is also geometrical. *Helictites* is a term introduced by Moore (1954) but it is usually used for formations on dripstones and walls, while those eccentric formations which grow on the floor upward are logically called *heligmites*. These terms are gradually being taken up into the German language.

Eccentrics grow on rock as well as on flowstone and dripstone and especially on stalactites, but more rarely on floor flowstone. They show a remarkable independence of the earth's force of gravity which indicates forces in their formation that are stronger. They are small and their forms are very thin (from 1 to a few millimeters), twisted, and often dendritic. They turn away from obstacles, also growing around other eccentric formations and sinter tubes without touching them. A systematic description of their forms will not be given here; for more details refer to Gèze (1957) and Trimmel (1965).

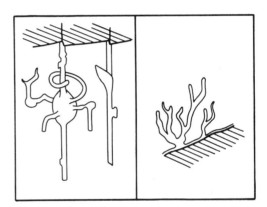

Fig. 13.20. Helictites and heligmites (eccentrics)

Eccentric formations have a fine, almost microscopic canal running through them into which the water is sucked in capillarily. At the tip it evaporates again before a drop can form. The emerging mass of water is so small that the force of gravitation is less than the effect of the surface tension. This is the explanation accepted today. If the amount of water emerging increases, e.g., as a result of an increase in pressure, so that drops can form, a normal sinter tube begins to grow. Moore's investigations (1954) show that the axis of the crystals runs subparallel to the axis of the eccentric formation, thus it does not coincide exactly with the direction of growth. Gèze (1957) mentions monocrystalline eccentric formations, also.

For a long time the origin of these forms was a matter of dispute. Certainly Prinz explained their formation by capillarity in 1909, however without recognizing the other connections (evaporation for example). Lobeck (1929) and McGill (1933) found they were the result of impurities in the water and Corbel (1947) showed that insoluble particles in the central canal encourage their creation on the surface of the sinter tube. Huff (1940) and McGrain (1942) explain them by water oversaturated with lime out of which

$CaCO_3$ crystallizes in the central canal of a sinter tube, thus hindering or obstructing the water's circulation. It was said that the water then emerged from the tube at a weak spot and formed helictites. An old explanation refers to the wind and the evaporation caused by it. Glory (1936) was of the opinion that a slight wind made eccentrics grow against the wind, while stronger winds on the other hand forced them to grow in the opposite direction. There are indeed wind eccentrics, so-called *anemolites,* but they are rare. Anemolites in Demänovské jaskiňe (ČSSR) grow in fine, feather-like eccentric formations against the wind, the same in La Clamouse (southern France). There are still more theories, e.g., electrically charged water particles which have become so from being reduced to spray are said to cause eccentric formations; this may happen as an exception but by no means could it be generally true.

D.2 Coral Formations

Of the many names only coral is appropriate as the name of this type of formation (Salzer, 1954). The surface can be cauliflower-like, or show small, connected balls of calcite, sometimes arranged like a clump of grapes, single-stemmed fungoid formations, or branched stems (coral-like formation). These forms are explained as deposits due to the evaporation of water ascending capillarily. It is striking that coral formations are frequently found on lofty ledges where evaporation reaches a maximum. Moore (1952) also places evaporation in the foreground, which implies the capillary rising of water. In cave entrances and in deep semi-caves a soft coral formation is to be found; organisms, especially algae, play a part in its creation.

D.3 Shields

Shields are semi-circular leaf formations which grow freely out into space; they have a diameter of 1 dm up to 1 m and are 1-2 cm thick. There are no known places in Central Europe where these formations can be found except in southern Slovakia (ČSSR) and in Jeskyňe Domica (Domitza Cave). They also occur in southern France, e.g., in Aven d'Orgnac. Kundert (1952) describes them in Lehman Cave (NY, USA).

Shields consists of two layers of sinter which are separated by an interstice (Kunsky, 1958). Water rises capillarily in this interstice according to Gèze (1957) until it reaches the edge where it evaporates, forming strips of growth. Kunsky also assumes that there is a hydrostatic pressure because frequently stalactites form on the edge and can scarcely be observed on earlier strips of growth.

The leaves of palm-stem stalagmites (see Chap. 13.3.1.3.Cb) are not shields, but they grow, like them, obliquely upward. Their circumference is not continuously smooth either, but rather lobed or even fringed. The surface is very rough and the water creeps up capillarily in it, evaporating only when it reaches the tips and edges; the further the lobes, the stronger evaporation is.

E. Cave Pearls

Cave pearls are small pieces of calcareous sinter which are usually completely round, however, some are oval or irregularly rounded; but they are always loose and have a smooth surface. Their diameter is about 1 cm, yet there are extreme exceptions up to 15 cm. In the cross-section there is a nucleus of foreign matter, usually rock or mineral particles; this is enclosed in concentric layers of calcite where the axis of the crystals is radially oriented (Hahne et al., 1968). These spherulites have their origin in pools of moving water rich in lime which causes the deposition of calcite. Constant motion is required so that the individual pearls do not stick together when sinter is deposited, if they did they would lose their main characteristic, namely that of being rounded and loose.

According to Hahne et al. (1968) the time necessary for the formation of cave pearls varies greatly. In old mines on the Ruhr there were some which proved to be 3 months old but also some 30 years old. Other authors (see Bibliography in Hahne et al., 1968 and Kirchmayer, 1964) state a period of between 25 and 170 years for their creation.

Dripstone particles, held in motion by water dripping into small pools, grind against one another until they are similar to cave pearls and are therefore also classed as such. A trivial name for these in German is Teufelskonfekt (devil's sweets).

13.3.1.4 Moonmilk

Moonmilk is a white mixture of microscopic crystals of carbonates with 35%-70% water. Ninety percent of the dry substances consist of calcite, frequently as lamellar, rhombohedric crystals, less frequently as lublinite, a thinly needled calcite (see also Král, 1975). In dolomitic rocks moonmilk contains 0.05%-10% Mg-hydrocarbonates (Bernasconi, 1959, 1961; Trimmel 1962), among them hydromagnesite (Davies and Moore, 1957; Gèze and Pobequin, 1962) and huntite (Gèze and Pobequin, 1962). In addition there are impurities, especially clay, which sits on the crystal lamellae in colloid particles (up to 2.5%) or which is embedded between the crystals (up to 8%; Bernasconi, 1961, 1975), and some organic stuffs (Caumartin and Renault, 1958). With over 35% H_2O moonmilk is plastic and with over 70% it is oversaturated with water, which it then loses. When it is dried, friable crusts and powdery residues are created.

Moonmilk is formed on one hand from deposits, on the other hand by the weathering of sinter. The latter takes place according to Caumartin et al. (1958) with the aid of biological processes. According to Mason (1959) bacteria play a part in its creation through sedimentation, also; moreover clay minerals seem to stimulate the process. The origin of moonmilk has by no means been explained yet; further intensive research is necessary.

The name moonmilk was derived from a deposit in Monloch = Mondloch on Pilatus Mountain (Switzerland). It appears for the first time in Gessner as lac lunae in 1455.

13.3.2 Cave Sulphates, Gypsum

Gypsum occurs as a new formation from oversaturated water containing gypsum. It is, however, not deposited in the same way as calcareous sinter by the loss of a component, but by a physical process:

$$Ca^{2+} + SO_4^{2-} + 2 H_2O \rightarrow CaSO_4 \cdot 2 H_2O.$$
gypsum in solution   solid gypsum

Sedimentation takes place only as a result of evaporation, except for those rare cases where a decrease in temperature is the cause. Gypsum deposits created by evaporation in caves are small. The voluminous accumulations of gypsum in the Carlsbad Caverns (NM, USA) are an exception, the manner in which they were created, also. There water containing gypsum out of Permian gypsum deposits on the surface, which was very heated by the subtropical sun, penetrated into the earth's depths. When it cooled off the excess gypsum was deposited, forming several meters of gypsum (Bretz, 1949; Good, 1957). Palmer et al. (1977) arrive at another interpretation. They have found numerous indications that the gypsum deposits in the caves of Carlsbad Caverns National Park were created before the caves by metasomatosis in the limestone. The fact which gives most support to this theory is that the textures of this gypsum correspond to those of the adjoining limestones. The displacement of CO_3^{2-} by SO_4^{2-} took place in the zone where formerly the body of groundwater, hypersaline and therefore rich in SO_4^{2-}, mixed with the freshwater flowing in. More precise details concerning the mechanics of this metasomatosis are lacking. The caves were created in a later phase during which especially the region containing the gypsum and adjoining limestones was dissolved.

Where there is no gypsum nearby within a wide radius, this material is a secondary formation in caves. As Pohl and White (1965) show, pyrite, FeS_2, is the initial material in the Carboniferous limestone of the Paleozoic tableland in the USA. It is however, to be assumed that the less stable markasite, the rhombic modification of FeS_2, is the origin in many cases (personal information from Pohl). Dark limestones frequently show fine grains of FeS_2 in thin sections; such is the case in the Carboniferous limestone of Mammoth Cave National Park and in the Lower Cretaceous Schrattenkalk of Hölloch. The SO_4^{2-} which reacts with limestone to form gypsum is created from FeS_2 by oxidation. This process of weathering takes place quickly under the influence of bacteria (*thiobacillus thioxidans* and *th. ferrooxidans*), but slowly without organisms.

$$4\,FeS_2 + 15\,O_2 + 14\,H_2O \rightarrow 4\,Fe(OH)_3 + 8\,H_2SO_4$$
$$8\,H_2SO_4 + 8\,CaCO_3 + 8\,H_2O \rightarrow 8\,CaSO_4 \cdot 2\,H_2O + 8\,CO_2$$
$$8\,CO_2 + 8\,CaCO_3 + 8\,H_2O \rightarrow 8\,Ca^{2+} + 16\,HCO_3^-$$

$$4\,FeS_2 + 15\,O_2 + 30\,H_2O + 16\,CaCO_3 \rightarrow 4\,Fe(OH)_3 + 8\,CaSO_4 \cdot 2\,H_2O + 8\,Ca^{2+} + 16\,HCO_3^-.$$

Limonite, FeO(OH) [simplified $Fe(OH)_3$], and gypsum are left behind while dissolved limestone is carried away. Limonite often occurs in caves as pseudomorphoses from pyrite and markasite lumps; gypsum is often found as two-dimensional radiating aggregates on cave walls, so-called gypsum starbursts, as needle-like or worm-like, bent monocrystals and gypsum cave flowers (oulopholites), as crusts, rarely as gypsum stalactites, frequently as powdery efflorescences, as an exception also as snowballs (Flint Ridge Cave). The crystals sit on the rock, growing from the base where the FeS_2 is located. Gypsum is deposited independently of the rock in cave clay by shifting solutions, usually as crystal arrows, rarely as complete forms.

1 mol of FeS_2 (120 g of pyrite) dissolves 4 mol of $CaCO_3$ (400 g of pure limestone). The former has an average density of 5.0 g/cm^3, the latter of 2.75 g/cm^3. Thus 1 cm^3 of pyrite dissolves 6 cm^3 of limestone (Bögli, 1972).

Gypsum created from pyrite has a volume 6.23 times greater than pyrite. This has morphological consequences. If a layer rich in pyrite is covered by a layer low in pyrite, gypsum forms in the interstices, putting pressure on the covering layer with its forces of crystallization. The limestone, which is otherwise so brittle, slowly bends away without cracking. This movement takes place along the cleavage planes of the calcite — inner crystalline sliding. Measurements are being taken in Berome Moore Cave (MO, USA) to determine the velocity of the process in a 30-cm-thick layer of limestone. In the caves of Mammoth Cave National Park processes of detachment caused by gypsum are common; they are an important factor in the formation of the caves there (Bögli, 1972).

Cave gypsum is common in the Mediterranean zone where the high temperatures considerably accelerate the process of its creation. The Grotte la Cigalère (Ariège, France) is famous for its gypsum flowers. In the Grotta del Vento in the Province of Ancona (Italy) gypsum was formed by the reaction of H_2S with limestone. Occasionally fine selenite needles (translucent gypsum) are found in artificial tunnels, also, where H_2S emanates from sulphur water.

In dolomitic limestones and in dolomites *epsomite* forms in the presence of SO_4^{2-}, $MgSO_4 \cdot 7 H_2O$ (epsom salts). Since this mineral is very soluble in water, it is only found in caves of arid and semi-arid, warm climatic regions, in the USA for example in New Mexico, and as an exception in Wyandotte Cave (IN, USA; Blatchley, 1897).

Alkali sulphates are rare in caves. *Thenardite*, Na_2SO_4 is created by the reaction of sulphates with the remaining content of Na^+ in marine sediments. In the USA this was discovered in Mammoth Cave (Kentucky) and in Wind Cave (South Dakota). *Mirabilite*, $Na_2SO_4 \cdot 10 H_2O$ is found in Flint Ridge Cave. It occurs in long, sinuous monocrystals up to 0.5 m in length, in hair-like cave flowers, in crystal slabs, rarely in stalactitic form. The sphere of existence of mirabilite is very limited, for it liquifies at more than 80% atmospheric humidity and at less than 60% it loses its crystal water and becomes thenardite.

13.3.3 Cave Minerals

Cave minerals have always been created after the formation of the cavities, thus they are new formations. They are formed partly from the rock by means of dissolution and recrystallization which is the way calcite and aragonite grow, partly out of cave sediments by means of weathering. If there are ore minerals in the rock or in the sediments washed in, the result is a multiplicity of secondary minerals. The creation of phosphates was already discussed in Chapter 13.2.3. Primary rock components and those of the sediments carried unchanged into the cave do not belong to the cave minerals. Clay minerals are usually referred to as cave minerals, which often hardly corresponds to the definition. Monroe (1975) mentions autochthonous clays created by the reactions between sediments and substances dissolved in seeping water.

Hicks (1950) lists 76 different cave minerals, whereas 69 are named in Moore's list (1970). Hill (1976) mentions 75 minerals. The following list ist based on the authors just referred to, on Klockmann (1923, 1978), Warwick (1962), Gèze (1965), Urbani (1969), Fischbeck and Müller (1971), Strasser (1970), Diaconu (1973), Diaconu et al. (1975) and Bertolani and Rossi (1975). For literature refer to Moore (1970) and Hill (1976) who name Anglo-American authors especially.

Table 13.8. Cave Minerals

Anglesite, $PbSO_4$, Ahumada Mine Cave, Mexico
Apatite, also Fluorapatite, Hydroxylapatite, Carbonate apatite; Skelskoy Caves, Krim; Javořićko Caves, ČSSR (Pfeiferová et al., 1975)
Aragonite, $CaCO_3$ widespread
Ardealite, $CaHPO_4 \cdot CaSO_4 \cdot H_2O$ – also with $4H_2O$ –, Csoklovina Cave, Rumania
Attapulgite, $Mg_5 (Si_4O_{10})_2(OH)_2 \cdot 8H_2O$, Carlsbad Caverns, NM, USA
Azurite, $2 CuCO_3 \cdot Cu(OH)_2$, Copper Queen Cave, AZ, USA
Barite, $BaSO_4$, Youlgrave Cave, GB
Beudantite, $PbSO_4 \cdot FeAsO_4 \cdot 2Fe(OH)_3$, Island Ford Cave, VA, USA
Biphosphammite, $(NH_4)H_2PO_4$, in bat guano, W. Australian Caves
Birnessite, $(Mg, Ca, K) Mn_7O_{14} \cdot 2H_2O$, Webers Cave, IA, USA
Bloedite, $Na_2SO_4 \cdot MgSO_4 \cdot 4H_2O$, Lee Cave, KY, USA
Brochantite, $CuSO_4 \cdot 3Cu(OH)_2$, Blanchard Mines Cave, NM, USA
Brushite, $CaHPO_4 \cdot 2H_2O$, Bärenfalle, Tennengebirge, Austria
Calcite, $CaCO_3$, common in caves
Carbonate apatite, $3 Ca_3(PO_4)_2 \cdot CaCO_3 \cdot H_2O$, El Capote Cave, Nueva Leon, Mexico
Cerussite, $PbCO_3$, Herman Smith Cave, IL, USA
Christobalite, SiO_2, Wind Cave, SD, USA
Cimolite, $Al_4(SiO_3)_6 \cdot 3H_2SiO_3 \cdot 3H_2O$, Lookout Cave, WA, USA
Coelestine, $Sr (SO_4)$, Miller Cave, TX, USA
Crandallite, $2 AlPO_4 \cdot Al(OH)_3 \cdot Ca(OH)_2 \cdot H_2O$, Pajares Cave, Puerto Rico
Cyanotrichite, $CuSO_4 \cdot 3 Cu(OH)_2 \cdot 2Al(OH)_3 \cdot 2H_2O$, Blanchard Mines Cave, NM, USA
Diadochite, $4 FePO_4 \cdot Fe_2(SO_4)_3 \cdot Fe(OH)_3 \cdot 24H_2O$, Feengrotten, Saalfeld, GDR
Dolomite, $CaMg(CO_3)_2$, Lehman Caves, NV, USA
Elaterite, elastic bitumen, Matlock, GB
Endellite, $Al(OH) (Si_2O_5) \cdot Al(OH)_3 \cdot 2H_2O$, Carlsbad Caverns, NM, USA
Epsomite, $Mg(SO_4) \cdot 7H_2O$, Wyandotte Cave, IN, USA
Ferghanite, $U_3(VO_4)_2 \cdot 6H_2O$, Tyuya Cave, USSR
Fluorapatite, $3Ca(PO_4)_2 \cdot CaF_2$, Cueva el Indio, Venezuela
Fluorite, CaF_2, Dürschrennen, Säntis, Switzerland
Francoanellite, $K_3Al_5H_6(PO_4)_8 \cdot 12H_2O$, Castellana Grotte, Italy
Galenite, PbS, Herman Smith Cave, IL, USA
Giobertite, $MgCO_3$-gel, dolomite caves, Southern France
Geothite, $FeO(OH)$, Hölloch, Switzerland
Guanite, s. Struvite
Guanovulite, $KNH_4(SO_4) \cdot 2H_2O$, Piedras Cave, Honduras
Gypsum, $CaSO_4 \cdot 2 H_2O$, widespread, La Cigalère, Ariège, France
Halloysite, $Al(OH)Si_2O_5 \cdot Al(OH)_3 \cdot x H_2O$, Grotte di Frasassi, Prov. Ancona, Italy (see endellite)
Halotrichite, $FeAl_2(SO_4)_4 \cdot 22H_2O$, Diana Cave, Banat Mountains, Rumania
Hematite, Fe_2O_3, Wind Cave, SD, USA
Hemimorphite, $Zn_3Si_2O_7 \cdot Zn(OH)_2 \cdot H_2O$, Broken Hill Cave, Rhodesia
Hexahydrite, $MgSO_4 \cdot 6H_2O$, Lee Cave, KY, USA
Hibbenite, $2 Zn_3 (PO_4)_2 \cdot Zn (OH)_2 \cdot 7H_2O$, Hudson Bay Cave, B.C. Canada
Hopeite, $Zn_3 (PO_4)_2 \cdot 4H_2O$, Broken Hill Cave, Rhodesia
Huntite, $3 MgCO_3 \cdot CaCO_3$, frequent in moonmilk
Hydrocalcite, $CaCO_3 \cdot 6H_2O$, common in moonmilk under $5°C$
Hydromagnesite, $4 MgCO_3 \cdot Mg(OH)_2 \cdot 4 H_2O$, common in moonmilk in dolomite caves
Hydroxylapatite, $3 Ca_3(PO_4)_2 \cdot Ca(OH)_2$, Negra Cave, Puerto Rico
Hydrozincite, $2 ZnCO_3 \cdot 3 Zn(OH)_2$, Island Ford Cave, VA, USA
Ice, H_2O, ice caves, e.g., Eisriesenwelt, Salzburg, Austria
Jarosite, $KFe_3 (SO_4)_2 (OH)_6$, Tintic Cave, UT, USA
Lecontite, $NaNH_4 (SO_4) \cdot 2H_2O$, Piedras Cave, Honduras
Limonite, $FeO (OH)$, frequent in caves

Table 13.8 (Continued)

Lithophorite, $LiMn_3Al_2O_9 \cdot 3H_2O$, Cueva de los Cristales, Venezuela
Magnesite, $MgCO_3$, Titus Canyon Cave, CA, USA
Malachite, $CuCO_3 \cdot Cu(OH)_2$, Lilburn's Cave, CA, USA
Martinite, $Ca_3(PO_4)_2 \cdot CaCO_3 \cdot Ca(OH)_2$, Negra Cave, Puerto Rico
Melanterite, $FeSO_4 \cdot 7H_2O$, Wilson Cave, NV, USA
Minervite, $AlPO_4 \cdot nH_2O$, Grotte de Minerve, Hérault, France
Mirabilite, $Na_2SO_4 \cdot 10H_2O$, Flint Ridge Cave, KY, USA
Misenite, $KHSO_4$, cave on Cape Miseno, Naples, Italy
Monetite, $CaHPO_4$, Aggtelek, Hungary
Monohydrocalcite, $CaCO_3 \cdot H_2O$, in coral formations and moonmilk, rare; Franconian Switzerland, Western Germany
Montmorillonite, $(Al, Mg)_8 (Si_2O_5)(OH)_{10} \cdot 12 H_2O$, Calif. Cave, USA
Nesquehonite, $MgCO_3 \cdot 3 H_2O$, Le Moulis, France
Newberyite, $MgHPO_4 \cdot 3 H_2O$, Skipton Cave, Australia
Nitrammite, $NH_4(NO_3)$, Nicajack Cave, TN, USA
Nitrocalcite, $Ca(NO_3)_2 \cdot H_2O$, Mammoth Cave, KY, USA
Nitromagnesite, $Mg(NO_3)_2 \cdot 6 H_2O$, Great Cave, KY, USA
Olivenite, $Cu_2(AsO_4)(OH)$, Tintic Cave, UT, USA
Opal, $SiO_2 \cdot 2 H_2O$, Grotte des Grès, Sahara, S. Algeria
Palygorskite, $(Mg_3, Al_2) Si_4O_{10}(OH)_2 \cdot H_2O$, Cueva de los Ursulas, Venezuela
Parahopeite, $Zn_3(PO_4)_2 \cdot H_2O$, Hudson Bay Cave, B.C., Canada
Pickeringite, $MgAl_2(SO_4)_4 \cdot 22H_2O$, Diana Cave, Banatgebirge, Rumania
Pigotite, Organic Mineral, Sea-Cave, Porthcurnow, Cornwall, GB
Psilomelane, MnO_2-gel, Jasper Cave, WY, USA
Pyrite, FeS_2, Dachstein-Mammut Cave, Austria
Pyrrhotine, FeS, Herman Smith Cave, IL, USA
Quartz, SiO_2, Wind Cave, SD, USA
Salmoite, Hudson Bay Cave, B.C., Canada
Salt, $NaCl$, Welbeck Colliery, GB
Scholzite, $Ca_3(PO_4)_2 \cdot Zn(OH)_2 \cdot H_2O$, Island Ford Cave, VA, USA
Selenite, s. Gypsum
Sepiolite, meerschaum, $2 Mg_2Si_2O_5 \cdot Mg(OH)_2 \cdot 4H_2O$, Zbražow Cave, ČSSR
Smithsonite, $ZnCO_3$, Herman Smith Cave, Il, USA
Soda niter, KNO_3, Leonhardscave near Homberg, W. Germany
Spangolite, $Cu_6Al(SO_4)Cl(OH)_{12} \cdot 3 H_2O$, Blanchard Mines Cave, NM, USA
Spencerite, $Zn_3(PO_4)_2 \cdot Zn(OH)_2 \cdot 3 H_2O$, Hudson Bay Cave, B.C., Canada
Sphalerite, ZnS, Herman Smith Cave, IL, USA
Struvite, $MgNH_4(PO_4) \cdot 6H_2O$, common in guano caves
Sulphur, S, Shor-Su-Cave, Usbekistan, USSR
Syngenite, $K_2Ca(SO_4)_2 \cdot H_2O$, un-named cave in Western Australia
Taranakite, $6 Al(PO_4) \cdot 2K(OH) \cdot 18 H_2O$, Pig Hole Cave, VA, USA
Tarbuttite, $Zn_3(PO_4)_2 \cdot Zn(OH)_2$, Broken Hill Cave, Rhodesia
Tenorite, CuO, Calumet Cave, AZ, USA
Thenardite, Na_2SO_4, Mammoth Cave, KY, USA
Tinticite, $2 FePO_4 \cdot Fe(OH)_3 \cdot 3H_2O$, Tintic Cave, UT, USA
Trihydrocalcite, $CaCO_3 \cdot 3H_2O$, in moonmilk
Turanite, $Cu_3(VO_4)_2 \cdot 2 Cu(OH)_2$, Tyuya-Muyun Cave, USSR
Tyuyamunite, $Ca(UO_2)_2(VO_4)_2 \cdot nH_2O$, Tyuya-Muyun Cave, USSR
Vanadinite, $3 Pb_3(VO_4)_2 \cdot PbCl_2$, Havasu Canyon Cave, AZ, USA
Variscite, $AlPO_4 \cdot 2 H_2O$, Drachenhöhle, Mixnitz, Austria
Whitlockite, $Ca_3(PO_4)_2$, El Capote Cave, Nueva Leon, Mexico

14 Speleogenetics

Speleogenetics are defined as the totality of all processes which effect the creation and development of natural underground cavities. These comprise corrosion, erosion, and incasion as have already been discussed. They are influenced by lithology, tectonics, and climate from which springs the morphological and karst-hydrological diversity of karst regions. The result is that speleogenetic theories are heavily dependent on the region studied by the author in question. Yet it can be generally stated that in karst interstices and joints are widened to form caves by the dissolution of rocks. D.C. Ford (1970) rightly questions the validity of this statement as a definition of speleogenetics, considering it to be much too general and to express too little. Other authors, mostly American, have also expressed negative opinions of general speleogenetic theories, such as White and Longyear (1962) who observed: "... the multitudinous theories are neither correct nor incorrect in the general case, they are irrelevant" (quoted by Ewers, 1972). Howard (1963) said: "... an universally applicable origin of caves is impossible unless one speaks in the vaguest and most inconsequential terms," (p. 54). Halliday (1960) expresses it a little more mildly: "Only in the broadest terms can it be said that all limestone caves develop in the same way, and terminology which suggests that this is true should be replaced by the description of individual speleogenetic sequences," (p. 23, Abstract).

Karst-hydrological and speleogenetic theories overlap in many cases so that, for example, Thraikill's paper on the elements of karst hydrology (1968) applies equally to the formation of caves. Reference should be made to Chapters 6.3 and 7, but repetitions cannot be avoided in the following.

14.1 The Role of Joints and Bedding Interstices in Speleogenetics

Interstices and open joints are the starting points for speleogenetic processes.

Bedding interstices (a) and joints (b) differ in several points which are of hydrological importance:

1a) Bedding interstices are always subcapillary and of similar width. There are exceptions when the bedding plane has served as a plane of tectonic movement in folded regions. In capillary interstices water moves only under pressure, that means only under phreatic conditions.

b) Joints can occur as capillary interstices, yet as tension joints they can be open, showing a width of up to a few decimeters. At the escarpments of a plateau the rock pushes toward the space whereby capillary interstices become widened to open joints or tension cracks are formed. Open joints have water flowing through them under vadose conditions, and are, moreover, prerequisite to the formation of purely vadose caves (see Fig. 11.1).

2a) Bedding interstices stretch out in all directions (see Chap. 14.2). The distances possible amount to many kilometers.

b) Joint interstices, especially tension joints, cannot as a rule be followed for farther than a few hundred meters; faults can be followed for a few hundred meters up to a few kilometers.

c) Overthrust planes of nappes are extensive but there is no effect on cave formation. At best caves in basal beds are cut into the overthrust plane by erosion. Even in cases where limestone occurs on both sides of the overthrust plane there are hardly any karst-hydrologically active connections, unless the surface was crossed by faults which were created later. The cause might be that it was sealed off as a result of mylonitization or by means of clayey rock material thrusted by the nappe.

3a) Bedding interstices extend far, especially in carbonate rocks, and touch large portions of the ground surface which can supply a great quantity of water. Folds provide no obstacle to this as long as the impermeable rocks in the culmination of the fold lie below the karst water level. If they project up above it, the result is a canalization of the water in the syncline such as was the case in the second phase of development in Butler Cave (VA, USA).

b) Joints usually stand at a steep angle or vertically, providing water passages which are more or less straight and short. The areas they drain directly are small — apart from their function as swallow-holes at the surface. Conditions for the drainage of large portions of surface are considerably better in a network of joints.

The optimum is a combination of bedding and joint interstices as this not only creates favorable conditions for drainage but also promotes corrosion by mixing waters.

14.2 The Development from Interstice to Cave Passage Under Phreatic Conditions

Interstices and open joints stand at the beginning of all underground karstification. Open joints are hydrologically previous. In an experiment with a block of limestone (90% $CaCO_3$) 40 x 60 cm^2 in surface area and 30 cm thick no interstices could be found with a magnifying-glass. A circle of 10 cm was cut into this block and a pressure-box was set on it. The pressure of a 40-m water column was applied to the area within the circle. After about 1 h a capillary interstice became visible as water seeped out and large drops gradually formed. With a 50-m water column there was noticeable dripping and new wet places were observed (Bögli, 1964b, 1969d).

This experiment proves the permeability of even subcapillary and capillary interstices when the pressure gradient is sufficient. The interstice which responded first showed a speed of seepage v_s of 0.30 m/h. Differences in pressure of a 40-m water column are not rare in mountain karst. The pressure gradient h/l is, however, approx. 10^2-10^4 times smaller because of the much greater distances and the resulting v_s equals 0.1-0.01 x v_s [see Chap. 5.5.6, Eqs. (5.6) and (5.8)]. For the first active interstice there would be a v_s of 0.030 m/h if the distance of seepage was 30 m so that the length of time of seepage would be 1000 h or 41.7 d. In capillary interstices the water covers several tens or hundreds of meters in a year. Therefore the initial phase of underground karstification lasts very long. This fact and the resultant widening during this period has been the great speleological problem — and is in part still today.

Capillary interstices must be hydrologically pervious from the beginning of karstification, but they are at a disadvantage in comparison to open interstices. The question therefore arises as to why the open joints of at least 2 mm width postulated as necessary by O. Lehmann (1932) are so often not effective for the formation of caves. As shown empirically water not infrequently follows narrow interstices although open joints are there at its disposal.

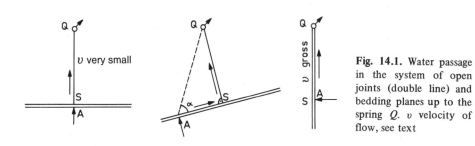

Fig. 14.1. Water passage in the system of open joints (double line) and bedding planes up to the spring Q. v velocity of flow, see text

If water flowed out of the rock at A into the open joint which is a few centimeters in width, at S it would seep into a bedding plane and it would reach the spring at Q. S is the foot of the perpendicular out of Q to the open joint and thus $\overline{SQ}$ is the shortest connection. There are now three resulting possibilities:

a) the open joint runs perpendicular to the pressure gradient A. The water follows this gradient and seeps into the bedding plane directly opposite A; this is so because every other point is farther away from Q and therefore has a smaller pressure gradient (Fig. 14.1, left). The open joint does not become a karst water passage.

b) the open joint runs at the angle α to $\overline{AQ}$, which corresponds to the usual case. Since the water does not experience any loss of pressure, moving at a minimal velocity in the wide open joint, it seeps into the interstice at S, where the greatest pressure gradient occurs. It thereby chooses the hydrologically shortest way from A to Q by way of S (Fig. 14.1, middle). The open joint becomes a water passage for a short distance:

$$\overline{AS} = AQ \cdot \cos \alpha$$
$$\overline{SQ} = AQ \cdot \sin \alpha$$
$$\overline{ASQ} = AQ (\sin \alpha + \cos \alpha).$$

c) If the open joint runs directly to the spring, the use of it as a water-passage is obvious and is moreover also still the case under vadose conditions (Fig. 14.3, right).

The equation in (b) is the basic equation for each possible case, even where the diverting bedding plane does not run through a point opposite A.

Several examples in Hölloch with $\alpha = 90° \pm 15°$ show that open joints are not accepted by water as water passages, for instance in the commercial part at the Kapelle where the throw of the fault is 5 m, or in the Orgelwand-Seilwand-fault with a throw of 60 m (see Fig. 5.16). The water passage running almost perpendicularly to this open joints lies 60 m deeper than the overflow. Of the dozens of bedding planes cut open by the Orgelwand only one has become karst-hydrologically active. During the phase of folding it had served as a plane of movement and was consequently a little wider, yet still decidedly capillary.

The distance $\overline{SQ}$ is presumably straight under the ideal conditions of an interstice which is equally wide in all directions, whose walls are equally rough, when the composition of the rock is the same everywhere. In this straight line maximum velocities of flow are found. On all other courses the velocities are the lower, the greater the distance is from the straight line. The expression of this decrease in speed is the elliptical — more precisely lenticular — cross-section of the passage (see Chap. 12.2.1).

In nature there are no ideal conditions. First in an open joint the movement of the water is not a matter of one bedding plane but of many and of joint interstices as well. They are not of equal value nor are they individually of equal width along their whole length, nor homogenous in their material, as there are impurities, calcite veins, and fossils. As a result the water moves in sinuous, criss-crossed courses in the interstices. $\overline{SQ}$ is then at best the center line of all statistically possible water-courses. Yet the rule that water follows the hydrologically shortest way is always valid. The heavier the flow, the more the interstice widens there, and the more the discharge diminishes in the neighboring water-courses, partly until they are completely in disuse. If an initial network forms a main passage early, a simple, hardly branched passage is created. When there are water-courses of almost similar value, they all grow to become cave passages which are connected with one another like a network: passage network. If the connections are locally intensified, it is called a labyrinth.

The initial phase lasts relatively long. In this period the future passage network is established; it will be reduced to a few or one single water-course as the other individual ones are left in disuse.

The descriptions up to now have presumed that the water-courses were laid on one and the same bedding plane. Such is frequently the case when the direction to the local base level runs almost parallel to its strike. Hölloch partly represents this type. Less frequently the local base level lies in a direction perpendicular to them. The water then crosses the folds and can reach the surface in, for example, the rising flank of a fold. That was originally the case in Butler and Breathing Cave in VA, USA (Deike, 1960). If necessary the water reaches the surface by means of open joints from the bedding interstice. However, more frequently the strata dip down and it must search for a way to the spring by means of a system of joints. D.C. Ford described such examples (1965, 1968, 1970) in the central Mendips. This leads necessarily to a much longer course of flow than corresponds to the distance between the point of seeping-in and the spring.

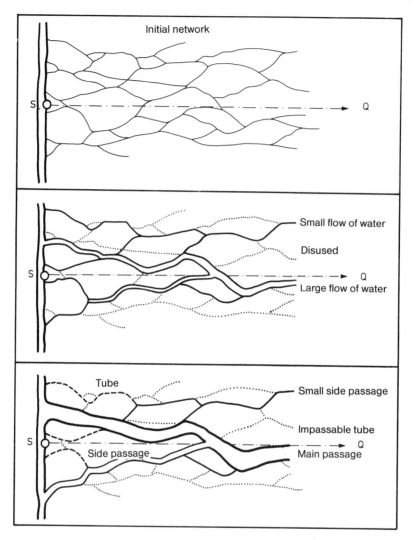

Fig. 14.2. From the initial network to the cave passage or network of passages. *Q* spring; *S* foot of the perpendicular from Q to the open joint

14.3 The Development to a Cave Level

The further fate of the cave's development depends on the behavior of the karst water surface. If it is constant for a longer time, passage cross-sections grow until they reach the breaking limit of the rock — beyond that point incasion dominates the shaping of the cave. Such conditions are found in tectonically and morphologically quiet regions, e.g., in the peneplains of North America; here Flint Mammoth Cave represents the prototype. For this reason American speleogenetic theories assume a quasi-stationary karst water surface in flat-lying limestones. Over longer periods of time, however, there are changes in the position of the local base level even in such regions and thus the development of new underground karst water levels, so-called evolution levels according to Sawicki (1909). They correlate to terrasses on the surface (Deike, 1967; Miotke and Palmer, 1972). Since in general only slight high-water levels of a few meters occur in the peneplains, the passages run almost horizontally. Therefore the different evolution levels can be well distinguished although they are in part lying only 10-20 m above one another.

In regions which suffer stronger tectonic movements developments take a somewhat different course (Davies, 1960). Deike (1960) describes Breathing Cave (VA, USA) which developed in the deep phreatic zone. Its folded limestone strata lie embedded between quartz sandstones. The water followed the network of joints and thus a cave system was created that was an extremely geometrical network with a difference in elevation of over 100 m. When later the local base level was laid deeper this cave with its original drainage perpendicular to the strike was transformed into a cave with run-off parallel to it and it now follows the axial dip of the syncline, along with Butler Cave (Fig. 14.3).

Where valleys cut down quickly as in the Alps, the formation of evolution levels occurs only during the phase in which there is a pause in this activity.

The still small diameter of the cave passages and the breakdown debris in them lead to losses of pressure (see Chap. 5.5.7); these can only be compensated for by a great head or by a strong high-water zone. The evolution levels fluctuate around a median position with large amplitudes, in Hölloch ± 50 m. The water courses which were formed in the first phase of development remain fixed on it and enlarge. As the deepening of the valley recommences karst hydrology must adapt to the changing circumstances; a deeper system of passages is established in the zone of the new local base level. The upper system, which at first is still phreatic, constantly loses an increasing amount of water to the lower one; the permanent karst water surface sinks. Gradually the upper system comes under the conditions of the high-water zone. Further development is now limited to periods of high water, when phreatic conditions dominate. In periods of normal water (vadose, air-filled) the feeders cut into the passage floor, creating channels and canyons. If, in the course of on-going underground karstification, also the high-water level sinks below the upper evolution level, the latter becomes inactive – the expansion of the cavities is completed, but not their shaping by breakdown and the deposition of calcareous sinter.

If the pauses in the deepening of the valley do not last too long, the water leaves the upper evolution level before it has reached maturity of form. The original forms, from which the conditions of creation can still be easily recognized, remain preserved. If these periods last a longer time, however, the cross-sections of the passages become so large that there is general breakdown.

This is true of many cave systems in the eastern Alps, e.g., Tantal Cave, Eisriesenwelt and the Dachstein Caves. The early forms are completely destroyed by breakdown or buried under debris from it (see Chap. 11).

For the clarification of cave genetics it is thoroughly relevant that in the case of breakdown especially joint surfaces are laid open, which suggests that they are the basis of the origin of the passage. Thus by means of breakdown a bedding-plane passage

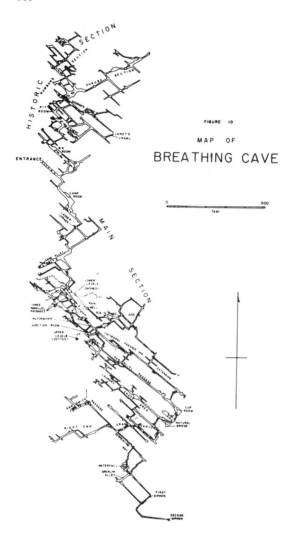

Fig. 14.3. Deep phreatic network of passages in Breathing Cave (Deike, 1960)

can be changed to have the appearance of a fictitious joint passage. Since in addition joint surfaces become exposed which have just as little to do with karstification as with cavity formation, serious false conclusions are unavoidable. For this reason most of the longer alpine caves especially in the eastern Alps have been classed as joint-controlled caves.

Signs of phreatic origin
a) Network caverns — key forms of the phreatic zone
b) Bedding-plane passages — key forms of the phreatic zone
c) Symmetrical elliptical (lenticular) cross-sections — key forms for phreatic conditions (phreatic zone + high-water area of the vadose zone, larger siphons)
d) Inverse potholes from corrosion by mixing waters — key forms for phreatic conditions
e) Descents and ascents alternating in a passage — phreatic zone
f) Hydrically formed passage ceilings — phreatic conditions

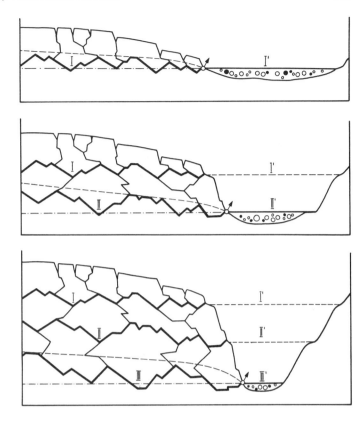

Fig. 14.4. Cross-sections showing the development of an alpine cave of the Hölloch type. *I, II, III:* evolution levels; *I', II', III':* correlating valley floors or local base levels, respectively

14.4 Primary and Secondary Vadose Cave Formation

Water that flows in the underground under vadose conditions has an open water surface and moves pressure-free in a gravitational channel. It behaves like water on the surface. Therefore passages originating in the vadose zone descend continuously. The runway must be able to accept the discharge; this presupposes a sufficiently wide connection to the spring, a case of pronounced karst-hydrological activity. Only in the back-water region (high-water zone) do temporary phreatic conditions occur.

Primary, vadose cave formation by water occurs without a previous phreatic phase. Only open joints 1 mm in width and more come into question for this. It is a condition which O. Lehmann (1932) assumed to be necessary for cave formation in general. This type is, however, comparatively rare. It is characterized by a rather straight, little-branched course. There is no network. The gradient is uniform. The passages are narrow and high, frequently steep or shaft-formed. They are the one-cycle caverns named by Davis (1930); to these Pohl's (1955) "vertical shafts" must also be counted.

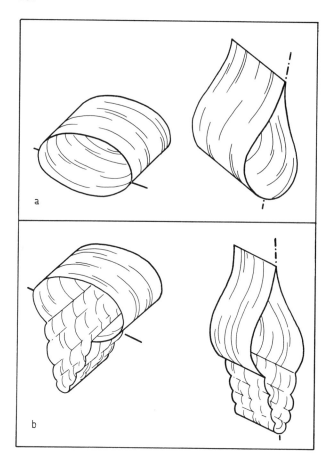

Fig. 14.5 a, b. Cross-sections in two-cycle caverns. **a** Cross-sections which were always inactive in the vadose phase; only the phreatic phase is recognizable; **b** cross-sections in feeders; above the phreatic portion, below it the part of vadose origin

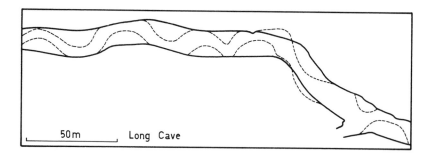

Fig. 14.6. Canyon (*broken lines*) in Long Cave (KY, USA) according to Deike (1967, p. 91)

Characteristics of primary vadose caves
a) The cave systems are either simple or branched, connections between them are rare and purely accidental
b) The passages follow open joints
c) The passage's form is high and narrow, with the shape of open joints and canyons
d) The gradient is uniform and without rocky ascents opposed to the general inclination

In the case of *secondary vadose cave formation* there is a preceding phreatic phase. In the following vadose phase the deepest possible passage in a network is always taken as the discharge conduit; the many other passages become dry, inactive, and remain preserved as phreatic remnants. The feeders run less steeply than is the case in a primary vadose system. Even though the gradient is essentially uniform, ascents remaining from the phreatic phase are not rare, nor are the siphons which are thereby formed (see Chap. 5.4 with Fig. 5.7). The passage cross-sections have either the old, phreatic shape still, if the passage became inactive right at the beginning of the vadose phase, or otherwise a combination of shapes made up of remainders of the phreatic, lenticular cross-sections lying above (shoulder of the canyon) and the vadose canyon cut into the floor of the same. Davis (1930) called this type a two-cycle cavern.

When canyons are formed in passages with broad, flat floors and a slight incline, meanders often occur (Figs. 12.3 and 14.6). Such forms were described by Deike (1967) as found in caves in Mammoth Cave Plateau (KY, USA); together with White he also pursued the problem mathematically (1969).

As the loops of the meanders move forward simultaneously with the deepening of the channel, the deeper parts are advanced, in comparison to the upper ones, in the direction of the flow. This occurrence is to be found fairly frequently in deep canyons — up to 10 m and more.

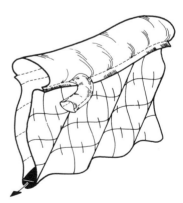

Fig. 14.7. Meandering canyon according to Ewers (1972, Fig. 24)

14.5 Widening of Interstices

For capillary interstices to become passable for water phreatic conditions must dominate. Problems of corrosion result from this which have already been pointed out several times.

In the laboratory Ewers (1972) investigated the development and widening of water conduits in physically soluble rocks. He pressed a block of salt with a polished surface on to a water-filled, transparent, plastic cushion. Between the two lay a capillary interstice. At one end of the plane he introduced water into this interstice, at the other he allowed for drainage. Between the two points an even pressure gradient developed. After a few minutes a conduit began to form at the point of inflow which developed in the direction of the point of outflow. At the same time thinner side branches formed. When the conduit reached the outflow the first phase was completed (Fig. 14.8a). The main canal continued to grow as the cross-section enlarged, while the side branches simultaneously reduced their growth and finally ceased it altogether.

Ewers' experiments are based on physical processes of dissolution which take place without the introduction of any substances. Therefore they cannot be directly applied to the dissolution of carbonate rocks. They only resemble dissolution by aggressive water, but its ability to dissolve is quickly exhausted. In the initial stage this type of dissolution cannot be reckoned with in limestone.

The main problem does indeed not lie in the determination of the direction of flow of the water, but in corrosion itself. As opposed to Ewers' experiment, moreover, the water flows in previously determined courses which it only has to widen. This would correspond to about phase two in Ewers' experiments.

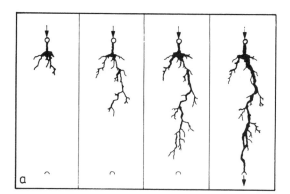

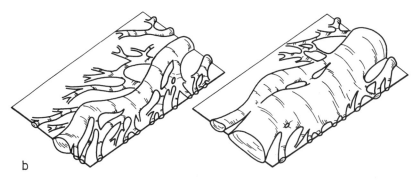

Fig. 14.8. a, b. Development of water conduits acc. to Ewers (1972, Fig. 6 and 12). **a** Development up to the beginning of phase 2; **b** sections showing the further development in phase 2

Since corrosion by mixing waters plays an important role especially in the initial phase, foreign water must be introduced on the bedding plane. This takes place by means of joints. Wherever they cut the bedding plane a mixture results and hence corrosion by mixing waters and an expansion of the interstice in the direction of the outflow. If the joints run almost parallel to the pressure gradient, the bedding-plane passage will continue to develop in the direction of the joint but on the bedding plane. The resulting conduit is bedding-plane and joint-controlled. If, however, they run obliquely to the pressure gradient, widening continues away from the joint plane, which can have no further influence on the direction; the conduits are only bedding-plane-controlled.

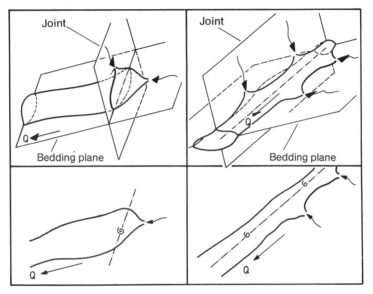

Fig. 14.9. Bedding-plane passages (*spirals* inverse potholes due to mixing corrosion). *Left* only bedding-plane-controlled; *right* bedding-plane and joint-controlled

It becomes clear that statistically the probability of corrosion by mixing waters is greatest near the surface of the karst water-body. The shallow phreatic zone thereby becomes favored for the formation of cave passages. However, it must be emphasized that water passages through the deep phreatic zone are thoroughly possible, especially when interstices and open joints are scarce. The smaller the possibility of the meeting of two waters, the deeper in the phreatic zone the meeting will take place, according to statistics.

14.6 Phases in the Development of Cavities

In the following, terms are used which are common in the Davis theory of cycles, but which show no relationship to it, neither to a development cycle of caves nor to a karst cycle. Rather they mean phases in development. These do not depend only on the time which has passed but also on the state of the rock and on other factors. In a cave passage where the parts are of the same age, various stages of development can be observed. In thick-layered limestone such a passage can show the early stage of maturity without incasion, while in rock that has been under heavy tectonic stress old age may already have begun with the destruction of early forms by breakdown.

The schematic organization of cave development should be functionally divided into a speleogenetic and a karst-hydrological phase. The *speleogenetic* one for a two-cycle cavern comprises:

a) previous phase: the rock persists in the given state; its cavities are water-filled, and nothing changes

b) initial phase: the local base level has reached such a position that a pressure gradient is created in the rock's interstices. The water moves so that the widening of the interstices begins. Predominant corrosion by mixing waters, phreatic zone

c) youth: the water flows faster, erosion joins corrosion (Plate 8.3). Underground cavities develop to the size of caves in the sense of the definition of a cave. Bumps occur but other forms of breakdown are lacking mostly phreatic

d) maturity: cross-sections continue to increase until the first signs of incasion with occasional ceiling breakdown. Phreatic and/or vadose (high-water zone, feeders, perhaps inactive)

e) old age: incasion, especially ceiling breakdown occurs generally. Covering and destruction of the forms of maturity. Physiognomically joints appear in the foreground (fictitious joint passages). Usually vadose and generally inactive, rarely phreatic

f) senility: the cave becomes destroyed

The *karst-hydrological* phase runs parallel to the development of the cave. The two are closely linked

a) previous phase: all interstices and open joints are filled with water. There is no pressure gradient, thus also no flow

b) initial phase: a pressure gradient is created because the local base level has been laid deeper. The water at first moves slowly, however with increasing velocity. The karst water surface forms. Deep phreatic

c) phase of karstification: when karst-hydrological activity has been started this phase begins. The permanent karst water level sinks until close to the local base level. Karst-hydrological activity advances into the deep phreatic zone

d) phase of cessation of sinking of the karst water level. Underground karstification reaches its maximum, karst-hydrological development its zenith. A renewed deepening of the local base level leads to a revival of the phase of karstification

In conclusion it should be pointed out that each cave presents an individuality to which speleogenetic theories cannot always do justice. Single phases may have been omitted, special circumstances may have led to an unexpected development or even to a unique situation. In semi-arid and arid regions cave development takes a different course — or perhaps it unexpectedly follows theory precisely as if water were present in vast amounts as in humid climates; this could, by the way, be explained by the Pluvial Period. In short, the systems of development presented cannot claim to be of complete general validity.

Phases in the Development of Cavities

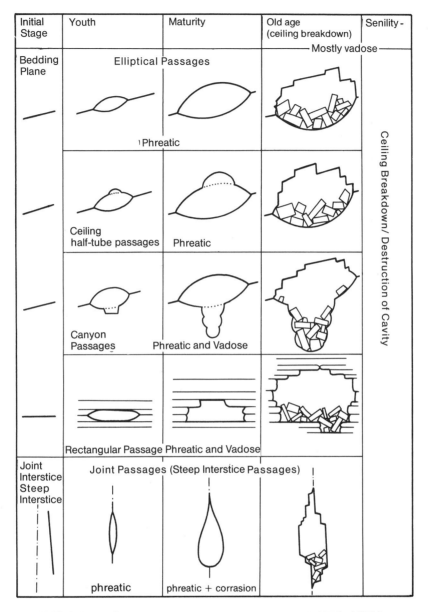

Fig. 14.10. Schema of the development of passages (acc. to Bögli, 1956a, 1976a)

15 Speleometeorology – Speleoclimatology

Speleometeorology is concerned with instantaneous processes in the atmosphere of caves, speleoclimatology with their average states. However, with the exception of wind, variations in the parameters are so small that there is only a slight difference between instantaneous and average values. Therefore speleometeorology and speleoclimatology are usually used with the same meaning, apart from their differing points of view. The former is preferred in English and in German usuage (Kyrle, 1923; Myers, 1962; Wigley and Brown, 1976), the latter in French (Gèze, 1965). In Germany and Austria mining terms such as Höhlenwetter, Bewetterung and Wetterwechsel (change in wind direction) are frequently used, thus in Kyrle who defines Höhlenwetter as "the total contents of the cave in gas-form". He supports the use of these expressions with the argument that "cave meteorology is very similar to that in mines" (p. 203).

Cave meteorology is essentially different from conditions on the earth's surface. The behavior of a cave's atmosphere is the result of pronounced local causes; local should be understood here to mean interconnected cavities and their openings to the surface. The absolute lack of any sunshine is the most important difference. On the other hand the earth's warmth is of some importance although its effects are usually overestimated. Moreover a cave's atmosphere lacks a free expanse of space, thus the air can move only linearly. In large cavities convection currents occur in addition. Then there are static states which are permanent in caves whereas their equivalent on the surface occurs only occasionally for a short time, for example constantly high air humidity near 100%, minimal variations in temperature, slight or no movement of air locally.

15.1 Movement of Air in Caves

The air in underground cavities is set in motion by differences in pressure on the one hand, and by flowing water on the other (Cigna, 1967, 1968). Trombe (1947, 1952) assumes that the effect of flowing water on the air is limited to waterfalls and is therefore only local: "Aucun mouvement d'air ne se produit en général dans l'ensemble du réseau par l'effet d'une chute d'eau" (1947, p. 99). On the other hand Myers understands it as an important motive force for the movement of subterranean air (cave wind): "In an active cave the most obvious agent promoting the circulation of the air is the freely running stream" (1962, p. 226). The varying conditions in the caves investigated are reflected in the different hypotheses. There is a wide span of possibilities for the velocity of the wind. Although it is so small in large caverns that it is unnoticeable, it occasionally rises in narrow passages to the strength of a storm. In a narrow opening (0.6 m²) of Pinargözü Cave (Turkey) even a velocity of 46.2 m/s (166.3 km/h) was measured (Bakalowicz, 1972).

Differences in air pressure are partly of exogenous origin, i.e., the result of exterior causes which are therefore foreign to the cave, partly of endogenous origin, and partly due to the contrast between the open atmosphere and the cave's air.

To what degree the differences in pressure express themselves as cave wind is a question of the form of the subterranean cave system. One must on principle distinguish between cave systems with several openings to the exterior and those with only one. In the first case exogenous forces can set the air within the earth into motion. This type of cave is called *dynamic*. Cave systems with only one opening to the outside scarcely react to exogenous factors, apart from the weak disturbances caused by variations in density: *static caves*. Where there is a change in type, in *statodynamic caves*, the main entrance counts for the larger part of static conditions while smaller openings either become active or inactive depending on the meteorological circumstances, e.g., Dobšinská ladová jaskyňa (Dobšinská Ice Cave) in Slovakia (ČSSR).

Bock (1913) was the first to attempt to express the circulation of cave air in equations, but their application failed because of the large number of parameters which can hardly be determined.

15.1.1 Exogenous Factors of Pressure Differences

An underground cave system with various openings is exposed to a wind on the surface (Fig. 15.1). In general under these circumstances the following is valid for the outside openings:

$$p_1 \neq p_2 \neq p_3 \neq \ldots$$

In this way pressure gradients are created which cause an air draught from the place of higher to the place of lower pressure. In through-caves the windward side has the higher, the lee-side the lower pressure.

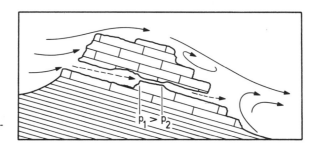

Fig. 15.1. Cave wind in a through-cave (see text)

In wide cave passages the pressure decrease is slight. It is concentrated on the narrow passes, e.g., between p_1 and p_2 in Fig. 15.1. In this stretch there is also the strongest draught. If a side passage took off at this point, air would be sucked up it according to the Bernoulli equation (Chap. 5.5.3).

A difference in pressure in the open atmosphere works in a similar way but it is only slight even when the distances between the openings are kilometers long. The cave winds caused by this are weak and are usually overlaid by winds of other origins and thus unrecognizable.

Even differences in temperature outside the openings can cause differences in barometric pressure. If one entrance lies in a cool wood or on the shaded side of a mountain, the other on a warm, sunny slope, the pressure on the cool side is higher than on the sunny. Generally this can be proved only when the openings are approximately at the same height.

15.1.2 Endogenous Causes of Pressure Differences

Pressure differences within underground cavities can only be caused by different densities of the atmosphere which are the result of varying temperature, humidity or CO_2 content. Air movements which have this origin are convection currents:

$$\text{liter weight (pond)} = \frac{1.293}{1 + 0.00367 \cdot t} \cdot \frac{p}{760}$$

$$\text{air density} = \frac{0.001293}{1 + 0.00367 \cdot t} \cdot \frac{p}{760}$$

These formulas are valid for dry air and can give useful approximate values for moist air, when the value for the given moisture content is set in place of 1.293 at 0°C or 0.001293 respectively. p is the actual air pressure.

Table 15.1. Liter weight of the atmosphere at 760 mm Hg dependent on the temperature t and the humidity H (according to Trombe, 1947). The liter weight x 10^{-3} corresponds numerically to the density

t (°C)	H (%)					
	50	60	70	80	90	100
−10	1.341	1.341	1.341	1.341	1.341	1.341
− 5	1.316	1.316	1.316	1.315	1.315	1.315
0	1.291	1.291	1.291	1.291	1.290	1.290
5	1.268	1.267	1.267	1.266	1.266	1.266
10	1.244	1.244	1.243	1.243	1.242	1.242
15	1.222	1.221	1.220	1.219	1.219	1.218
20	1.199	1.198	1.197	1.196	1.195	1.194
25	1.177	1.176	1.175	1.173	1.172	1.170
30	1.156	1.154	1.152	1.150	1.149	1.147

Endogenous Causes of Pressure Differences

According to this a column of air 1 cm² in area at its base and 10 m in height has a weight of 1.291 pond or a pressure of 1.291 pond/cm² at 0°C and 70% humidity; at 100% it is 1.290 pond/cm², still referring to 760 mm Hg. This difference in humidity results in a pressure difference of 0.001 pond/cm² or a water-column (h) of 0.001 cm.

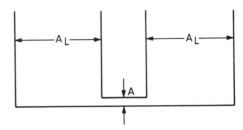

Fig. 15.2. see text

If the cross-section A of the connecting tube is small in comparison to A_L of the air-columns, the pressure gradient concentrates on the narrow passes so that Torricelli's theory can be applied here:

$$v = \sqrt{2gh'}.$$

h' is the height of the air column which can be calculated from the water column:

$$h' = \frac{h}{|\rho|}.$$

In the above example h would equal 0.001×10^{-2} m and

$$v = \sqrt{2g \frac{0{,}001 \cdot 10^{-2}}{0{,}001291}} = 0{,}39 \text{ m/s}.$$

An air movement of 0.39 m/s is scarcely perceptible to our senses. According to Table 15.1 differences in temperature of 5°C, e.g., from 10° to 15°C, result in a difference in pressure of 0.024 pond/cm² at 100% humidity, 24 times the value resulting from a 30% difference in humidity. That corresponds to an air velocity of 1.90 m/s. But a difference in temperature of 5°C is very rare in connected shafts with no outside opening. Normally the difference ranges around 1°C so that air movements caused thereby are far less than 1 m/s.

In conclusion it can be said that air movements caused by endogenous differences in temperature and humidity are close to the limits of perceptibility. Their importance lies in the fact that they intermix cave air, balancing differences as well as transporting humidity in the parts of the cave closed to the exterior: convection currents. The humidity can precipitate as condensed water and thus participate in the widening of the cavities by corrosion.

15.1.3 Cave Winds as a Result of Temperature Contrast Between Open Atmosphere and Underground Cavities

When an underground cave system possesses two or more outside openings at different heights, cave winds of varying strength occur when there is a difference in temperature between the cave and the open atmosphere. Apart from a few deviations, e.g., in the case of ice caves, the temperatures inside the mountain correspond approximately to the annual average temperature on the surface or are slightly lower as a result of the water from melting snow in early summer (high Alps). This constancy of temperature stands in contrast to the variations of the open atmosphere where the Central European extremes lie at $-30°$ and $+38°C$, the Irish at $-7°$ and $28°C$. The differences between the medians of January and July decreases from $22°C$ on the western border of the USSR to $9°C$ in Ireland. This results in differences in temperature of up to $20°C$ and more between the endokarst and the outside atmosphere. In the north of the USA the difference between the extreme temperatures of January and July varies from $50°C$ on the Atlantic Coast to $70°C$ near St. Paul to $36°C$ on the Pacific Coast. In the south it is $40°C$ on the Atlantic Coast (southern Georgia), $50°C$ in Arkansas, and $24°C$ on the Pacific Coast.

With the aid of Table 15.1 the masses of the air columns can be calculated if t, H, the height of the air column H_L and the air pressure p are known. Since approximate values are sufficient, the average p at a given height can be determined from Table 15.2.

Table 15.2. Average value of the air pressure dependent on the altitude. a) according to Kohlrausch (1930). $t = 0°C$, $\varphi = 50°N$; b) according to Thommen (?), gliding temperature according to CINA, $\varphi = 45°N$

H (m above sea level)	a) Pressure (mm Hg)	Conversion factor	b) Pressure (mm Hg)	Conversion factor
0	762.0		760.0	
200	743.2	0.975	742.1	0.976
400	724.9	0.951	724.6	0.953
600	707.1	0.928	707.5	0.931
800	689.7	0.905	690.6	0.909
1000	672.7	0.883	674.1	0.887
1200	656.2	0.861	657.9	0.866
1400	640.0	0.840	642.0	0.845
1600	624.2	0.819	626.4	0.824
1800	608.8	0.799	611.2	0.804
2000	593.7	0.779	596.2	0.784
2200	579.1	0.760	581.7	0.765
2400	564.8	0.741	567.2	0.746
2600	551.0	0.723	553.2	0.728
2800	537.5	0.705	539.3	0.710
3000	524.2	0.688	525.8	0.692

Example: the lower opening of a cave lies 730 m above sea level, the upper one 500 m higher — such are the conditions of the western section of the Hölloch system.

The temperatures in the cave at 980 m above sea level, the center of the air column, are between 4 1/2° and 5°C. The arrival of water from melted snow was taken into account so the temperatures are converted to the monthly average. The exterior temperatures were taken from neighboring stations.

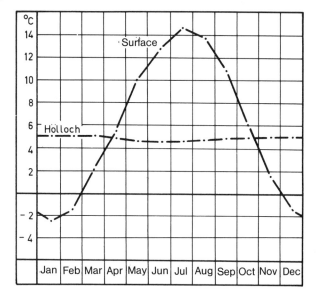

Fig. 15.3. Temperature curves of the monthly average for Hölloch and the surface at 1000 m above sea level

From Table 15.1 the liter weight or the density (liter weight x 10^{-3}) is determined for the air pressure at 1000 m above sea level with the aid of the conversion factor.

Table 15.3. t°C (1), liter weight: pond/l (2), G of the 500-m-high air column (area of cross-section 1 cm²) in pond (3), Diff. between the G of the two corresponding air columns in the atmosphere G_f and in the cave G_H (4). H: humidity of the atmosphere

Month	Open atmosphere (H: 80%)			Hölloch (H: 100%)			$G_g - G_H$
	(1)	(2)	(3)	(1)	(2)	(3)	(4)
1	−2.4	1.156	57.80	5.0	1.123	56.15	1.65
2	−1.5	1.151	57.55	5.0	1.123	56.15	1.40
3	2.0	1.136	56.81	5.0	1.123	56.15	0.66
4	5.5	1.121	56.05	4.8	1.124	56.19	−0.14
5	10.0	1.103	55.12	4.5	1.125	56.25	−1.13
6	12.8	1.091	54.53	4.5	1.125	56.25	−1.72
7	14.5	1.083	54.17	4.5	1.125	56.25	−2.08
8	13.8	1.086	54.32	4.6	1.125	56.21	−1.91
9	10.7	1.100	54.98	4.7	1.124	56.21	−1.23
10	6.0	1.119	55.95	4.8	1.124	56.19	−0.24
11	1.5	1.138	56.92	4.9	1.123	56.17	0.75
12	−1.5	1.151	57.55	5.0	1.123	56.15	1.40
Extreme values							
	25	1.040	52.02	4.5	1.125	56.25	−4.23
	20	1.061	53.04	4.5	1.125	56.25	−3.21
	−10	1.189	59.47	5.0	1.123	56.15	3.32
	−20	1.236	61.79	5.0	1.123	56.15	5.64
	−25	1.261	63.04	5.0	1.123	56.15	6.89

($G_f - G_H$) can be tested with an altimeter as the difference in height or in pressure, respectively. In Hölloch differences in pressure of a 1.14 cm water column (10 m difference in height according to the Thommen altimeter) were measured at temperatures slightly under 0°C in front of and behind the Wettertüre in the commercial part of the cave. These differences in pressure can be converted into wind velocities according to Chapter 15.1.2.

$$v \, [m/s] = \sqrt{2g \frac{h}{\rho_L} 10^{-2}}, \text{ h in cm of water column}$$

Table 15.4. The monthly average wind-velocity v (4) dependent on (G_f-G_H) (3), t°C (1) and liter weight (2) and ρ resp. (Lg · 10^{-3}) at the Wettertüre in Hölloch; calculated

Month	(1)	(2)	(3)	(4)	Month	(1)	(2)	(3)	(4)
1	6.0	1.119	1.65	17.01	7	5.5	1.121	−2.08	−19.08
2	6.0	1.119	1.40	15.67	8	5.6	1.120	−1.91	−18.29
3	6.0	1.119	0.66	10.76	9	5.7	1.120	−1.23	−14.68
4	5.8	1.120	−0.14	− 4.78	10	5.8	1.120	−0.24	− 6.48
5	5.5	1.121	−1.13	−13.54	11	5.9	1.119	0.75	11.47
6	5.5	1.121	−1.72	−17.35	12	6.0	1.119	1.40	15.67

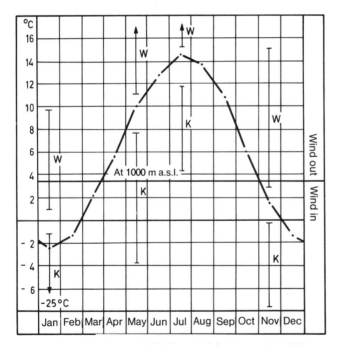

Fig. 15.4. Directions of ventilation at the entrance to Hölloch during the passage of warm and cold air respectively. (*K, W* corresponding extent of temperature). *Line of dots and dashes* the annual periodicity of the temperature on the surface (1000 m above sea level). *Heavy line* temperature of the cave at 1000 m above sea level, 4.8°C

Negative values for (3) and (4) mean that the colder cave air is flowing out of the entrance, positive mean the cave wind is blowing inward. According to Table 15.3 there is calm (equilibrium) in the cave when the outside temperature is approx. 4.8°C at 1000 m above sea level and the weather is calm. With a west wind it sinks to 3°C. A change of wind direction in winter from inward to outward means warm air in the atmosphere and with it rain, commencement of snow-melting and danger of high water.

The measurements taken at the Wettertüre show that the wind's velocity measured at 9 m/s lies 5 m/s under the velocity calculated. That was to be expected as there must be other connections which pursue paths still unknown. But it is true here, too, that physically, natural cavities and cave systems can hardly be apprehended precisely, however the values calculated do offer approximations or hints of conditions which have not yet been recognized.

Besides the annual periodicity of the average velocities of the wind, daily periodicity occurs also. In addition there are aperiodic variations of temperature during the passage of warm and cold air-masses with corresponding behavior of the cave winds.

Only in midsummer is there uninterrupted ventilation outward at the entrance. In the other seasons changes in the direction of the air current are normal, even though in winter the air moves predominantly inward. Concerning calculations on ventilation see McElroy (1966) who bases his work on mining conditions.

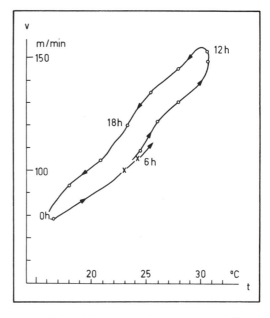

Fig. 15.5. Daily periodicity of wind's velocity v and the outside temperature t in the cave of Pinargözü according to Bakalowicz (1972). v as function of t (11th Aug. 1970, 06:00 to 12th Aug., 1970, 06:00)

15.1.4 Air Movements Caused by Flowing Water

As a result of friction with flowing water, air is moved forward. If the water course has narrow passes or siphons inserted in it, local air circulation takes place in front of them (a), the air skimming back along the ceiling to the point of departure. Rapidly flowing water sweeps air along with it at the transition to a pressure flow (b), or sucks it into the center of the whirl (c; Fig. 15.6).

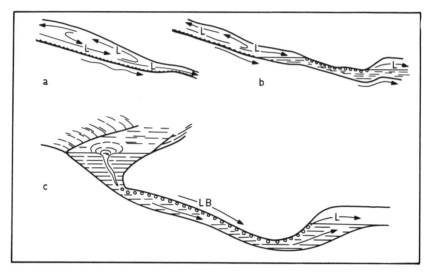

Fig. 15.6. Explanation see text. *L* air in motion; *LB* transport of air bubbles

Because of its buoyancy the air swept along is carried under the ceiling. The steeper the passage, the higher the velocity of flow necessary to be able to move the air downward against its buoyancy (see Chap. 12.2.2, ceiling channel passage).

Air carried in this way in narrow passages under phreatic conditions effects an air movement by means of suction before the swallowing point; when the water re-emerges air movement is caused by an increase in pressure. Even this air movement is too slow as a rule for it to be perceived by the senses but evidence can easily be found by the movement of swaths of smoke.

Waterfalls are especially effective, above all those in narrow shafts where the water touches the walls. In this case the effect of a water-jet air pump occurs with suction at the upper end and a considerable increase in pressure in the air at the lower end (see Fig. 15.7 a-d).

a) waterfall in a wide shaft which is closed at the bottom: closed circulation of air

b) waterfall in a wide shaft which is open at the bottom: partly closed circulation of air, partly cave wind in the direction of the water flowing away

c) waterfall in a narrow shaft: closed circulation of air prevented, cave wind in side passage below

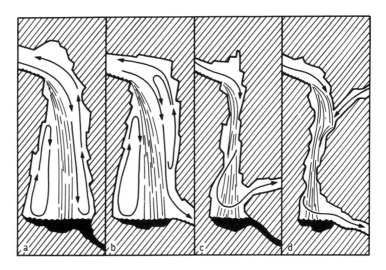

Fig. 15.7 a-d. Movement of air in waterfalls (see text). *Arrows* direction of wind's movement

d) waterfall in narrow shaft with side passage: air is sucked in through the waterfall's shaft but also through the side passage and is blown away below as a strong cave wind. Closed circulation of air prevented.

15.2 Cave Temperatures

On one hand cave temperatures are determined essentially by the climatic mean values, on the other by the water flowing in the underground, which usually has a cooling effect. The cold waters of melting snow are of great importance in the mountains because of their long duration — in the Alps, March to July.

The thermal relationships between water, air, and rock are determined by their specific heat capacity and their density.

Table 15.5. Specific heat capacity and ρ (density) of water, rock and air (1 cal = 4.1868 J)

	Spec. h.c. (cal/g)	ρ (g/cm^3)	(cal/cm^3)	(cm^3/cal)
Water	1	1	1	1
Limestone	0.210	2.7	0.567	1.76
Air	0.241	0.001293	0.0003116	3209

Since the heat content of a volume of air is 3200 times lower than that of the same volume of water and 1820 times lower than that of limestone, the temperature of the air quickly adapts to that of the water or limestone — provided there is intensive contact between the media. They warm up and cool off respectively so slightly per day that they can generally be neglected. Longer periods of one-sided temperature divergence from the mean value are, however, clearly recognizable, especially in ice caves (see Chap. 16).

Wagner (1960) names a depth of 20 m for the zone of neutral temperature in Central Europe. That does not apply to karst because air and water can circulate almost freely as opposed to the case in an underground of a different nature. Hauser and Oedl (1926, pp 91/92) report concerning Eisriesenwelt (Austria): "In winter, on the other hand, we still found ice formations at the first connecting side tunnel, a sure sign that the zero-isotherm has advanced 1 km further into the mountain." That is approx. 1 1/2 km distant from the entrance and does not yet lie in the neutral-temperature zone. In Hölloch a decline in temperature of 5°C was noticed over 1000 m inside the cave in February 1956; moreover the place lies approx. 500 m vertically below the surface. In active caves the cold water from melting snow cools off the air and the rock until well into summer. As a consequence the zone of neutral temperature must lie beneath the karst region which has a through-flow of water; for Hölloch this means about 1000 m beneath the surface.

In dynamic caves the circulation of air creates a temperature equilibrium. In the Dachstein Ice Cave (Austria) Saar (1954) noticed that the temperature sank with increasing altitude at a ratio of 1°C/200 m; this value is approximately the same in Hölloch. The decrease in temperature with increasing altitude corresponds approximately to that in the open atmosphere.

In active endokarst the geothermal gradient is of no noticeable importance as the heat supplied from the earth is carried away by water. Hölloch can again serve as an example. The passages of this system lie between 50 and 900 m beneath the earth's surface. At least 2000 l of water seep in through every square meter of the earth's surface annually. The flow of heat from the interior of the earth amounts to 400 kcal/m^2/y (Toperczer, 1960). If the water flowing through the cave is warmed by 0.2°C, it is sufficient to transport the total amount of this heat away or to prevent the karstified zone above from being geothermally heated up. During the Ice Age precipitation in the cold periods was carried away out into the foreland as glacial ice and only a small portion found its way into the endokarst. With 20% of today's drainage, i.e., 400 l/m^2 · y, an increase in temperature of 1°C would suffice to carry away the geothermal heat, with 200 l/m^2 an increase of 2°C. This leads to the paradox that during glaciation Hölloch, and with it surely all deep alpine caves, was warmer than today. This also explains why the worm, *Octolasium transpadanum Rosa,* which today is found only south of the Alps, found refuge north of the alpine barrier in Hölloch, where it is frequently found in the high-water zone. It is a relict from a Pleistocene warm period which cannot be more closely identified — the cold periods caused its death on the earth's surface and in caves situated at slight depth (Bögli, 1973a).

Ford (1975) and Ford et al. (1976) describe temperature developments in 9-km-long Castleguard Cave under Columbia Icefield (Canada, B.C./Alberta) which at present shows neither a through-flow of water nor a movement of air which is of dynamic origin. Glacial water flows by unknown paths on a deeper level through limestone, carrying away part of the geothermal heat (Castleguard Springs: 2.2°C). The flow of heat to the overlaying glacier is between 10% and 40% of the expected value. The geothermal gradient amounts to 45 m/°C and more in the region of the cave so that 3.6°C is found in the center of the cave, 0°C at the upper end under the glacier, and -3.8°C at the lower end, 320 m deeper.

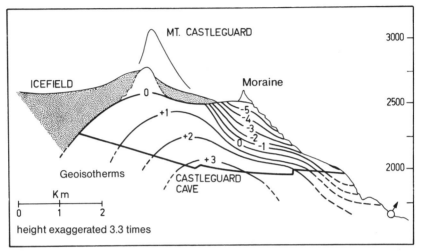

Fig. 15.8. Geoisotherms in Castleguard Cave at Columbia Icefield (Canada) according to Ford et al., (1976)

In static caves the atmosphere is stratified depending on its temperature or on its density. Rising, static caves are traps for warm air and are relatively warm, descending caves are traps for cold air so with increasing depth the temperature sinks, occasionally as low as to form perennial ice (see Chap. 16).

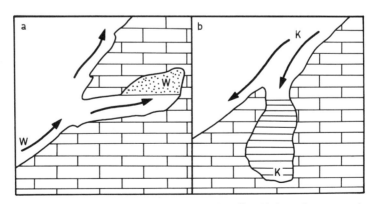

Fig. 15.9 a, b. Static caves; *a* summer; *b* winter. *W* warm-air pocket; *K* cold-air pocket; *arrow* air movement

Table 15.6. Temperatures in static caves

Abisso Enrico Revel (Apuanic Alps)		Grotta Gigante (Triestine karst) (Polli, 1953, 1958)	
Depth (m)	t (°C)	Depth (m)	t (°C)
0	18.7	0	12.5
25	8.0	24	10.4
50	5.6	100	9.5
85	3.4		
182	2.2		
304	1.3		

15.3 Humidity of the Air

The relative humidity is generally high, between 90% and 100% (Gèze, 1965) and variations are slight, apart from the entrance region when the cave wind blows inward. In the Grotta di Padriciano (Triestine karst) degrees of humidity between 34% and 97% were measured on the surface during 10 d in July, but already 30 m within the cave it was constant at 98% and 330 m from the entrance at 100% (Tommasini and Candotti, 1968).

A wind blowing upward in cave passages rapidly reaches the saturation limit. Condensation takes place on the walls, normally as drops of condensed water, in ice caves as hoar-frost and ice-flowers. If there are sufficient condensation nuclei, fog forms. If there are not, the air can be oversaturated and as soon as such a passage is entered there is a spontaneous formation of fog.

The air aspirated in winter warms up, the relative humidity sinks, occasionally even as far as below 70%. The cave dries and the walls become light. The front of dry walls advances gradually into the cave and in Hölloch can reach a distance from the entrance of almost 2 km by the end of winter.

In arid and semi-arid climates the humidity in the caves often sinks under 80% again; sodium or potassium minerals serve as indicators of this (see Chap. 13.3.2).

16 Ice Caves

According to definition ice caves are caves with permanent ice, even when the ice remaining in autumn is not larger than a felt hat. Caves in ice are called glacier caves. Although it may seem inconsistent, the distinction is made in this way. The primary factor is the formation of ice within the cave yet only its preservation is decisive in making an ice cave out of an ordinary cave. For the former cold winters are required, for the latter, however, cool summers in humid climates. That limits their occurrence in Central and Western Europe to the high mountains and highlands over 700 m above sea level.

Three theories have been found to explain the existence of permanent ice. One assumes that the ice is a relic from the Pleistocene Period (Dawkins, 1876). Providing evidence against this theory are the facts that the ice forms anew even today and that periods when the ice decreases, even thaws completely, alternate with periods when it increases. The oldest ice found up to the present (3000 years old) is in the Romanian ice cave Ghetarul de la Scarisvara. It looks as if European ice caves were completely free of ice in the post-glacial climate optimum (Trimmel, 1968).

The summer-ice theory is considerably older. Fugger (1891-1893) refers to places in literature from the 17th and 18th centuries. At that time the opinion held was that the nature of the cave "is so fantastic that when winter freezes most intensively outside it has mild air inside, but cold, yes, even ice-cold when the sun shines most warmly" (quoted by Kyrle, 1923, p. 110). These subjective findings did not stand up to measurement. Yet the summer-ice theory found support until far into the 19th century.

The third, the winter-ice theory, is the oldest but was only recognized generally at the end of the 19th century. Prévost found concerning the Glacière near Besancon in 1792 that the static ice cave preserved winter's cold even in summer, which was why the ice remained preserved. Thury (1861) observed that the formation of ice is favored to take place in spring and autumn when water and cold temperatures coincide in the underground. Moreover he also already distinguished between static and dynamic ice caves. Browne (1865) used this division to classify the ice caves he described. Fugger (1891-1893), who investigated the various theories, especially with his series of temperature measurements, subsequently substantiated the winter theory and helped it to become accepted. However, he did deal especially with static ice caves. Primarily Austrian scientists have been participating in the investigation of ice caves since, for indeed the largest in the world are in Austria, for example Eisriesenwelt in Salzach Valley and Dachstein-Rieseneishöhlen (Dachstein Giant Ice Caves) in the Salzkammergut. (Bock, 1913b; Hauser and Oedl, 1926; Saar, 1954).

The formation of ice by evaporation was also observed in caves with seeping water and a strong air-draught with temperatures little above 0°C, e.g., in Geldloch in Oetscher (Kyrle, 1923). However, only fine ice coverings are created in this way.

In *static ice caves* cold winter air becomes caught in descending pocket caves. They frequently begin on the bottom of dolines; these work like funnels to capture the air. In the course of the winter the cold air fills the cavities (Fig. 16.1: *hatched*), cooling off the rock to under 0°C. When in spring the snow begins to melt the water penetrating the cave freezes to form considerable masses of ice (*cross-hatched*). If the cave entrance is not too wide and sufficiently shaded (shady side of a valley or mountain, trees), and the cave deep enough, the cold air warms up in its upper portion – part of the ice melts. In late autumn and early winter the cave is again filled with cold air. The summer's water seeping out of the rock, which is still warmer than the air, freezes: the second period of ice formation begins. When the influx of water ceases in mid-winter the formation of ice stagnates until spring – only snow blown in increases the mass of ice formed by water. Static ice caves occur in deeper situations, e.g., the Glacière near Belfort (France) at 700 m above sea level and Grotte de la Glacière at 630 m a.s.l. (Sweeting, 1972).

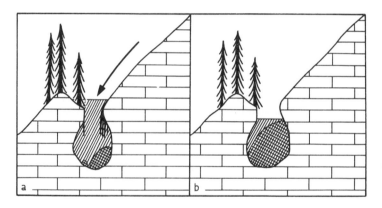

Fig. 16.1 a, b. Static ice caves (see text). **a** late autumn; **b** early summer

In high mountains there are static ice caves which can hardly be called genuine ice caves for the ice found in them is névé, firn, i.e., created from snow blown in. A better name would be firn cave. Kyrle, however, counts them among the ice caves since they usually also have genuine cave ice in them.

Dynamic ice caves have several outside openings at various heights. As a rule they belong to extensive underground cave systems. Air movements of exogenous-endogenous origin cause the formation of ice and also preserve it. In the winter half-year temperatures inside the cave are higher than outside; they are between 0.5° and 3°C. As in a chimney the air rises up to the upper openings. Cold air far below 0°C is drawn in from outside through the lower openings. This cools the rock off, creating undercooled regions around the lower entrances. Because of the air's low capacity for heat the 0°C-front moves into the cave only slowly, depending on conditions, a few hundred meters during the winter, as an exception also as far as 1500 m (Eisriesenwelt, Hauser et al., 1926).

In summer the cave air is considerably colder than the outside air. It sinks downward. The replacement comes in from the high outside openings; upon entrance it is additionally cooled by the remainders of snow in dolines, shafts, and open joints. Therefore it reaches the frost zone around the lower entrances with temperatures only slightly above 0°C. Usually it continues to cool off to frost temperatures. Only in August and September, occasionally earlier, does the air become a little warmer than 0°C. This is the period of melting ice. Already in October the outside temperature at the outside openings sinks to below 0°C and the air current changes back to that of the winter half-year.

In dynamic caves, too, there are two phases for the formation of ice, and one of melting plus one of stagnation. In late autumn summer's water, still seeping through the rock, freezes upon reaching the cavity with the cooled air. As the cold season advances, the flow of water ceases and the period of stagnation begins, even if temperatures continue to fall. Evaporation can even cause a slight decrease in the ice. In spring and early summer water from melting snow reaches the cave and freezes in the frost zone around the lower outside openings: the amount of ice reaches its maximum. When the direction of the air current changes, the temperature gradually rises above 0°C: the period of melting begins.

Saar's investigations (1954) in Dachstein-Rieseneishöhle offer good insight into the course temperatures and air-current direction take during a year. For several years this was considered to be the largest ice cave in the world (1910-1919) until Eisriesenwelt in Salzachquertal surpassed it (Czoernig-Czernhausen, 1920, 1926; Hauser et al., 1926).

Table 16.1. Average temperatures 1920-1924 in Dachstein Riesenhöhle according to Saar (1954). M: month; A: entrance 1458 m above sea level; B: Tristandom, 100 m from entrance; C: Artusdom, 400 m from entrance; W: direction of air current, e: inward, a: outward

M:	1	2	3	4	5	6	7	8	9	10	11	12
A:	−14.7	−8.7	−0.5	0.7	10.5	12.8	10.8	8.6	10.9	5.9	−2.6	−1.8
B:	−8.8	−6.1	−2.4	−1.4	−1.0	−0.1	0.4	2.1	1.2	1.3	−3.0	−2.3
C:	0.9	0.9	1.2	1.4	1.7	1.9	2.3	2.6	2.3	1.6	1.2	1.2
W:	e	e	e	e	e/a	e/a	a	a	a	a/e	e	e

For the higher ice-free portion (upper Artusdom, point of measurement) there is a corresponding glaciated portion lying deeper (lower Artusdom, Tristandom, point of measurement; Bock et al., 1913).

Eisriesenwelt in Salzachtal is situated 200 m higher and shows an ice area of 20,000 m^2. The ice is 12 m thick in places and stretches from the entrance 650 m into the mountain without interruption; a further 650 m show even larger permanent ice formations. Here too investigations of temperature were made with numerous series of measurements. Two examples of these are presented in Table 16.2.

Table 16.2. Temperatures in Eisriesenwelt on 31.3. and 3.4.1921 (Hauser and Oedl, 1926)

Place	Altitude (m above sea level)	31.3.1921 Wind inward		3.4.1921 Wind outward		Wind's strength (m/s)
		Time	t(°C)	Time	t(°C)	
In front of cave	1641	9.30	+0.5	11.35	+9.5	
Entrance gate	1660	9.45	−0.8	11.45	−0.4	6.37
Hymir	1755	11.30	−1.0			
Sturmsee	1769	12.30	−1.4	12.45	−1.3	5.74
Mörkdom	1762			12.50	−0.3	
Eispalast	1759			12.55	−0.5	
End of floor ice	1755			13.00	−0.4	
U-tunnel, P. 23	1731	14.00	−1.0	13.25	0.0	
Midgard, P. 26	1770	15.30	−1.0	13.45	+1.0	
Midgard, P. 33	1781	16.20	−0.2	15.50	+0.3	
Dripstone wood	1740	18.00	+0.8			
Straight joint	1720	19.00	+1.0			

Statodynamic ice caves have a mixture of both types. One of the most significant of this kind is Dobšinská ladova jaskyňa (Dobšinská Ice Cave) in Slovakia (ČSSR; Steiner, 1922; Droppa, 1960, 1967). It consists of a static part which cannot cool down sufficiently, and a dynamic part between the main entrance and a doline; an exchange of air takes place with the main entrance through the incasion debris of the doline.

Table 16.3. Temperatures t(°C) in Dobšinská Ice Cave (acc. to Droppa, 1960) in the years 1950 and 1951

Place	28.9.50 t	23.1.51 t	6.5.51 t
Rim of the doline, 971 m above sea level	9.8	−6.4	8.4
Entrance, 964 m above sea level	1.2	−5.6	1.2
Mala sien, 956 m above sea level, 35 m from entrance	0.4	−5.0	−3.2
Organ, 929 m above sea level, 105 m from entrance	0.5	−3.8	0.1
Dno Pekla, 900 m above sea level, 180 m from entrance, ice-free, static part	1.2	1.2	1.2

The world of ice formations is very rich, although not as rich as that of calcareous sinter. They resemble one another in many ways, especially where the forms are the direct result of flowing water. However, where conditions of formation and removal differ from one another, there are no mutually corresponding formations. In the case of ice there are no forms which correspond to straw stalactites, eccentrics, or other capillary, sinter forms. However, there is the bizarre form of the dentritic ice which at the beginning of the warm period forms on ice walls from melted ice-water running down. In the vicinity of zero degrees the ice recrystallizes easily. Therefore during the melting period there is recrystallization during which large ice-crystals are formed. The thin, fine crystals lying in the spaces between them melt faster forming a network of engraved lines looking like a honey-comb (Kyrle, 1923).

17 Classification of Underground Cavities

The classification of underground cavities can be made on one hand according to their genesis or according to their size, or on the other hand according to striking characteristics, which is, however, quite arbitrary.

17.1 Definition of Cave

Not all underground cavities are called caves, for according to the ISU (International Speleological Union) the term cave is defined as a "natural, underground cavity, large enough to be entered by man," which can be partially or entirely filled with sediments, water, or ice (Fink, 1973, p. 34). The term cave thus presupposes a size large enough to allow a human being, even if only a child, to crawl inside. Accordingly the human is the measure of all caves. This definition is, however, not absolute as small passages and rooms which are connected to caves are included with them and so are called caves, too. The scientist is free to depart from the definition and name smaller cavities caves also if it seems necessary to him. In the same way animal constructions from the badger's hole to the mousehole are termed caves by biologists, the smaller ones microcaves.

As a synonym for cave grotto is also used. This word is derived from Italian – grotta – or French – grotte –, which can generally be translated as cave. A grotto is a small to medium-sized, natural or artificial cave, generally richly decorated with speleothems (Fink, 1973, p. 37). But in German this term should be used only for artificial cavities imitating caves.

Caves which are of more complicated construction, i.e., branched and criss-crossed caves, maze caves, or a combination of shafts and horizontal passages are termed cave systems.

17.2 Genetic Classification

If underground cavities are divided up according to the manner in which they were created, the result is two principal groups: primary caves and secondary caves.

17.2.1 Primary Caves

All underground cavities which are created simultaneously with the rock are called primary caves. Genetically they have nothing to do with underground drainage. These are caves in igneous rocks and sedimentary primary caves. To the former belong vesicles and lava tunnels, to the latter overcovering caves, reef caves, and tufa caves.

Primary Caves in Igneous Rocks

Vesicles are magmatic gas bubbles which are retained when volcanites harden. They rarely reach a diameter of a few decimeters and seldom ever the size demanded by the cave definition.

Lava tunnels form in the less viscous pahoehoe-lava of basalt's chemistry. They are created under the hardened lava-covering when the liquid rock beneath flows away. A tunnel remains behind; on its walls lava drops run down and harden. On the side walls lava ledges give evidence of former stages of filling. Lumps of lava accumulate on the floor. In Lavabeds National Monument (Northern California) 296 such tunnels have been counted, some of them miles long. In adjacent Oregon and in the State of Washington there are innumerable others, winding their way under the endless forests. On Lanzarote they reach a total length of almost 16 km; among them the Cueva de los Verdes is 6.1 km long according to Trimmel (1968). In Iceland the Raufarholshellir stretches 3.2 km in a lava stream from the turn of the last millenium. Hawaii and Japan also have lava tunnels.

Primary Caves in Sediments

Cavities of a larger size are not possible in sediments which have a loose stage previous to their diagenesis, unless the components had a corresponding size. In landslide debris small, irregularly formed caves occur between huge boulders, so-called *overcovering caves*. Kyrle (1923) even classes them as secondary caves but landslide debris should be understood as a new deposit, as one gigantic mass of debris; during its creation cavities were enclosed in it.

If sediments are deposited in a solid state the formation of sedimentary primary caves is quite possible. In coral and sponge reefs the animal colonies, proliferating abundantly in the warm water which is rich in nourishment, enclose cavities: *reef caves*. They can be found in every size. However, as a rule they do not last long as they fill with coral or reef sand respectively.

Tufa caves belong in the same category. They form where calcareous tufa grows over rock and tufa walls. They are found in the tufa barriers of Yugoslavia's rivers, thus near Jaice where the Pliva flows into the Vrbas. Höllgrotte near Baar (Ct. Zug, Switzerland) was created in tufa deposits below the rock wall of Lorze Gorge; it has been commercialized because of its attractive dripstone formations. It allows an especially good study of the conditions of tufa cave formation (see Fig. 13.15).

17.2.2 Secondary Caves

Secondary caves are created after diagenesis. They can be formed by forces working from the exterior; exogenous caves, or by forces within the mountain itself: endogenous caves.

Exogenous Caves

Weathering and erosion from the exterior work on the rock. Generally the result is a *rock-shelter* or a *shelter cave*; the former is a shallow cave with a more or less flat bottom under an overhanging rock ledge, the latter shows a maximum horizontal extension without the depth exceeding the width of the mouth.

Wind caves are created by the action of the wind which in a storm tears away components of rock loosened by weathering or which grinds down the rock with the sand it carries. Wind caves are characteristic of arid regions.

River-bank caves are hollowed out by the moving water of rivers and streams. In the initial stage they are potholes. However, they can develop into genuine caves. Lettenmayer Cave near Kremsmünster (Austria) is approx. 25 m deep (Schadler, 1920).

Surf or wave-cut caves are usually shelter caves. Open joints can be widened up to the size of caves by the pressure of the breaking waves and by the currents caused by them. If corrosion plays a larger part in the process of widening, such caves should be classed as karst caves, e.g., Grotta Smeralda close to Amalfi or the Blue Grotto on Capri.

Endogenous Caves

Endogenous caves are created by agents active inside the rock. Tectonic processes and corrosion are the initiators of the creation of a cavity; when the size is sufficient, erosion and incasion set in also.

Endogenous caves are divided into tectonic caves, open joint caves, and karst caves.

In *tectonic caves* the cavities were created by tectonic processes; the subsequent changes they have undergone by corrosion and erosion are only negligible; occasionally changes by incasion have been of a larger dimension. Alpine tension joints in crystalline rocks represent this type. Not infrequently they show a rich content of minerals created hydrothermally, particularly in the case of rock-crystal. Tension joints created later, especially in sediments, are rich in calcite. Tension joints frequently form in folded rock; they occur on the outside of the folds where tractive forces are at work (Fig. 11.1b).

Alpine tectonics cause open joints to be torn open by the sliding away of sections of nappes or of sedimentary complexes.

Open joint caves have been counted among tectonic caves up to now. However, they form a category of their own as their cavity is created not by tectonic but by morphological processes, such as by mass movements, e.g., by the slipping away of limestone layers on a more or less plastic base, e.g., marl, toward the adjacent valley (see Fig. 11.1c). Nothing is changed by the fact that the cracks may have been tectonically prescribed for the cavity created has not been formed tectonically. Frequently they are open above.

Karst caves (corrosion caves), one of the two main themes of this book, occur much more frequently than all the other types of cave together; they are also much larger and are created by subterranean water. Karst caves are classified according to the manner of flow: vertical cave is the term used when the water flows downward to the karst water body or to an impermeable base, horizontal cave when it flows sideways to the local base level. This classification is meaningful hydrologically but vertical caves often run in every direction but vertically, and horizontal caves in mountains are by no means horizontal. Morphographically the two names apply approximately only in cave systems in tablelands and under peneplains.

Karst caves can be divided into:

bedding caves — which run between two layers of similar, soluble rock,

contact caves — which run between two different kinds of rock of which the underlying one is insoluble, and

joint caves — which are created along joints and faults.

17.3 Geological-Petrographical Classification

A classification of caves according to the type of rock suggests itself but the number of rocks with endogenous caves is relatively small — and such a classification would refer only to these. The lithological criteria are solubility, geomorphological hardness and resistance to weathering.

Chalk caves — in general rare because of their geomorphological softness; an exception is the Parisian Basin.

Limestone and dolomite caves (carbonate rock caves) are the caves common in carbonate karst; corrosion caves. Because of the high degree of geomorphological hardness of the rock there are also karst caves which already began to form in the Mesozoic Period.

Gypsum caves — widespread in gypsum karst but because of the softness of the rock they originated in the Quaternary.

Sandstone caves — created by the dissolution of the matrix and subsequent washing away of the sand (erosion), but also by the dissolution of lime sandstones. The proportion of erosion is high. Quartzite caves are included in this group for the present.

Conglomerate caves are created by the dissolution of the pure limestone components and/or the matrix as well as by erosion.

Granite caves are formed by tectonic or mass movements. Examples have become known in Sweden.

Glacier caves in glacier ice or in ice in general.

17.4 Classification According to Size

The size of a cave is of scientific consequence because every meter wider, longer, or deeper can be confirmation of an existent thesis — or else evidence for its repudiation; it may also be the source of new knowledge whether it be only of local speleogenetic or of more general importance. The objective of cave surveying is on the one hand to prepare a cave map, which is indispensable as a basis for scientific work, and on the other hand to understand the entire network of underground cavities as far as they are accessible to humans. Therefore the extent of a cave largely depends on the stage of exploration. The result is the disadvantage that caves slip from one category into the next higher as the stage of their exploration advances.

The following categories are distinguished according to the length measured:

small caves	up to 50 m
medium caves	50-500 m
large caves	500-5000 m
giant caves/caverns	over 5000 m

In the last ten years the exploration of caves has been accelerated; this has been due in part to a sense of competition. In 1965 there were still only five caves more than 25 km in length, among them Hölloch (Switzerland) the longest with 81 km. Early in 1980 there were already 33; Hölloch had slipped to third place with 140 km after Flint Mammoth Cave (341 km) and Optimističeskaja Peschtschera (142.5 km).

Also the difference in altitude within the mountain is of significance; it gives a karst profile with speleogenetic and speleomorphological insights from the surface down to the deepest point of a cave system, frequently in the karst water surface.

This manner of classifying caves (Tables 17.1 and 17.2) could be extended by a list of the largest verticals in the caves; according to Courbon (1973) they measure 337 m in Puits du Pot II (France), 333 m in Sotano de las Golondrinas (Mexico), 328 m in the Gouffre d'Aphanié (France), and 320 m in Puits Lépineux. But the largest cavities are also worthy of mention. The largest is to be found in La Torca del Carlista (Spain) with $500 \cdot 230 \cdot 125$ m^3 (L $\cdot$ W $\cdot$ H). The Salle de la Verna (France) attains $230 \cdot 180 \cdot 150$ m^3, the Grotta Gigante near Triest (Italy) $240 \cdot 180 \cdot 138$ m^3 and the Sotano de las Golondrinas (Mexico) $305 \cdot 125 \cdot 333$ m^3. Beside them the Schwarze Dom in Hölloch seems rather small with its approx. 400,000 m^3 volume.

Table 17.1. Caves with a measured length of over 26 km according to Courbon et al. (1975), Chabert (1977), British Caver (vol. 77) concerning Russian Caverns, and personal communications, as of 1979/80

1.	Flint Mammoth Cave System (KY, USA)	341.2 km
2.	*) Optimističeskaja Peschtschera (Podolia, Ukraine, USSR)	142.5 km
3.	Hölloch (Muota Valley, Switzerland)	139.8 km
4.	*) Ozernaja Peschtschera (Podolia, Ukraine, USSR)	104 km
5.	Jewel Cave (SD, USA)	100 km
6.	Ojo Guareña (Burgos, Spain)	61 km
7.	Réseau Felix Trombe / Réseau de la Coume-di-Quarnède (Hte-Garonne, France)	60 km
8.	Greenbrier Caverns / Organ Cave (WV, USA)	58.2 km
9.	Friar Hole (WV, USA)	50.4 km
10.	Wind Cave (SD, USA)	49.2 km
11.	Easgill Caverns (Lanc., U.K.)	45.4 km
12.	Cumberland Caverns (TN, USA)	43.8 km
13.	Crevace Cave (MO, USA)	43.4 km
14.	Eisriesenwelt (Salzburg, Austria)	42 km
15.	Zolužka Peschtschera (Ukraine, USSR)	40 km

Table 17.1. Continued

16.	Ogof Ffynnon Ddu (S. Wales, U.K.)	38.5 km
17.	Réseau de la Pierre St. Martin (Pyr.-Atl., France)	38.4 km
18.	Sloans Valley Cave System (KY, USA)	36.7 km
19.	Réseau de la Dent de Crolles (Isère, France)	36.3 km
20.	The Hole (WV, USA)	35.8 km
21.	Dachstein Mammoth Cave (Oberösterreich, Austria)	35 km
22.	Siebenhengste Cave System (BE, Switzerland)	ca. 35 km
23.	Binkley's Cave System (IN, USA)	34.2 km
24.	Carlsbad Caverns (NM, USA)	33.2 km
25.	Blue Spring Cave (IN, USA)	31.5 km
26.	Tantal Cave (Salzburg, Austria)	30.5 km
27.	Butler Sinking Creek Cave System (V, USA)	28.8 km
28.	Ogof Agen Allwed (S. Wales, U.K.)	27.7 km
29.	Culverson Creek Cave System (WV, USA)	27 km
30.	Systema Cavernario de los perdidos (Cuba)	26 km

*) In 1979 the connection of the two Russian Gypsum Caves was erroneously reported, but they "have not been connected" (British Caver, vol. 77, p. 26).

Table 17.2. Caves with a difference in altitude of more than 840 m a.s.l. (Authors as in Table 17.1) as of 1979/80

1.	Gouffre Jean Bernard (Hte-Savoye, France)	1358 m
2.	Gouffre de la Pierre St. Martin (Pyr.-Atl., France)	1350 m
3.	Avenc B.15 (Huesca, Spain)	1150 m
4.	Gouffre Berger (Vercors, Isère, France)	1141 m
5.	Schneeloch (Tennengebirge, Salzburg, Austria)	1111 m
6.	Hoyos de Pilar / Cima G.E.S.M. (Malaga, Spain)	1098 m
7.	Lamprechtsofen (Salzburg, Austria)	1028 m
8.	Réseau Felix Trombe / Réseau de la Coume-di-Quarnède (Hte-Garonne, France)	1018 m
9.	Snežnaja (Caucasus, USSR)	ca. 1000 m
10.	Réseau des Aiguilles (Dévoluy, Hts-Alpes, France)	980 m
11.	Garma Ciega / Cellagua (Santander, Spain)	970 m
12.	Gouffre Touya de Liet (Pyr.-Atl., France)	966 m
13.	Kievskaja ou Kilsi (Pamir-Alai)	964 m
14.	Antro di Corchia (Toscana, Italy)	950 m
15.	Gouffre du Cambou de Liard (Pyr.-Atl., France)	926 m
16.	Grotta di Monte Cucco (Umbria, Italy)	922 m
17.	Abisso Michele Gortani (Friuli, Italy)	920 m
18.	Feuertal Cave System (Totes Gebirge, Austria)	913 m
19.	Berger-Plattenecksystem (Tennengebirge, Salzburg, Austria)	900 m
20.	Hochleckengrosshöhle (Höllengebirge, Austria)	896 m
21.	Sistema Purificaciòn (Mexico)	893 m
22.	Gouffre de la Fromagère / Cialet d'Engins (Isère, France)	870 m
23.	Sotano de San Augustin (Mexico)	861 m
24.	Trunkenboldschacht (Totes Gebirge)	859 m
25.	Hölloch (Muota Valley, Switzerland)	856 m
26.	Gruberhornhöhle (Salzburg, Austria)	854 m
27.	Siebenhengste Cave System (BE, Switzerland)	ca. 840 m

17.5 Classification According to Prominent Characteristics

To be precise it is not a case of classification but of accentuating one or the other group from the totality. This is neither scientific nor systematic, but it can be very useful for certain purposes. For instance water flows through active cave sections and cave visitors must heed the danger of high water, whereas inactive caves have no threat of water, yet frequently have attractive dripstone formations. The differentiation between wet and dry caves is unclear since it can indicate inactive caves with much or little moisture as well as active and inactive caves.

Swelling caves (see Chap. 1.2.1), dripstone caves, aragonite caves, and phosphate caves are self-explanatory, breakdown caves also. Gypsum caves are formed in gypsum but ice caves are not formed in ice — they contain permanent ice.

This short list could be extended practically at will, but its purpose is only to illustrate how random such a classification is.

Appendix (A)

Conventional Cave Signs

Even today karsthydrologic, speleomorphologic, and speleogenetic research, as far as it can be pursued by cavers, still consists of the surveying of caves, the recording of hydrological and geological data and all other facts of the cave, and of the production of cave maps in which the accumulated facts are entered. Signs are required which allow the simplest possible recording of data but with a multiplicity of details sometimes makes the process difficult. The idea of complete consistency in the signs was abandoned and allowance made for the fact that an author must create his own sign for unusual forms.

The UIS (Union Internationale de Spéléologie) decided concerning the selection of signs in the Committee for the Standardization of Cave Signs to set up three groups of signs:

a) cave signs for the earth's surface (Fig. A.1)
b) signs for large cave systems with plans on a small scale (Table A.1 and Fig. A.2)
c) signs for smaller caves with large-scale plans (Table A.1).

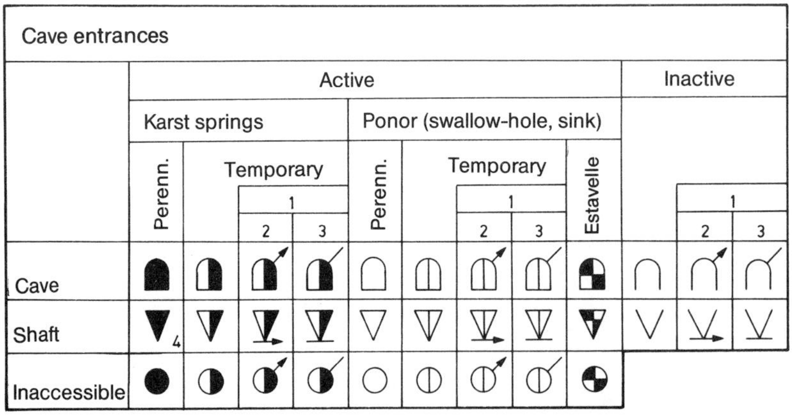

Cave entrances												
	Active									Inactive		
	Karst springs				Ponor (swallow-hole, sink)							
	Perenn.	Temporary			Perenn.	Temporary			Estavelle			
			1				1				1	
			2	3			2	3			2	3
Cave	◐	◠	◠	◠	◠	◠	◠	◠	◐	⌒	⌒	⌒
Shaft	▼₄	▽	▽	▽	▽	▽	▽	▽	▼	∨	∨	∨
Inaccessible	●	◐	◐	◐	○	○	○	○	⊕			

Fig. A.1. Signs for cave entrances according to B.R.G.M. (Trimmel et al., 1966). *1* cutting into an underground water course; *2* perennial underground water course; *3* temporary underground water course; *4* vauclusian spring, entrance possible with diving equipment

Conventional Cave Signs

a) Cave signs for the earth's surface were taken over from the B.R.G.M. (Bureau des Recherches Géologiques et Minières, Paris), which set up a system concerning the objects of karst hydrology on the surface and at cave entrances for its karst-hydrological maps. The list of sings for the forms of surface karst which was drawn up by the "Commission des Phénomènes karstiques" of the "Comité national de Géographie de France" under P. Fénelon was thereby completed. The signs set up by the B.R.G.M. have been tested many times, especially in France, but also in numerous other countries in general usage.

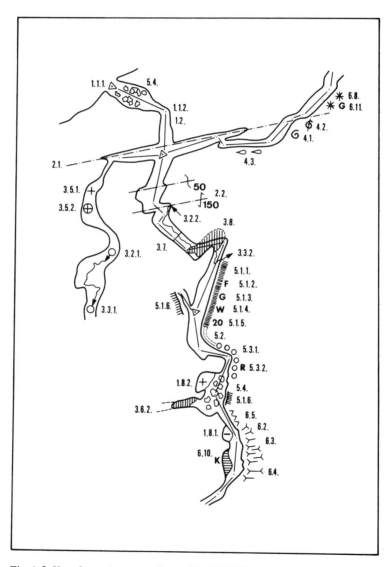

Fig. A.2. Use of cave signs according to Bögli (1970)

b) Maps for large cave systems demand other signs than plans for small caves. When the scale is small, 1:500, 1:1000 and smaller, there is very often very little or no space remaining between the boundaries of cavities for the sighting lines of the surveyor. Adapted to this difficulty, the sings have been created so that they can be drawn in outside the boundary lines. These signs were created by Bögli in 1952 for the maps of Hölloch and have been tested by the AGH (Arbeitsgemeinschaft Höllochforschung); they have been in use since then (Bögli, 1956c, 1970, 1976a).

c) The large scales, 1:100, 1:200, which are usual for smaller caves, provide the possibility of placing the signs for objects in their right position on the map. The proposal was submitted to the International Congress for Speleology in Ljubljana in 1965 by M.H. Fink as commissioned by the "Kommission für die Vereinheitlichung der Höhlensignaturen" of the UIS (Union Internationale de Spéléologie).

In the post-war years when speleology received fresh impetus, not only as a sport but especially as a science, individual groups of cave explorers created their own lists of signs. They were different from group to group. At the first International Congress for Speleology in Paris (1953) the attempt was made to standardize them. At the third congress in Vienna-Salzburg in 1961 A. Bögli's proposal for group (b) was accepted; it was confirmed in Ljubljana in 1965 and at the same time that of the B.R.G.M. and of Fink was accepted for group (c) (Trimmel and Audetat, 1966).

In the table the signs for group (b) have been up-dated by the author. The possibility exists of supplementing the signs of one group by those of the others.

It is in the interest of speleology as a science, of speleologists, and of cave explorers that the legibility of cave maps be increased by the use of the standardized lists and that their meaning be completely comprehensible even in the case of texts in foreign languages.

Table A.1. Cave signs for the groups (b) and (c)

1.	Surveying and Relief	Bögli Group b)	Fink Group c)
1.1.1.	Survey point 1st order	△	△
1.1.2.	Survey point 2nd order	•	○ •
1.1.3.	Altitude		⌖620
1.1.4.	Height in reference to entrance		+4 -25
1.2	Line of sighting	_._	-----
1.3.	Isohypses, direction of inclination		/600
1.4.	Slope		30°→

Conventional Cave Signs 241

Table A.1. Continued

1.5.	Passage with the position of a cross-section, unexplored branch, passage height	
1.6.	Crossing passages	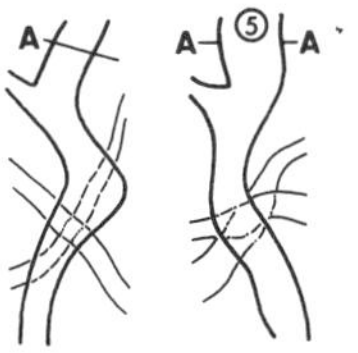
1.7.	Step with meter data	
1.8.1.	Abyss	
1.8.2.	Chimney, dome	
1.8.3.	Domepit	
2.	**Geological details**	
2.1.	Fracture, open joint	
2.2.	Fault with dip separation and direction of the separation, height of a step caused by fault	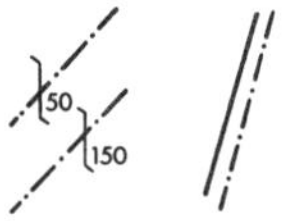
2.3.1.	Strike and dip, in general	
2.3.2	horizontal	
2.3.3.	slightly inclined	
2.3.4.	strongly inclined	
2.3.5.	vertical	
3.	**Hydrological details**	
3.1.1.	Underground water course	
3.1.2.	Cave river	
3.1.3.	Cave river, temporary	

Table A.1. Continued

3.2.1. Spring on passage floor

3.2.2. Spring on the side

3.3.1. Swallow-hole in the passage floor

3.3.2. Swallow-hole on the side

3.4. Seepage through sand and gravel

3.5.1. Sound of water in the rock or out of open joints

3.5.2. Sound of water not always audible

3.6.1. Siphon

3.6.2. End siphon

3.7. Waterfall, cascade

3.8. Underground lake partly touching ceiling

3.9. Oozing into the cave

3.10. Snow, ice with date

4. **Forms due to erosion and corrosion**

4.1. Erosional pothole (passage floor)

4.2. Solution pockets, ceiling pockets

4.3. Scallops

5. **Clastic Deposits**

5.1. Cave clay

5.1.1. " dry

5.1.2. " moist

5.1.3. " with gypsum crystals

Conventional Cave Signs 243

Table A.1. Continued

5.1.4. "	with worm heaps	W
5.1.5. "	20 cm thick	20
5.1.6.	Rock with clay covering	
5.2.	Sand	
5.3.1.	Gravel, angular	
5.3.2.	Gravel, rounded	R
5.4.	Boulders	
6.	**Chemical deposits**	
6.1.	Calcareous sinter in general, flowstone	
6.2.	Stalactites	
6.3.	Stalagmites	
6.4.	Dripstone-column	
6.5.	Flowstone	
6.6.	Rimstone pools	
6.7.	Moonmilk	M
6.8.	Crystals	
6.9.	Eccentrics, helictites, heligmites	
6.10.	Calcite crystals	K
6.11.	Gypsum crystals	G
7.	**Speleometeorology**	
7.1.	Direction of wind with date, hour	
7.2.	Frequently changing winds	

References

Abel, O.: Urweltliche Höhlentiere. Höhlenkundl. Vortr. *4*, Wien (1922)
AGH, Arbeitsgemeinschaft Höllochforschung: Höllochplan 1:10'000, Kopie des Archivplanes 1:1000 (1970)
Alfirević, S.: Contribution à la connaissance de la morphologie des sources sousmarines. Congr. Yougosl. Spéléol. *II*, Split, 49-57 (1958)
Allen, J.R.L.: On the origin of cave flutes and scallops by the enlargement of inhomogeneities. Atti Rass. speleol. Ital. 3-19 (1972)
Allen, J.R.L.: Transverse erosional marks of mud and rock: Their physical basis and geological significance. In: Sedimentary Geology, Intern. J. applied regional sedimentology, Elsevier, Amsterdam, 1971
American Geological Institute: Dictionary of geological terms. Dolphin Reference Book C 360 (1962)
Andrieux, C.: Contribution à l'étude du climat des cavités naturelles des massifs karstiques. III. Evapo-condensation souterraine. Ann. Spéléol. *25*, 531-559 (1970)
Arnberger, E.: Höhlen und Niveaus. Die Höhle *6/1*, 1-4 (1955)
Aubert, D.: Structure, activité et évolution d'une doline. Bull. Soc. Neuchâteloise Sci. Nat. 113-120 (1966)
Avias, J.: Karst of France. In: Herak, M., Stringfield, V.T. (ed.): Karst. London: Elsevier 1972, 129-188
Avias, J., Dubois, P.: Sur la nappe aquifère des karst barrés par la faille du Bas-Languedoc. Spelunca, Mém. *3*, 68-72 (1963)
Bakalowicz, M.: La rivière souterraine de Pinargözü (Taurus, Turquie). Ann. Spéléol. *27/1*, 93-103 (1972)
Bakalowicz, M.: Les grandes manifestations hydrologiques dans le monde. Spelunca, Bull. *2*, 38-40 (1973)
Balász, D.: Karst regions in Indonesia. Karst-Es Barlang-kutatás *5*, 3-61 (1968)
Ballif, Ph.: Wasserbauten in Bosnien und der Herzegowina I, Wien 1896
Barsch, D.: Studien zur Geomorphogenese des zentralen Berner Jura. Basler Beitr. Geogr., Basel 1969
Basset, W.A., Basset, A.M.: Hexagonal stalactite from Rushmore Cave. S.D. Natl. Speleol. Soc. Bull. *24*, 88-94 (1962)
Batsche, H., Bauer, F., et al.: Kombinierte Karstwasseruntersuchungen im Gebiet der Donauversikkerung in den Jahren 1967-1969. Steir. Beitr. Hydrogeol. 5-176 (1970)
Bauer, E.W.: Höhlen, Welt ohne Sonne. Int. Libr., Esslingen: J.F. Schreiber 1971a
Bauer, E.W.: The Mysterious World of Caves. Int. Libr., London: Collins Publ. 1971b
Bauer, F.: Nacheiszeitliche Karstformen in den österreichischen Kalkhochalpen. 2e Congr. Int. Spéléol. *I*, 299-328 (1963)
Bauer, F.: Kalkablagerungen in den österreichischen Kalkhochalpen. Erdkunde *18*, 95-102 (1964)
Bauer, F.: Kalkabtragungsmessungen in den österreichischen Kalkhochalpen. Erdkunde *18*, 95-102 (1964)
Bauer, F., Zötl, J.: Karst of Austria. In: Karst. Herak und Stringfield (eds.). London: Elsevier 1972, 225-265
Bauer, F., Zötl, J., Mayr, A.: Neue karsthydrographische Forschungen und ihre Bedeutung für Wasserwirtschaft und Quellschutz. Wasser, Abwasser. 280-297 (1958)
Bauer, M.: Beiträge zur Geologie der Seychellen. N. Jahrb. Mineral. *II* (1898)

Bernasconi, R.: Contributo allo studio del Mondmilch. Stud. Stor. Rass. Speleol. Ital. Mem., Como, 39-56 (1959)
Bernasconi, R.: 4ème contribution à l'étude du Mondmilch: L'évolution physico-chimique du Mondmilch. Rass. Speleol. Ital. Mém. *5*, 75-100 (1961)
Bernasconi, R.: Le Mondmilch calcitique et ses formes cirstallines. Stalactite *25/2*, 6-10 (1975)
Bertolani, M., Rossi, A.: The speleological complex "Grotta grande del Vento − Grotta del Fiume" in the Frasassi Canyon (Ancona, Italy). Proc. 6th Int. Congr. Speleol., Olomouc, 357-366 (1975)
Biese, W.: Über Höhlenbildung: I. Teil, Entstehung der Gipshöhlen am südl. Harzrand und am Kyffhäuser. Abh. Preuss. Geol. Landesamt, N.F.H. 137 (1931)
Binder, H.: Der Hungerbrunnen. Mitt. Ver. Natw. Math., Ulm, No. 25, 231-277 (1957)
Binder, H.: Niederschlag, Abfluß und Verdunstung im Gebiet des Blautopfes. Ib. Mitt. Oberrhein. Geol. Ver. *42*, 63-75 (1960a)
Binder, H.: Die Wasserführung der Lone. Jahresh. Karst-Höhlenkd. *21*, 211-248 (1960b)
Birot, P.: Le relief calcaire. Cent. Doc. Univ. Paris, 1966
Birot, P., Corbel, J., Muxart, R.: Morphologie des régions calcaires à la Jamaique et à Puerto Rico. Paris: CNRS 1968, 335-392
Bixio, R.: Les grottes tectoniques en roches karstifiables. Proc. 7th Int. Speleol. Congr. Sheffield 1977, 47-50
Blatchley, W.S.: Indiana caves and their fauna. IN. Dept. Geol. Natl. Res., Annu. Rep. *21*, 121-212 (1897)
Blessing, H.M.: Karstmorphologische Studien in den Berner Alpen. Tübinger Geogr. Stud., No. 65 (1976)
Blumberg, P.N.: Flutes: A study of stable, periodic dissolution profiles. Ph.D. thesis (Dept. Chem. Eng.), Univ. Michigan, 1970, 170 pp.
Bock, H.: Der Karst und seine Gewässer. Mitt. Höhlenkd. *6/3*, (1913a)
Bock, H.: Mathematisch-physikalische Untersuchungen der Eishöhlen und Windröhren. In: Die Höhlen im Dachstein. Graz 1913b, 102-144
Bock, H., Lahner, G.: Die Dachsteinrieseneishöhle. In: Die Höhlen im Dachstein. Graz 1913, 15-34
Bögli, A.: Probleme der Karrenbildung. Geogr. Helv. *3*, 191-204 (1951)
Bögli, A.: Grundformen von Karsthöhlenquerschnitten. Stalactite, Z. Schweiz. Ges. Höhlenforsch. 56-62 (1956a)
Bögli, A.: Der Chemismus der Lösungsprozesse und der Einfluß der Gesteinsbeschaffenheit auf die Entstehung des Karstes. Rep. Comm. Karst Phenomena IGU, New York 1956b, 7-17
Bögli, A.: Die Erforschung des Höllochs. Stalactite 1956/4, 69-75 (1956c)
Bögli, A.: Kalklösung und Karrenbildung. Z. Geomorph., Suppl. *2*, 4-21 (1960a)
Bögli, A.: Karsthydrographische Untersuchungen im Muotatal. Regio Basiliensis 68-79 (1960b)
Bögli, A.: Der Höhlenlehm. Atti Symp. Int. Speleol., RSI, Mem. *5*, 11-29 (1961a)
Bögli, A.: Karrentische. Z. Geomorph. 185-193 (1961b)
Bögli, A.: Höhlenkarren. 3. Int. Kongr. Spelaöl. Wien 1961, 25-27 (1963a)
Bögli, A.: Korrosive Bildungsbedingungen von Höhlenräumen. 3. Int. Kongr. Spelaöl. Wien 1961, 29-33 (1963b)
Bögli, A.: Beitrag zur Entstehung von Karsthöhlen. Die Höhle *14*, 63-68 (1963c)
Bögli, A.: Mischungskorrosion, ein Beitrag zum Verkarstungsproblem. Erdkunde *18/2*, 83-92 (1964a)
Bögli, A.: Die Kalkkorrosion, das zentrale Problem der Verkarstung. Steir. Beitr. Hydrogeol. 1963/64, 75-90 (1964b)
Bögli, A.: La corrosion par mélange des eaux. Int. J. Speleol. *1*, 61-70 (1964c)
Bögli, A.: Le Schichttreppenkarst, un exemple de complexe glacio-karstique. Rev. Belg. Géogr. *1/2*, 64-82 (1964d)
Bögli, A.: The role of corrosion by mixed water in cave forming. In: Problems of Speleological Research, Proc. Int. Speleol. Conf., Brno, ČSSR, 1964, 125-131 (1965)
Bögli, A.: Karstwasserfläche und unterirdische Karstniveaus. Erdkunde *20*, 11-19 (1966)
Bögli, A.: Höhlenniveaus und Höllochniveaus. 4ème Congr. Int. Spéléol., Ljubljana 1965, 23-27 (1968a)

Bögli, A.: Präglazial und präglaziale Verkarstung im hinteren Muotatal. Regio Basiliensis, Basel, 135-153 (1968b)
Bögli, A.: CO_2-Gehalte der Luft in alpinen Karstböden und Höhlen. 5. Int. Kongr. Speläol., Stuttgart, S. 28/1-9 (1969a)
Bögli, A.: Inkasion. Laichinger Höhlenfreund, No. 7 (1969b)
Bögli, A.: Shafts. Actes 3ème Congr. Suisse Spéléol. *1967*, 17-18 (1969c)
Bögli, A.: Neue Anschauungen über die Rolle von Schichtfugen und Klüften in der karsthydrographischen Entwicklung. Geol. Rundsch. *58*, 395-408 (1969d)
Bögli, A.: Diskussionsbeitrag zu Ek: Abondance du gaz carbonique dans les fissures des grottes. 5. Int. Kongr. Speläol., Stuttgart, p. 14/2 (1969e)
Bögli, A.: Le Hölloch et son karst. Ed. la Baconnière, Neuchâtel (Suisse). French and German version (1970)
Bögli, A.: Corrosion by mixing of karst waters. Trans. Cave Res. Group G.B. *13/2*, 109-114 (1971a)
Bögli, A.: Karstdenudation – das Ausmaß des korrosiven Kalkabtrages. Regio Basiliensis *12/2*, 352-361 (1971b)
Bögli, A.: Gips in Höhlen. Urner Min. Freund *10*, No. 6, 77-84 (1972)
Bögli, A.: Neue Ergebnisse der Höllochforschung. 6th Int. Congr. Speleol. Olomouc, ČSSR, Proc. III, 19-24 (1973a)
Bögli, A.: Studie zur Hydrographie der Poljen. Geogr. Z., Beih., 84-88 (1973b)
Bögli, A.: Zauber der Höhlen. Zürich: Silva 1976a
Bögli, A.: CO_2-Gehalte der Luft und Kalkgehalte von Wässern im unterirdischen Karst. Z. Geomorphol. Suppl. *26*, 153-162 (1976b)
Bourgin, A.: La Luire et la Vernaison souterraine. Ann. Spéléol. 31 (1946)
Brandt, A., Kempe, S., et al.: Geochemie, Hydrographie und Morphogenese des Gipskarstgebietes von Düna/Südharz. Geol. Jahrb.ser. C, H. 15 (1976)
Bretz, J.H.: Vadose and phreatic features of limestone caverns. J. Geol. *50*, 675-811 (1942)
Bretz, J.H.: Carlsbad Caverns and other caves of the Guadalupe block, N.M. J. Geol. *57/5*, 477-490 (1949)
Bretz, J.H.: Caves of Missouri. Geol. Surv. and water resources, Rolla (1956)
Bretz, J.H.: Bermuda, a partially drowned late mature, pleistocene karst. Bull. Geol. Soc. Am. *71*, 1729-1754 (1960)
Breznik, M.: Akumulacija na cerkniškem in Planinskem Polju. Geologija 7, Geološki zavod, 119-149 (1962)
Bridge, J.: Ebb and flow springs in the Ozarks. School of Mines and Metallurgy, Rolla, 17-26 (1924)
Brinkmann, R.: Abriß der Geologie I, 10. ed. Stuttgart: Enke 1967
Broeker, W.S., Olson, E.Z.: ^{14}C-dating of cave formations. Natl. Speleol. Soc. Bull. *21*, 43 (1959)
Browne, G.F.: Ice-caves from France and Switzerland (cit. from Kyrle, 1932) (1865)
Buckland, W.: Reliquia Diluvianae. London 1823, 303 pp.
Bull, P.A.: Lamination or warves? Processes and mechanisms of fine-grained sediment deposition in caves. Proc. 7th Int. Speleol. Congr. Sheffield, 86-89 (1977a)
Bull, P.A.: Surge marks in caves. Proc. 7th Int. Speleol. Congr. Sheffield, 89-92 (1977b)
Burger, A.: Hydrogéologie du Bassin de l'Areuse. Bull. Soc. Neuchâteloise Géogr., Suisse (1959)
Caumartin, V., Renault, Ph.: La corrosion biochimique dans un réseau karstique et la genèse du Mondmilch. Notes Biospéléol. *13*, 87-109 (1958)
Cayeux, L.: Roches carbonates, calcaires et dolomies. Paris: Masson 1935
Chabert, C.: Les grandes cavités mondiales. Spelunca 1977/2, Suppl. (1977)
Chabot, G.: Les plateaux du Jura central: étude morphogénique. Publ. Fac. Lett. Univ. Strasbourg (1927)
Chaptal, L.: La lutte contre la sécheresse. La Nature *289*, 449-454 (1932)
Choppy, J.: Vermiculures d'argile sur une coulée stalagmitique. Bull. Comm. Natl. Spéléol. 5, 6 (1955)
Cigna, A.A.: An analytical study of air circulation in caves. Int. J. Speleol. *3/1+2*, 41 ff (1967)
Cigna, A.A.: Air circulation in caves. Proc. 4th Int. Congr. Speleol. Ljubljana 1965, 43-49 (1968)
Cigna, A.A.: Considerazioni sulle teorie speleogenetiche. In: Le Grotte d'Italia. Riv. Ist. Ital. Speleol. 391-408 (1975)

References

Cigna, A.A., Cigna, L., Vido, L.: Quelques considérations sur l'effet sel dans la solubilité des calcaires. Ann. Spéléol. *18*, 185-191 (1963)
Ciry, R.: Une catégorie spéciale de cavités souterraines: les grottes cutanées. Ann. Spéléol. *14*, 23-30 (1959)
Clarke, F.W.: The data of geochemistry. U.S. Geol. Surv. Bull. 770 (1924)
Coleman, J.C.: Stalactite growth in the New Cave, Mitchellstown. Br. Caver *13*, 23/24 (1945)
Colvee, P.: Cueva en cuarzitas en el Cerro Autana. Territ. federal Amazonas. Bol. Soc. Venez. Espeleol. *4/1*, 5-13 (1973)
Corbel, J.: Observations sur le karst couvert de Belgique. Bull. Soc. Belg. Etud. Géogr. 95-105 (1947)
Corbel, J.: Drei Beiträge in: H. Lehmann, Das Karstphänomen in den verschiedenen Klimazonen. Erdkunde VIII, 119/120 (1954)
Corbel, J.: Les Karsts du nord-ouest de l'Europe. Rev. Géogr. Lyon, Mém. Doc. *12* (1957)
Corbel, J.: Les karsts de l'est canadien. Cahiers Géogr. Québec, 193-216 (1958)
Corbel, J.: Vitesse de l'érosion. Z. Geomorphol. 1-28 (1959a)
Corbel, J.: Erosion en terrain calcaire. Ann. Géogr. *366*, 97-120 (1959b)
Courbon, P.: Atlas des grands gouffres du monde. Vioud et Coumes, Apt en Prov. (1973)
Courbon, P., Chabert, C.: Les grandes cavités mondiales. Spelunca *4*, 5-8 (1975)
Cramer, H.: Die Systematik der Karstdolinen. N. Jahrb. Mineral., Geol., Paläont. *85B*, 293-382 (1941)
CRF, Cave Research Foundation: Map of the Flint Ridge Cave System, 1st ed. 1964
Curl, R.L.: The aragonite-calcite problem. Natl. Speleol. Soc. Bull. *24*, 57-73 (1962)
Curl, R.L.: Scallops and flutes. Trans. Cave Res. Group G.B. 121-160 (1966)
Curl, R.L.: Minimum diameter stalactites. Natl. Speleol. Soc. Bull. *34*, 129-136 (1972)
Curl, R.L.: Minimum diameter stalagmites. Natl. Speleol. Soc. Bull. *35*, 1-9 (1973)
Curl, R.L.: Deducing flow velocity in cave conduits from scallops. Natl. Speleol. Soc. Bull. *36/2*, 1-5 (1974)
Curl, R.L.: Ableitung der Fließgeschwindigkeit in Höhlen aus den Fließfazetten. Mitt. Verb. Dtsch. Höhlen- und Karstforscher *21/3*, 49-55 (1975)
Cvijić, J.: Das Karstphänomen. Geogr. Abh. 215-319 (1893)
Cvijić, J.: Morphologische und glaziale Studien aus Bosnien, der Herzegowina und Montenegro: 2. Karstpolien. Abh. Geogr. Ges. *3*, 1-85 (1901)
Cvijić, J.: Hydrographie souterraine et évolution morphologique de karst. Rec. Trav. Inst. Géogr. Alpine *6*, 376-420 (1918)
Cvijić, J.: The evolution of lapiés. Geogr. Rev. *14*, 26-49 (1924)
Cvijić, J.: La geographie des terrains calcaires. Acad. Serbe Sci. Arts, Monographies CCCXLI, Beograd (1960)
Czoernig-Czernhausen, W.: Die Eisriesenwelt. Mitt. Dtsch.-Österr. Alpen-Ver. No. 9-12 (1920)
Czoernig-Czernhausen, W.: Die Höhlen Salzburgs. Ver. Höhlenkde. Salzburg 1926
Davies, W.E.: Features of cavern break-down. Bull. Natl. Speleol. Soc. Am. *11*, 34-35 (1949)
Davies, W.E.: Mechanics of cavern break-down. Natl. Speleol. Soc. Bull. *13*, 36-42 (1951)
Davies, W.E.: Origin of caves in folded limestone. Natl. Speleol. Soc. Bull. *22/1*, 5-18 (1960)
Davies, W.E., Chao, C.C.T.: Report on sediments in Mammoth Cave, Kentucky. U.S. Geol. Surv. (1959)
Davies, W.E., Legrand, H.E.: Karst of the United States. In: Karst. Herak and Stringfield (eds.): London 1972, 467-505
Davies, W.E., Moore, G.W.: Endellite and Hydromagnesite from Carlsbad Caverns. Natl. Speleol. Soc. Bull. *19*, 24-27 (1957)
Davis, W.M.: Origin of limestone caverns. Geol. Soc. Am. Bull. *41*, 475-628 (1930)
Dawkins, W.B.: Observations on the rate at which stalagmite is being accumulated in the Ingleborough Cave. Rep. Br. Assoc. Adv. Sci., 80 (1874)
Dawkins, W.B.: Die Höhlen und Ureinwohner Europas (deutsch von Sprengl, I.W.). (from Kyrle, Theoretische Speläologie, 1923, p. 110), 1876
Defaure, M.: Le karst et les poljés du Peloponnes. In: Birot, Le relief calcaire, 1966, 174-192
Deike, G.H.: Origin and geologic relations of Breathing Cave, Virginia. Natl. Speleol. Soc. Bull. *22/1*, 30-42 (1960)

Deike, G.H.: The development of caverns of the Mammoth Cave region. Thesis, PA State Univ. (1967)
Deike, G.H., White, W.B.: Sinuosity in limestone solution conduits. Am. J. Sci. *267*, 230-241 (1969)
Demangeot, J.: Géomorphologie des Abruzzes adriatiques. CNRS, Mém. Doc. *3* (1965)
Demangeot, J.: Sur une courbe de dissolution des calcaires en montagne méditerranéenne. CNRS, Mém. Doc. *4*, 185-193 (1968)
Diaconu, G.: Presence of the Pickeringite in the Diana Cave, Baile Herculane. 6th Int. Congr. Speleol., Abstr., 18/19 (1973)
Diaconu, G., Medesan, A.: Sur la présence du Pickeringite dans la grotte de Diana. 6th Int. Congr. Speleol., Tome I, 231-239 (1975)
Dongus, H.: Die Oberflächenformen der mittleren Schwäbischen Alb. Jahresh. Karst-Höhlenkd. *4*, 21-44 (1963)
Douglas, I.: An evaluation of some techniques used in the study of landforms with special reference to limestone areas. Unpublished B. Litt. thesis, Univ. of Oxford, 1962
Douglas, I.: Some hydrologic factors in the denudation of limestone terrains. Z. Geomorphol. *12*, 241-255, 1968
Droppa, A.: Die Höhlen Demänovské Jaskyné. Slovak. Acad. Sci., Bratislava (1957)
Droppa, A.: Dobšinská L'adová jaskyňa. Bratislava (1960)
Droppa, A.: Jaskyna na Slovensku. Obzor ČSSR (1967)
Dubois, P.: Les circulations souterraines dans les calcaires de la région de Montpellier. Bull. Bur. Rech. Géol. Min., 31 (1961)
Dubois, P.: Etudes des réseaux souterrains des Rivières Buèges et Virenque. 2ème Congr. Int. Spéléol. 1958, 167-175 (1962)
Dubois, P.: Les circulations souterraines dans les karsts barrés de Bas-Languedoc. 3ème Congr. Int. Spéléol. 1961, H. 2, 167-174 (1964)
Dubois, P., Griosel, Y.: La source sousmarine de l'Abyss (Etang de Thau – Bas-Languedoc). Spelunca, Mém. *3*, 87-91 (1963)
Eckert, M.: Das Karrenproblem: Die Geschichte seiner Lösung. Z. Naturw. *5*, 321-432 (1895)
Eckert, M.: Das Gottesackerplateau. Z. Dtsch.-Oesterr. Alpenver. *31*, 52-60 (1902)
Egli, P.: Beitrag zur Kenntnis der Höhlen in der Schweiz. Inaug.-Diss. Phil. Fak. II, Univ. Zürich (1904)
Eisenstuck, M.: Die Kalktuffe der mittleren Schwäbischen Alb. Diss. Univ. Tübingen (1949)
Eissele, K.: Eine Karst-„Röhre" im obern Weissen Jura der Schwäbischen Alb. In: Eissele, K., Groschopf, P., Zur Karsthydrologie der Schwäbischen Alb. Jahresh. Karst-Höhlenkd. 1963. München 1963, 81-85
Eissele, K., Groschopf, P.: Zur Karsthydrologie der Schwäbischen Alb. Jahresh. Karst-Höhlenkd. München 1963, 81-92
Ek, C.: Conduits souterrains en relation avec les terraces fluviales. Ann. Soc. Géol. Belg. *84*, 313-340 (1961)
Ek, C.: Abondance du gaz carbonique dans les fissures des grottes. 5. Int. Kongr. Speläol. Stuttgart, Vol. 2, S 14/1-3 (1969)
Ek, C., Gilewska, S., Kaszowski, L., et al.: Some analyses of the CO_2-content of the air in five Polish caves. Z. Geomorphol. *13/3*, 226-286 (1969)
Ek, C., Mathieu, L.: La daïa Chiker (Moyen Atlas, Maroc). Ann. Soc. Géol. Belg. 65-103 (1964)
Ernst, L.: Zur Frage der Mischungskorrosion. Die Höhle *15*, 71-75 (1964)
Ewers, R.O.: A model for the development of subsurface drainage routes along bedding planes. M.D.-Thesis, Univ. Cincinnati, Dept. Geol. (1972)
Faust, B.: Salpeter Mining in Mammoth Cave, KY. Filson Club (1967)
Feitknecht, W.: Allgemeine und physikalische Chemie. Basel: E. Reinhardt 1949
Fink, M.H.: Höhlen ohne natürliche Eingänge. Actes 4ème Congr. Int. Spéléol. 1965, 435-440 (1968)
Fink, M.H.: Mehrsprachiges Lexikon der Karst- und Höhlenkunde. Entwurf. ISU, Int. Speleol. Union (1973)
Fischbeck, R., Müller, G.: Monohydrocalcite, Hydromagnesite, in speleothems of Fränkische Schweiz, Western Germany. Contrib. Mineral. Petrol. *33*, 87-92 (1971)

Fischer, W.: Gesteins- und Lagerstättenbildung im Wandel der wissenschaftlichen Anschauung. Stuttgart: Schweizerbart 1961
Flandrin, J., Paloc, H.: Etude d'une source de karst: la fontaine de Vaucluse (France). 5. Int. Kongr. Speläol. Stuttgart, Abh. 5, Hy 13/1-15 (1969)
Folk, R.L.: Practical petrographic classification of limestone. Bull. Am. Assoc. Petrol. Geol. *43*, 1-38 (1959)
Ford, D.C.: The origin of limestone caverns: a model from the central Mendip Hills, England. Nat. Speleol. Soc. Bull. *27*, 109-132 (1965)
Ford, D.C.: Features of cavern development in central Mendip. Trans. Cave Res. Group, G.B. *10/1*, 11-25 (1968)
Ford, D.C.: Geologic structures and theories of limestone cavern genesis. Brit. Speleol. Assoc., Settle, Yorkshire, 35-45 (1970)
Ford, D.C.: Geologic structure and a new explanation of limestone cavern genesis. Trans. Cave Res. Group, G.B. *13*, 81-94 (1971)
Ford, D.C.: Castleguard Cave, an alpine cave in the Canadian Rockies. Stud. Speleol. *2/7-8*, 299-310 (1975)
Ford, D.C., Harmon, R.S., et al.: Geohydrologic and thermometric observations in the vicinity of the Columbia Icefield, Alberta and British Columbia, Canada. J. Glaciol. *16*, No. 74, 219-230 (1976)
Ford, D.C., Schwarcz, H.P.: Radiometric age studies of speleothem. Geol. Surv. Can., Paper 76-1B (1976)
Ford, T.D.: Structures in limestone affecting the initiation of caves. Trans. Cave Res. Group, G.B. *13/2*, 65-71 (1971)
Frank, R.: The clastic sediments of Douglas Cave, Stuart Town, N.S. Wales. Helictite 7, 3-13 (1969)
Frank, R.: The effect of non-climatic factors on flowstone deposition. Proc. 6th Int. Congr. Speleol. I, 413-417 (1975)
Franke, H.W.: Altersbestimmung mit radioaktivem Kohlenstoff. Naturwissenschaften 38, 527-528 (1951)
Franke, H.W.: Der schichtweise Aufbau der Bodenzapfen. Die Höhle *12*, 8-12 (1961a)
Franke, H.W.: Formgesetze des Höhlensinters. Atti Symp. Int. Speleol. Varenna, 1960 (Mem. RSI), 185-209 (1961b)
Franke, H.W.: Formgesetze der Korrosion. Jahresh. Karst-Höhlenkd. *18/3*, 1962, 207-224 (1963a)
Franke, H.W.: Formprinzipien der Tropfsteine. 3. Int. Kongr. Speläol. Wien, 1964, No. 2, 63-72 (1963b)
Franke, H.W.: Mischungskorrosion in Haarrissen. Die Höhle *16*, 61-64 (1965a)
Franke, H.W.: The theory behind stalagmite shapes. Stud. Speleol. *1*, 89-95 (1965b)
Franke, H.W.: Ein speläochronologischer Beitrag zur postglazialen Klimageschichte. Eiszeitalter und Gegenwart *17*, 149-152 (1966)
Franke, H.W.: Das Wachstum der Tropfsteine. 4th Int. Congr. Speleol. 1965, *3*, 97-103 (1968)
Franke, H.W.: Sub-minimum stalagmites. Natl. Speleol. Soc. Bull. *37*, 17 (1975)
Franke, H.W., Geyh, M.A.: Radiokohlenstoffanalysen an Tropfsteinen. Umschau, 91-92 (1971)
Franz, H.: Feldbodenkunde. Wien: Georg Fromme 1960
Fränzle, O.: Die Opferkessel im quarzitischen Sandstein von Fontainbleau. Z. Geomorphol. *15/2*, 212-235 (1971)
Frear, G.L., Johnston, J.: The solubility of calcium carbonate in aqueous solutions at 25°C. J. Am. Chem. Soc. *51*, 2082 (1929)
Freien, R.K.: Wässrige Lösungen, Bd. 2: Ergänzungen. W. de Gruyter, Berlin, New York, 1978
Friese, H.: Die Karsthohlformen der Schwäbischen Alb. Stuttgarter Geogr. Stud. H 37/38 (1933)
Frodl, F.: Die Höhlen des mährischen Karstes als Lagerstätten von Düngephosphaten. Brünn 1923
Fuchs, F.: Studien zur Karst- und Glazialmorphologie in der Monte-Cavallo-Gruppe. Frankfurter Geogr. H. *47* (1970)
Fugger, E.: Eishöhlen und Windröhren. Jahrb. K.u.K. Oberrealschule Salzburg, 24-26 (1891-1893)
Gádoros, M.: Über die Wasserbewegungen im tiefen Karst. 5. Int. Kongr. Speläol., Vol. 5, Hy 14/1-3 (1969)

Gallocher, P.: Contribution à l'étude de l'émérgence sousmarine de Port-Miou. Ann. Spéléol. *9/3*, 169-181 (1954)
Gams, J.: H geomorfologiji kraškega korozije. Acta carsologica II, Acad. Sci. Art Slovenica. Ljubljana, 29-65 (1959)
Gams, J.: Measurements of corrosion intensity in Slovenia and their geomorphological significance. Geogr. Vestnik, Ljubljana *34*, 3-20 (1962)
Gams, J.: Types of accelerated karst corrosion. Probl. Speleol. Res., Proc. Int. Speleol. Conf., Brno, 1964, 133-139 (1965)
Gams, J.: Some morphological characteristics of Dinaric Karst. Geogr. J. *135*, 563-572 (1969)
Gams, J.: Die zweiphasige quartärzeitliche Flächenbildung in den Poljen und Blindtälern des nordwestlichen dinarischen Karstes. In: Ergebnisse der Karstforschung in den Tropen und im Mittelmeerraum. Geogr. Z., Beih. Wiesbaden: Franz Steiner 1973, 143-149
Gams, J.: Towards the terminology of the polje. Proc. 7th Int. Speleol. Congr. Sheffield 1977, 201-202
Gansser, A.: The Roraima Problem (S. Amer.). Verhandl. Naturf. Ges. Basel, *84*, 80-97 (1974)
Gardner, J.H.: Origin and development of limestone caverns. Bull. Geol. Soc. Am. *46* (1935)
Garrels, R.M., Christ, C.L.: Solutions, Minerals and Equilibria. New York: Harper and Row 1965
Garrels, R.M., Thompson, M.E., Siever, R.: Control of carbonate solubility by carbonate complexes. Am. J. Sci. *259*, 24-45 (1961)
Gerstenhauer, A.: Ein karstmorphologischer Vergleich zwischen Florida und Yucatan. Deutscher Geographentag Bad Godesberg, Wiss. Abh., 332-344 (1968)
Gerstenhauer, A.: Der Einfluß des CO_2-Gehaltes der Bodenluft auf die Kalklösung. Erdkunde *26/2*, 116-120 (1972)
Gerstenhauer, A., Pfeffer, K.-H.: Beiträge zur Frage der Lösungsfreudigkeit von Kalkgesteinen. Abh. Karst-Höhlenkd. A/2 (1966)
Geyh, M.A.: Diskussionsbeitrag zu Groschopf: Karsthydrographische Probleme der Schwäbischen Alb. 5. Int. Kongr. Speläol. Stuttgart, Hy 6/4 (1969)
Gèze, B.: L'origine des eaux souterraines. Ann. Spéléol. *2/1*, 3-10 (1947)
Gèze, B.: La genèse des gouffres. Ier Congr. Int. Spéléol., 1952, Paris, 11-23 (1953)
Gèze, B.: Les cristallisations excentriques de la grotte de Moulis. CNRS Paris (1957)
Gèze, B.: La spéléologie scientifique. Ed. du Seuil, Le Rayon de la science *22* (1965)
Gèze, B.: Lexique des termes français de spéléologie physique et de karstologie. Ann. Spéléol. *28/1*, 1-20 (1973)
Gèze, B., Pobequin, Th.: Contribution à l'étude des concrétions carbonatées. 2ème Congr. Int. Spéléol. 1958, Bari, Actes I, 396-414 (1962)
Gicquel, P., Renault, Ph.: L'Aven de Jean-Nouveau. Ann. Spéléol. *11/3*, 113-124 (1956)
Giessler, A.: Das unterirdische Wasser. Berlin: Deutscher Verlag Geisteswiss. 1957
Glanz, Th.: Das Phänomen der Meermühlen von Argostolion. Steir. Beitr. Hydrogeol. *17*, 113-128 (1965)
Glory, A.: Stalactites excentriques. Spelunca *7*, 93-102 (1936)
Gmelin's Handbuch der anorganischen Chemie, Calcium. 8. ed., Teil B. Verlag Chemie GmbH, Weinheim, 1961
Goguel, J.: Données techniques sur l'effondrement des cavités souterraines. Ann. Spéléol. *8*, 1-8 (1953)
Good, J.M.: Non-carbonate deposits of Carlsbad Caverns. Natl. Spéléol. Soc. *19*, 11-23 (1957)
Goodchild, M.F., Ford, D.C.: Analysis of scallop patterns by simulation under controlled conditions. J. Geol. *79*, 52-62 (1971)
Gortani, M.: Apunti sulla classificazione dei pozzi naturali. Ier Congr. Int. Spéléol. 1952, Paris, t. 2, 25-28 (1953)
Götzinger, G.: Die Phosphathöhle von Csoklovina in Siebenbürgen. Mitt. Geogr. Ges. Wien *62*, 304-333 (1919)
Götzinger, G.: Die Phosphatvorräte Österreichs. Speläol. Jahrb., 98-102 (1928)
Graf, D.L., Lamar, J.E.: Properties of Calcium and Magnesium Carbonates and their bearing on some uses of Carbonates. Econ. Geol. 639-713 (1955)
Gripp, K.: Über den Gipsberg in Segeberg. Jahrb. Hamburg. Wiss. Anst. *80* (1912)

Groner, U.: Untersuchungen an Höhlenlehmen (Hölloch, Schweiz). Unveröff. Diplomarbeit, Geogr. Inst. Univ. Zürich (1977)

Groom, G.E., Williams, V.: The solution of limestone in South Wales. Geogr. J. *131*, 37-41 (1965)

Groschopf, P.: Pollenanalytische Datierung württembergischer Kalktuffe und der postglaziale Klimaablauf. Jh. Geol. Abt. Württ. Staatl. Landesamtes *2*, 72-94 (1952)

Groschopf, P.: Beiträge zur Holozänstratigraphie SW-Deutschlands nach ^{14}C-Bestimmungen. Jahresh. geol. Landesamtes Baden-Württ. *4*, 137-143 (1961)

Groschopf, P.: Die geologischen Voraussetzungen für die Erschließung von Karstwasser im Blautal. Jahresh. Karst-Höhlenkd., 71-79 (1963)

Groschopf, P.: Karsthydrographische Probleme der Schwäbischen Alb. 5. Int. Kongr. Speläol., 1969, Stuttgart, Vol. 5, Hy 6/1-4 (1969)

Grund, A.: Die Karsthydrographie. Studien aus Westbosnien. Geogr. Abh. *7/3*, Wien (1903)

Guidebook of the Congress Excursions: 4th Int. Congr. Speleol., 1965, Ljubljana (1965)

Guide Michelin: Gorge de Tarn. Service de tourisme Michelin, Paris 1951

Hahne, C., Kirchmayer, M., et al.: Höhlenperlen (Cave pearls). Neues Jahrb. Geol. Paläontol. Abh. *130*, 1-46 (1968)

Halliday, W.R.: Changing concepts of speleogenesis. Natl. Speleol. Soc. Bull. *22*, 23-28 (1960)

Halliday, W.R.: Caves of California. Seattle: Selbstverlag (1962)

Harmon, R.S., Thompson, P., et al.: Uranium-series dating of speleothems. Natl. Speleol. Soc. Bull. *37*, 21-33 (1975)

Harned, H.S., Davis, R.: The ionisation constant of carbonic acid in water... Am. Chem. Soc. J. *65*, 2030-2037 (1943)

Harned, H.S., Scholes, S.R.: The ionisation constant of HCO_3^- from 0°C to 50°C. Am. Chem. Soc. J. *63*, 1706-1709 (1941)

Harned, H.S., Owen, B.B.: The physical chemistry of electrolytic solutions. 3rd ed., New York: Reinhold 1958

Haserodt, K.: Untersuchungen zur Höhen- und Altersgliederung der Karstformen in den nördlichen Kalkalpen, Inaug.-Diss., Münchner Geogr. H. 27 (1965)

Hauser, E., Oedl, R.: Eisbildungen und meteorologische Beobachtungen. In: Die Eisriesenwelt im Tennengebirge. Wien: Verl. Speläol. Inst. 1926, 1926, 77-105

Heim, A.: Geologie der Schweiz. Vol. 1. Leipzig: Chr. Herm. Tauchnitz 1919

Heinemann, U., Kaaden, K., Pfeffer, K.-H.: Neue Aspekte zum Phänomen der Rillenkarren. Abh. Karst-Höhlenkd, A, No. 15, Festschrift Alfred Bögli, 1977, 56-80

Herak, M., Stringfield, V.T.: Karst; Important Karst Regions of the Northern Hemisphere. London: Elsevier 1972

Hicks, F.L.: Formation and mineralogy of stalactites and stalagmites. Natl. Speleol. Soc. Bull. *11*, 63-72 (1950)

Hill, C.A.: Cave minerals. Natl. Speleol. Soc. Huntsville AL, USA (1976)

Hjulström, F.: Studies on the morphological activities of rivers. Bull. Geol. Inst. Uppsala *25*, 221-527 (1935)

Hohl, R.: ABC der Geologie. In: Die Entwicklungsgeschichte der Erde. Hanau a/M.: W. Dausien 1971, 677-852

Holland, H.D., Kirsipu, T.V., et al.: Chemical evolution of cave waters. J. Geol. *72*, 36-67 (1964)

Howard, A.D.: The development of karst features. Natl. Speleol. Soc. Bull. *25*, 45-65 (1963)

Howard, A.D.: Verification of the "Mischungskorrosion" effect. "Cave notes" 8/2, Dept. Geogr. Johns Hopkins Univ., Baltimore MD, 1966, 9-12

Hubbert, M.K.: The theory of ground water motion. J. Geol. *48*, 785-944 (1940)

Hudson, R.G.S., et al.: Geology of the Yorkshire dales. Proc. Geol. Assoc. *44*, 228-269 (1933)

Huff, C.C.: Artificial helictites and gypsum flowers. J. Geol. *58*, 641-659 (1940)

Jaeckli, H.: Kriterien zur Klassifikation von Grundwasservorkommen. Eclogae Geol. Helv. *63/2*, 389-446 (1970)

Jennings, J.N.: Syngenetic karst in Australia. In: Contributions to the study of karst. Dept. Geogr. Publ. G/5, Austral. Natl. Univ., Canberra 1968, 41-110

Jennings, J.N.: Karst. London: M.I.T. Press 1971
Jennings, J.N., Sweeting, M.M.: The limestone ranges of the Fitzroy Basin, Western Australia, Geogr. Abh. *32* (1963)
Jiménez, A.N.: Carso profundisimo de Cuba. Proc. 6th Int. Congr. Speleol., III, Olomouc 1973, 225-227 (1976)
Johnston, J.: Determination of carbonic acid in natural waters. J. Am. Chem. Soc. 38/5, pp. 947-975 (1916)
Johnston, J., Williamson, E.D.: The complete solubility curve of calcium carbonate. Am. Chem. Soc. J. *38,* 975-983 (1916)
Johnston, J., et al.: The several forms of calcium carbonate. Am. J. Sci. *41,* 729-750 (1916)
Jones, R.J.: Aspects of the biological weathering of limestone pavement. Proc. Geol. Assoc. London *76,* 421-423 (1966)
Jordan, R.H.: An interpretation of Floridian karst. J. Geol. *58,* 261-268 (1950)
Käss, W.: Erfahrungen bei Färbeversuchen mit Uranin. Steir. Beitr. Hydrogeol. 21-65 (1965)
Käss, W.: Erfahrungen mit Uranin bei Färbeversuchen. Steir. Beitr. Hydrogeol. 123-134 (1967)
Katzer, F.: Karst und Karsthydrographie. Zur Kde. Balkanhalbinsel *8,* Serajevo (1909)
Kayser, K.: Morphologische Studien in Westmontenegro. 2. Die Rumpftreppe von Cetinje und der Formenschatz der Karstabtragung. Z. Ges. Erdkde. 26-49 und 81-102 (1934)
Kayser, K.: Karstrandebenen und Poljeboden. Zur Frage der Entstehung der Einebnungsflächen im Karst. Erdkde. *9,* 60-64 (1955)
Kayser, K.: Bemerkungen über den Pluralismus der Poljenentstehung und die Stellung des Poljes im Rahmen des Karstformenschatzes. In: Neue Ergebnisse der Karstforschung in den Tropen und im Mittelmeerraum. Wiesbaden: Franz Steiner 1973, 75-82
Keilhack, K.: Lehrbuch der Grundwasser- und Quellenkunde. Berlin 1917
Keller, M.: Unterwasserforschung im Blautopf bei Blaubeuren. Jahresh. Karst-Höhlenkd. 219-228 (1963)
Keller, R.: Gewässer und Wasserhaushalt des Festlandes. Leipzig: B.G. Teubner 1962
Kendall, P.F., Wroot, H.E.: The Geology of Yorkshire. Wien: Hollinek 1924
Kirchmayer, M.: Höhlenperlen (Cave pearls) aus Bergwerken. Sitzungsber. Oesterr. Acad. Wiss., Vol. 137 (1964)
Klaer, W.: Verwitterungsformen im Granit auf Korsika. Gotha: VEB Hermann Haack 1956
Klockmann, F.: Lehrbuch der Mineralogie. Stuttgart: Encke 1923, 1978
Klut: Untersuchung des Wassers an Ort und Stelle. Berlin: Springer 1943
Knebel, W.v.: Höhlenkunde. Braunschweig: Vieweg 1906
Knuchel, F.: Färbung des unterirdischen Abflusses der Schrattenfluh. Stalactite, Suppl. *7,* Soc. Suisse Spéléol. (1972)
Knutson, G.: Tracing groundwater flow in sand and gravel using radioactive isotopes. Steir. Beitr. Hydrodeol. 13-31 (1967)
Knutson, G., et al.: Field and laboratory tests of Chromium-51-EDTA and Tritium water as a double tracer for groundwater flow. "Radioisotopes in Hydrology", IAEA, Vienna, 347-363 (1963)
Kohlrausch, F.: Lehrbuch der praktischen Physik. 16. ed. Berlin: B.G. Teubner 1930
Köster, E.: Granulometrische und morphometrische Meßmethoden an Steinen, Körnern und sonstigen Stoffen. Stuttgart: Enke 1964
Král, Z.: Die Bedingungen der Farbigkeit von Tropfsteinformationen. Proc. 6th Int. Congr. Speleol. I. Olomouc 1973, 269-271 (1975)
Kraus, F.: Höhlenkunde. Wien: Carl Gerold's Sohn 1894
Krauskopf, K.: Dissolution and precipitation of silica at low temperatures. Geochim. Cosmochim. Acta *10,* 1-26 (1956)
Krebs, N.: Neue Forschungsergebnisse zur Karsthydrographie. Peterm. Geogr. Mitt. 166-168 (1908)
Krieg, W.: Höhlen und Niveaus. Die Höhle *5/1,* 1-4 (1954)
Krieg, W.: Zu "Höhlen und Niveaus". Die Höhle *6,* 74-77 (1955)
Krieg, W.: Exzessives Wachstum von Sinterröhrchen unter besonderen Bedingungen. Proc. 6th Int. Congr. Speleol. I, Olomouc 1973, 467-471 (1975)
Kunaver, J.: Guide through the high mountainous karst of the Julian Alps. 4th Int. Congr. Sepeol., Ljubljana (1965)

Kundert, C.J.: The origin of palettes, Lehmans Caves Natl. Mon., Baker, Nevada, Natl. Speleol. Soc. Bull. *14*, 30-33 (1952)
Kunsky, J.: Karst et grottes. Paris: Heintz 1958
Kyrle, G.: Theoretische Speläologie. Wien: Österr. Staatsdruckerei 1923
Lallemand, A., Paloc, H.: La méthode de détection au charbon actif pour les opérations des traçage à la fluorescéine. Paris: BRGM 1964
Lane, C.F.: Grassy Cove, an uvala in the Cumberland Plateau, Tennessee. J. Tenn. Acad. Sci. *27*, 291-295 (1952)
Lang: Diskussionsbeitrag zu Gádoros. In: 5. Int. Kongr. Speläol. Vol. 5, Hy 14/3 (1969)
Langelier, W.F.: The analytical control of anticorrosion water treatment. J. Am. Water Works Ass., 1500 (1936)
Laptev, F.F.: Aggressive Wirkungen der Wässer auf Karbonatgesteine, Gips und Beton (in Russian). Tr. Spetsgeo. *1*, Moskau (1939)
Lehmann, H.: Morphologische Studien auf Java. Geogr. Abh. 3, Stuttgart (1936)
Lehmann, H.: Karstentwicklung in den Tropen. Umschau in Wiss. Tech. *18*, 559-562 (1953)
Lehmann, H.: Studien über Poljen in den venezianischen Voralpen und im Hochapennin. Erdkunde *13/4*, 249-289 (1959)
Lehmann, H.: Polje. In: Westermanns Lexikon der Geographie, Vol. 3, 1970, 868
Lehmann, H.: Uvala. In: Westermanns Lexikon der Geographie, Vol. 4, 1970, 783
Lehmann, H., Krömmelbein, K., et al.: Karstmorphologische, geologische und botanische Studien in der Sierra de los Organos auf Cuba. Erdkunde *10/3*, 185-204 (1956)
Lehmann, O.: Über die Karstdolinen. Mitt. Geogr.-Ethnogr. Ges., Zürich, *31*, 43-71 (1931)
Lehmann, O.: Die Hydrographie des Karstes. Wien: Franz Deutike 1932
Leighton, M.W., Pendexter, C.: Carbonate rock types. Mem. Am. Assoc. Petrol. Geol. *1*, 33-61 (1962)
Libby, W.F.: Altersbestimmung mit der ^{14}C-Methode. Bibliogr. Inst. Mannheim (1969)
Liedtke, H.: Eisrand und Karstpoljen am Westrand der Lukavicahochfläche (Westmontenegro). Erdkunde *16/4*, 289-298 (1962)
Liszkowski, J.: Ist die Mischungskorrosion die einzige im phreatischen Bereich der Karstgrundwasserleiter wirksame Korrosionsform? Proc. 6th Int. Congr. Speleol. III, Olomouc 1973, 193-198 (1975)
Lobeck, A.K.: The geology and physiography of the Mammoth Cave National Park, KY, Geol. Surv. 31 (1929)
Löhnberg, A.: Zur Hydrographie des Zirknitzer Beckens. Diss. Georg August-Univ., Göttingen (1934)
Long, A., Martin, P.S.: Death of American ground sloths. Science *186*/4165, 638-640 (1974)
Louis, H.: Die Entstehung der Poljen und ihre Stellung in der Karstabtragung. Erdkunde *10*, 33-53 (1956)
Louis, H.: Allgemeine Geomorphologie. Berlin: Walter de Gruyter 1968
Lukas, G.: In: Maurin und Zötl, Die Untersuchung unterirdischer Wässer: B. Der chemische Nachweis des Salzdurchganges. D. Die Triftung von Bakterien. Steir. Beitr. Hydrogeol. 107-116 und 149-156 (1959)
Mais, K.: Vorläufige Beobachtungen über Kondenswasserkorrosion in der Schlenkendurchgangshöhle (Salzburg). Proc. 6th Int. Congr. Speleol. III, Olomouc 1973, 203-208 (1975)
Mallot, C.A.: Lost River at Wesley Chapel Gulf. Proc. IN. Acad. Sci. *41*, 285-316 (1932)
Mallot, C.A.: Karst valleys. Bull. Geol. Soc. Am. *50*, 1984 (1939)
Mallot, C.A.: Significant features of the Indiana Karst. Proc. IN. Acad. Sci. *54*, 8-24 (1945)
Mallot, C.A., Shrock, R.R.: Mud stalagmites. Am. J. Sci. *25*, 55-60 (1933)
Mangin, A.: Contribution à l'étude hydrodynamique des aquifères karstiques. Thèse, Univ. Dijon, Lab. souterrain du CNRS, Moulis (1975)
Maronny, G.: pH des mélanges de carbonates de lithium, bicarbonate de potassium et chlorure de potassium. Bull. Sci. Chim. Fr. *5*, 893 (1961)
Martel, E.A.: Les Abîmes. Paris: Delagrave 1894
Martel, E.A.: Nouveau traité des eaux souterraines. Paris: O. Doin 1921
Martel, E.A.: L'Aven Armand (Causse Méjean, Lozère). Millau: Artières 1948

Martin, C.: La résurgence sous-marine de Port-Miou. Spelunca *8/1* (1968)
Martin, J.: Quelques types de dépressions karstiques du Moyen Atlas central. Rev. Géogr. Maroc. *7*, 95-106 (1965)
Mason, W.M.A.: Biological aspects of calcite depositions. Rass. Speleol. Ital., Mem. *5/2*, 235-238 (1959)
Maucci, W.: L'ipotesi dell'erosione inversa... Le Grotte d'Italia. s. 4, Vol. IV, Bologna (1973)
Maull, O.: Vom Itatiaya zum Paraguay. Leipzig (1930)
Maull, O.: Handbuch der Geomorphologie, 2. ed. Wien: Franz Deutike 1958
Maurin, V., Zötl, J.: Die Untersuchung der Zusammenhänge unterirdischer Wässer mit besonderer Berücksichtigung der Karstverhältnisse. Steir. Beitr. Hydrogeol. (1959)
Maurin, V., Zötl, J.: Karsthydrographische Aufnahmen auf Kephallenia. Steir. Beitr. Hydrogeol. (1960)
Maurin, V., Zötl, J.: Karsthydrologische Untersuchungen auf Kephallenia. Österr. Hochschulz. 15.6.1963 Wien (1963)
Maurin, V., Zötl, J.: Karsthydrologische Untersuchungen im Toten Gebirge. Österr. Wasserwirtschaft *16*, 112-123 (1964)
Maxson, J.H.: Fluting and faceting of rock fragments. J. Geol. *48*, 717-751 (1940)
Mayr, A.: Blütenpollen und pflanzliche Sporen als Mittel zur Untersuchung von Quellen und Karstwasser. Anz. Österr. Akad. Wiss. (1953)
McElroy, G.E.: Mine ventilation. In: Mining Engineers Handbook I, 3rd ed. New York: John Wiley and Sons 1966
McGill, W.M.: Caverns of Virginia. VA. Geol. Surv. Bull. *35* (1933)
McGrain, P.: Helictites in the new discovery at Wyandotte Cave, Indiana. Proc. IN. Acad. Sci. *51*, 201-206 (1942)
Meinzer, O.E.: Outline of ground water hydrology with definitions. U.S. Geol. Surv., Water Supply Pap., 494, 1-67 (1923)
Mensching, H.: Beobachtungen zum Formenschatz des Küstenkarstes an der Kantabrischen Küste bei Santander und Llanes. Erdkunde *19*, 24-31 (1965)
Miotke, F.-D.: Karstmorphologische Studien in der glazialüberformten Höhenstufe der "Picos de Europa" in Nordspanien. Jahrb. Geogr. Ges. Hannover (1968)
Miotke, F.-D.: Die Messung des CO_2-Gehaltes der Bodenluft mit dem Drägergerät und die beschleunigte Kalklösung durch höhere Fließgeschwindigkeiten. Z. Geomorphol. N.F. *16*, 93-102 (1972)
Miotke, F.-D.: Carbon dioxide and the soil atmosphere. Abh. Karst-Höhlenkd. A/H 9 (1974)
Miotke, F.-D., Palmer, A.N.: Genetic relationship between caves and landforms in the Mammoth Cave National Park area. Veröff. Geogr. Inst. Tech. Hochsch. Hannover (1972)
Misik, M.: Anwendung der Schwerminerale in der paläogeographischen und stratigraphischen Forschung... in der Slowakei. Geol. Prace, Zošit *43*, 135-139 (1956)
Mitter, H.: Der physikalische Nachweis des Salzdurchganges und Messungen mit Radioisotopen. In: Maurin und Zötl: Die Untersuchungen der Zusammenhänge unterirdischer Wässer... Steir. Beitr. Hydrogeol., Graz 117-125 (1959)
Monroe, W.H.: Sinkholes and towers in the karst areas of North Central Puerto Rico. Geol. Surv. Res., Prof. Pap. *400* B (1960)
Monroe, W.H.: A glossary of karst terminology. Contr. to the hydrol. U.S. Geol. Surv. Water Supply Pap. *1899-K* (1970)
Monroe, W.H.: A possible origin of clay fills in cave. Proc. 6th Int. Congr. Speleol. I, Olomouc 1973, 509-512 (1973)
Monroe, W.H.: The Karst Landforms of Puerto Rico. Geol. Surv., Prof. Pap. *899*, Washington D.C., USA 1976
Moore, G.W.: Speleothem − a new cave term. Natl. Speleol. Soc. News 2 (1952)
Moore, G.W.: The origin of helictites. Natl. Speleol. Soc., Occ. Pap. 1 (1954)
Moore, G.W.: Aragonite speleothems as indicators of paleotemperature. Am. J. Sci. *254*, 746-753 (1956)
Moore, G.W.: Checklist of cave minerals. Natl. Speleol. Soc. News *28*, 9-10; 40 Bibliographical entries (1970)

Morawetz, S.: Zur Frage der Dolinenverteilung und Dolinenbildung im istrischen Karst. Peterm. Geogr. Mitt. *109*, 161-170 (1965)

Morton, F.: Die „Köhbrunnen" des Hallstätter Sees. Die Pyramide *11*, 70-72 (1963)

Moser, R.: Zur Abtragung im Dachsteingebiet. Jahrb. Oberösterr. Musealver. 305-307 (1956)

Moser, R.: Kalktische im Toten Gebirge und im Dachsteingebiet. Jahrb. Österr. Alpenver. 75-78 (1967)

Motts, W.S.: Geology and paleoenvironments of the northern segment, Capitan Shelf, New Mexico and West Texas. Geol. Soc. Am. Bull. *83*, 701-722 (1972)

Müller, R.: Hydraulik. In: Ingenieur Handbuch, 27. ed., Vol. 1, 1958, 496-598

Murawski, H.: Geologisches Wörterbuch, 5. ed. Stuttgart: Enke 1963, 1977

Murray, J.W.: The deposition of calcite and aragonite in caves. J. Geol. *62*, 481-492 (1954)

Myers, J.O.: Cave physics. In: British Caving, 2nd ed., Cullingford (ed.). London: Routledge and Kegan 1962, 226-251

Newell, N.D., Rigby, J.K., et al.: The Permian reef complex of Guadalupe Mountains Region, Texas and New Mexico. San Francisco: Freeman 1953

Nicod, J.: Recherches morphologiques en Basse-Provence calcaire. Thèse, Gap, Impr. Louis-Jean (1967)

Nicod, J.: Poljés karstiques de Provence, comparaison avec les poljés dinariques. Etud. Trav. "Méditerranée" *8*, Rev. Géogr. Pays Méditerr. (1969)

Nicod, J.: Essai sur les facteurs du régime des sources karstiques. Actes 93ème Congr. Natl. Soc. Savantes, Tours, 99-119 (1970)

Niggli, P.: Lehrbuch der Mineralogie II, 2. ed. Berlin: Gebr. Bornträger 1926

Oertli, H.: Karbonathärte von Karstgewässern. Stalactite *4*, Soc. Suisse Spéléol. 1-10 (1953)

Osinski, W.v.: Karst-windows. Proc. IN. Acad. Sci. *44*, 161-165 (1935)

Palmer, A.N., Palmer, M.V.: Landform development in the Mitchell Plain of southern Indiana. Z. Geomorphol. *19*, 1-39 (1975)

Palmer, A.N., Palmer, M.V., Queen, G.M.: Speleogenesis in the Guadalupe Mountains, New Mexico; Gypsum replacement of carbonate by brine mixing. Proc. 7th Int. Speleol. Congr. Sheffield, 333-336 (1977)

Pasini, G.: Sull'importanza speleogenetica dell'erosione antigravitativa. Le Grotte d'Italia, 4, Vol. IV, Bologna (1973)

Pauling, L.: Chemie − eine Einführung. 7. ed. Weinheim: Chemie 1967

Penk, A.: Über das Karstphänomen. Schr. Verbr. Naturw. Kennt. *44*, 1-38 (1904)

Penk, A.: Das unterirdische Karstphänomen. Recl. Trav. Offert à J. Cvijić, 175-197 (1924)

Peters, K.: Mechanochemische Reaktionen. 1. Eur. Symp. Zerkleinern, 1962, Frankfurt a/M. Weinheim: Chemie 1962, 78-98

Petránek, J., Pouba, Z.: Dating of the development of the Domica Cave, Sb. ústřed. Ústavu Dr. Radim Kettner, *18*, 254-272 (1951)

Petrik, M.: Measurements on submarine sources. Congr. Yugosl. Spéléol. II, Split, 49-57 (1958)

Pettijohn, F.J.: Sedimentary Rocks. 2nd ed. New York: Harper 1957

Pfeffer, K.-H.: Beiträge zur Geomorphologie der Karstbecken im Bereiche des Monte Velino (Zentralapennin). Frankfurter Geogr. H. 42 (1967)

Pfeffer, K.-H.: Flächenbildung in den Kalkgebieten. In: Ergebnisse der Karstforschung in den Tropen und im Mittelmeerraum. Geogr. Z., Beih. Wiesbaden: Franz Steiner, 1973, 111-132

Pfeffer, K.-H.: Zur Genese von Oberflächenformen in Gebieten mit flachlagernden Karbonatgesteinen. Wiesbaden: Franz Steiner 1975

Pfeffer, K.-H.: Probleme der Genese von Oberflächenformen auf Kalkgestein. Z. Geomorph. N.F., Suppl. *26*, 6-34 (1976)

Pfeiferová, A., Kvaĺek, M.: Vorkommen von Apatit in den Sinterfüllungen der Javoříčko Höhlen (ČSSR). Proc. 6th Int. Congr. Speleol., Olomouc I, 1973, 545-550 (1975)

Pia, J.: Theorien über die Löslichkeit des kohlensauren Kalkes. Mitt. Geol. Ges. Wien 46 (1953)

Picknett, R.G.: A study of calcite solutions at $10°C$. Trans. Cave Res. Group G.B. *7/1*, 39-62 (1964)

Picknett, R.G.: Saturated calcite solutions from $10°$ to $40°C$. Trans. Cave Res. Group G.B. *15/2*, 67-80 (1973)

Picknett, R.G.: Rejuvenation of aggressiveness in calcium carbonate solutions by means of magnesium carbonate. Proc. 7th Int. Congr. Speleol. Sheffield, 346-348 (1977)
Picknett, R.G., Bray, L.G., Stenner, R.D.: The chemistry of cave waters. In: Ford, T.D., Cullingford, C.H.D.: The Science of Speleology. London: Academic Press 1976
Pigott, C.D.: The structure of limestone surfaces in Derbyshire. Geogr. J. *131*, 41-44 (1965)
Pinchemel, P.: Les plaines de craie du nordouest de Paris. Paris: Colin 1954
Piper, A.M.: Groundwater in north central Tennessee. U.S. Geol. Surv. Water-Supply Pap. *640*, 1-238 (1932)
Pitty, A.F.: An approach to the study of karst water. Univ. Hull, Occ. Pap. Geogr. 5 (1966)
Playford, P.D., Lowry, D.C.: Devonian Reef complexes of the Canning Basin, Western Australia. Bull. Geol. Surv. W. Austral. 118 (1966)
Pobequin, T.: Contribution à l'étude des carbonates de calcium. Ann. Sci. Nat. Bot. *11*, 29-109 (1954)
Pohl, E.R.: Vertical shafts in limestone caves. Natl. Speleol. Soc., Occ. Pap. *2*, 3-24 (1955)
Pohl, E.R., White, W.B.: Sulfate minerals: their origin in the central Kentucky karst. Am. Mineral. *50*, 1461-1465 (1965)
Polli, S.: Meteorologia nella Grotta Gigante. 1er Congr. Int. Speleol. Paris *2*, 307-319 (1953)
Polli, S.: Cinque anni di meteorologia ipogea nella Grotta Gigante presso Trieste. Atti VIII, Congr. Naz. Speleol., Como, 1956, 166-179 (1958)
Polli, S.: Misure dell'accrescimento delle stalattiti. 2nd Congr. Int. Speleol., Bari, Actes I, 442-448 (1962)
Pommier, C., Garnier, J.J.: A propos des vermiculations argileuses. Bull. Comm. Natl. Spéléol. *5*, 7-8 (1955)
Popov, V., Gvodzdetskiy, N.A., et al.: Karst of the USSR. In: Karst. Herak and Stringfield (eds.). London: Elsevier 1972, 355-416
Potonié, H.: Entstehung der Steinkohle, 3. ed. Berlin 1905
Powell, R.L.: Alluviated springs of south central Indiana. Proc. IN. Acad. Sci. *72*, 182-189 (1963)
Powell, R.L.: NSS-Convention Guidebook. Natl. Speleol. Soc., Bloomington, Indiana (1965)
Prandtl, L.: Führer durch die Strömungslehre. 7. ed. Braunschweig: Vieweg 1969
Press, H.: 4. Wasserbau und Wasserwirtschaft, 5. Stau- und Kraftwerksanlagen. In: Hütte, Taschenbuch der Bautechnik, Vol. II, Berlin: W. Ernst 1969, 634-815
Prévost, P.: Recherches physico-mécaniques sur la chaleur. Paris 1972
Priesnitz, K.: Über die Vergleichbarkeit von Lösungsformen auf Chlorid-, Sulfat- und Karbonatgesteinen. Geol. Rundsch. *58*, 427-438 (1969)
Prinz, W.: Les cristallisations des grottes de Belgique. Nouv. Mém. Soc. Géol. Belg. 1-90 (1909)
Puri, H.S., Vernon, R.O.: Summary of the geology of Florida. Fla. Geol. Surv., Spec. Publ. *5*, Tallahassee (1964)
Rathjens, C.: Zur Frage der Karstrandebenen im Dinarischen Karst. Erdkunde *8*, 114 f. (1954)
Rathjens, C.: Beobachtungen an hochgelegenen Poljen im südlichen dinarischen Karst. Z. Geomorphol. 141-151 (1960)
Reclus, E.: Nouvelle Géographie universelle: France. Paris: Hachette 1881
Renault, Ph.: Dépots vermiculés d'argile de décalcification. 1er Congr. Int. Spéléol., Paris, 1952, *2*, 365-370 (1953)
Renault, Ph.: Une microforme spéléologique: vagues d'érosion. Spelunca *1*, 15-25 (1961a)
Renault, Ph.: Caractère des vagues d'érosion selon la morphologie des conduits karstiques. 3ème Congr. Int. Spéléol., Vienne, Vol. 2, 105-114 (1961b)
Renault, Ph.: Observations récentes sur les vermiculations argileuses. Spelunca *3/1*, 25-28 (1963)
Renault, Ph.: Contribution à l'étude des actions mécaniques et sédimentologiques dans la spéléogenèse. Ann. Spéléol. *22*, 5-21; 209-267; *23*, 259-307; 529-596 (1967/68)
Renault, Ph.: La formation des cavernes. Presse univ. France, Que sais-je? 1400 (1970)
Reuter, F.: Investigations in areas of salt and gypsum karstification. 6th Int. Congr. Speleol., Olomouc, ČSSR, Abstr. 13 (1973)
Reuter, F.: Untersuchungen in Salz- und Gipskarstgebirgen. Proc. 6th Int. Congr. Speleol. I, Olomouc, 1973, 313-318 (1975)

Rhoades, R.F., Sinacori, M.N.: Patterns of groundwater flow and solution. J. Geol. *49*, 785-794 (1941)

Riedl, H.: Grundsätzliche Bemerkungen zur feldmäßigen Untersuchung von Höhlensedimenten. Mem. Rass. Speleol. Ital., Como, 30-34 (1961)

Rinne, F.: Gesteinskunde. Leipzig: Max Jänecke 1928

Roglić, J.: Morphologie der Poljen von Kupres und Vukovo. Z. Ges. Erdkunde, Berlin, 299-316 (1939)

Roglić, J.: Quelques problèmes fondamentaux du karst. Inf. Géogr. *21/1*, 1-12 (1957)

Roglić, J.: Das Verhältnis der Flußerosion zum Karstprozeß. Z. Geomorphol. *4*, 116-128 (1960)

Roglić, J.: Les Poljés du karst dinarique et les modifications climatiques du quaternaire. Rev. Belg. Géogr. 105-125 (1964a)

Roglić, J.: Karst valleys in the Dinaric karst. Erdkunde *18*, 113-116 (1964b)

Roglić, J.: The depth of the fissure circulation of water and the evolution of subterranean cavities in the Dinaric karst. Proc. Int. Speleol. Conf. Brno, 25-36 (1965)

Roques, H.: Localisation conductométrique des émergences sousmarines de Port-Miou. Ann. Spéléol. *9/3*, 109-112 (1956)

Roques, H.: Considérations théoriques sur la chimie des carbonates. Ann. Spéléol. *17*, 463-467 (1962)

Roques, H.: Chimie des carbonates. Ann. Spéléol. *19*, 258-484 (1964)

Roques, H.: Problèmes de transferts de masse, posés par l'évolution des eaux souterraines. Ann. Spéléol. *24*, 455-494 (1969a)

Roques, H.: A review of present-day problems in the physical chemistry of carbonates in solution. Trans. Cave Res. Group G.B., 139-163 (1969b)

Rühl, A.: Studien in den Kalkmassiven des Apennin I. Z. Ges. Erdkunde, Berlin (1911)

Saar, R.: Geschichte und Aufbau der österreichischen Höhlendüngeraktion. Speleol. Mon. 7-9 (1931)

Saar, R.: Beiträge zur Meteorologie der dynamischen Wetterhöhlen. Mitt. Höhlenkomm. Wien, 1953, 5-25 (1954)

Salvayre, H.: Etude comparée des hydrogrammes de tarissement aux résurgences de la Sorgue, de la Vis et de l'Esperelle. 5. Int. Kongr. Speläol. Stuttgart, Vol. 5, Hy 9/1-8 (1969)

Salzer, H.: Beiträge zur Genese, Morphologie und Struktur von Tropfsteinformen in Naturhöhlen. Diss. Univ. Wien (1934)

Salzer, H.: Zauberwerk aus Stein. In: Karst und Höhlen in Niederösterreich und Wien. Wien: Jugend u. Volk, 53-60 (1954)

Sarvary, I.: How can the size of "cavernment" in karstic rocks be estimated? Proc. 6th Int. Congr. Speleol. III, Olomouc 1973, 267-272 (1975)

Sawicki, L.S.: Ein Beitrag zum geographischen Zyklus im Karst. Geogr. Z. *15*, 185-204, 259-281 (1909)

Schadler, J.: Die Phosphatablagerungen in der Lettenmayerhöhle bei Kremsmünster in Oberösterreich. Ber. Staatl. Höhlenkomm. Wien (1920)

Schauberger, O.: Über die vertikale Verteilung der nordalpinen Karsthöhlen. Mitt. Höhlenkomm. *1*, 21-28 (1955)

Scheu, E.: Die Entstehung von Trockentälern. Penk-Festschrift 1918

Schmid, E.: Höhlenforschung und Sedimentanalyse. Schr. Inst. Früh- Urgesch. Schweiz *13*, Basel (1958)

Schmidt, O.W.: Wörterbuch der Geologie, Mineralogie und Paläontologie. Berlin: Walter de Gruyter 1928

Schnitzer, W.A.: Eine neue Methode zur Markierung von Karstwasserwegen. Die Höhle *16/3*, 64-67 (1965)

Schnitzer, W.A.: Anwendung von Detergentien und verwandten Stoffen in der Hydrogeologie. Steir. Beitr. Hydrogeol. 231-234 (1967)

Schnitzer, W.A., Wagner, W.: Markierungsversuche mit Geruchsstoffen in der Karsthydrologie. Geol. Blatt NO-Bayern *17*, 179-194 (1967)

Schnitzer, W.A., Wagner, W.: Welche Anforderungen müssen Geruchsstoffe für Karstwassermarkierungen erfüllen? Vom Wasser *35*. Weinheim: Chemie 1969, 227-236

Seemann, R.: Pyritfunde in der Dachstein-Mammuthöhle (Oberösterr.). Die Höhle *21/2*, 83 (1970)
Seidel, K.: Dolomitisierung und Erzbildung in Karbonatgesteinen unter der Einwirkung von Salzsolen. N. Jahrb. Mineral., Geol., Paläont. Mh. 25-55 (1958)
Šerko, A., Mischler, I.: Die Grotte von Postojna. Ljubljana: Postojnska Jama 1958
Siever, R.: Silica solubility, 0-200°C, and the diagenesis of siliceous sediments. J. Geol. *70*, 127-150 (1962)
Simeoni, G.-P.: Etude de la région alimentaire de la nappe de la plaine du Bödeli. Rapport Centre hydrogéol. Univ. Neuchâtel (unveröffentlicht) (1973)
Sindowski, K.-H.: Mineralogische, petrographische und geochemische Untersuchungsmethoden, In: Bentz, Lehrbuch der angewandten Geologie, Vol. I, Allgemeine Methoden. Stuttgart 1961
Smith-D'Ans, A.: Einführung in die allgemeine und anorganische Chemie. Karlsruhe: G. Braun 1933
Stamp, L.D.: A glossary of geographical terms. Comm. Brit. Assoc. Adv. Sci. (1963)
Stchouzkoy-Muxart, T.: Solubilité de la calcite dans l'eau en présence de CO_2. Ann. Spéléol. *27*, 465-478 (1972)
Steinberg, F. v.: Gründliche Nachrichten von dem in Innerkrain liegenden Zirknitzer See. Graz (1781)
Steiner, L.: Die Temperaturverhältnisse in der Eishöhle von Dobšina. Met. Z. *19*, 193 (1922)
Stini, J.: Die Quellen. Wien: Springer 1933
Stini, J.: Tunnelbaugeologie. Wien: Springer 1950
Stini, J.: Randbemerkungen zur Frage der Entstehung von Höhlen. Geol. Bauwesen *18*, Wien (1951)
Stirn, A.: Kalktuffvorkommen und Kalktufftypen der Schwäbischen Alb. Abh. Karst-Höhlenkd., E/H *1* (1964)
Strasser, A.: Phosphatminerale aus einer Salzburger Höhle. Die Höhle *21/2*, 80-82 (1970)
Sweeting, M.M.: The karst lands of Jamaica. Geogr. J. 184-199 (1958)
Sweeting, M.M.: Some factors in the absolute denudation of limestone terrains. Erdkunde *18*, 92-95 (1964)
Sweeting, M.M.: Denudation in limestone regions, Introduction. Geogr. J. *131/I*, 34-37 (1965)
Sweeting, M.M.: The weathering of limestones with particular reference to the Carboniferous limestones of northern England. In: Essays in Geomorphology. London: Heinemann 1966, 177-210
Sweeting, M.M.: Karst Landforms. London: Macmillan 1972
Sweeting, M.M.: Some comments on the lithological basis of karst landform variations. Proc. 6th Int. Congr. Speleol. I, Olomouc 1973, 319-329 (1975)
Swinnerton, A.C.: The caves of Bermuda. Geol. Mag. *66*, 79-84 (1929)
Swinnerton, A.C.: Origin of limestone caverns. Bull. Geol. Soc. Am. *43*, 663-693 (1932)
Swinnerton, A.C.: Hydrology of limestone terrains. In: Meinzer, O.E., Physics of the Earth: IX, Hydrology. New York 1942
Szczerban, E., Urbani, F.: Formas carsicas en areniscas precambricas del territorio federal Amazonas y Estado Bolivar. In: Carsos de Venezuela, parte 4, Bol. Soc. Venez. Espeleol. *5*, 27-54 (1974)
Terzaghi, K.: Beiträge zur Hydrographie und Morphologie des kroatischen Karstes. Mitt. Jb. K. Ung. Geol. Reichsanstalt, *20*, 225-369 (1913)
Thomas, T.M.: Swallow holes on the millstone grit and Carboniferous limestone of the South Wales coalfield. Geogr. J. *120*, 468-475 (1954)
Thommen: Anleitung zum Gebrauch des Thommen-Altimeters
Thornbury, W.D.: Principles of Geomorphology, 2nd ed. New York: Wiley Int., 1969
Thrailkill, J.: Chemical and hydrologic factors in the excavation of limestone caves. Geol. Soc. Am. Bull. *79*, 19-46 (1968)
Thurber, D.L., Broeker, W.S., et al.: Uranium series ages of coral from pacific atolls. Science *149*, 55-58 (1965)
Thurner, A.: Hydrogeologie. Wien: Springer 1967
Thury, W.: Etudes sur les glacières naturelles. Diss. Bibl. Univ. Genève (1861)
Tillmans, J.: Die chemische Untersuchung von Wasser und Abwasser. Halle: W. Knapp 1932
Tintant, H.: La grotte de Bèze. "Sous le plancher", Spéléo-Club Dijon, 68-73 (1958)
Tommasini, T., Candotti, P.: Due campagne meteorologhiche nella Grotta di Padriciano. Atti Mem. Comm. Grotte. E. Boegan. Trieste, 59-78 (1968)

Toperczer, M.: Lehrbuch der allgemeinen Geophysik. Wien: Springer 1960
Tricart, J.: La partie orientale du bassin de Paris. Thèse, Univ. Paris (1949)
Trimmel, H.: Höhlen und Niveaus. Die Höhle *6/1*, 5-9 (1955)
Trimmel, H.: Die Arzberghöhle bei Wildalpen (Steiermark); ein Beitrag zu den Problemen der Höhlensedimente, der Bergmilchbildung und der Speläogenese. 2e Congr. Int. Spéléol. Bari, 1958, Actes I, 330-340 (1962)
Trimmel, H.: Speläologisches Fachwörterbuch. Landesver. Höhlenkd. Wien. Niederösterr., Wien (1965)
Trimmel, H.: Höhlenkunde. Braunschweig: Vieweg 1968
Trimmel, H., Audetat, M.: Signes conventionnels à l'usage des spéléologues. Stalactite *3*, 73-125 (1966)
Tripet, J.-P.: Etude hydrogéologique du bassin de la source de l'Areuse. Thèse Univ. Neuchâtel, Inst. Géol., Centre Hydrogéol. (1972)
Trombe, F.: Recherches souterraines dans les Pyrenées. II. Météorologie et hydrologie souterraine. Ann. Spéléol. *2*, 99-123 (1947)
Trombe, F.: Traité de Spéléol. Payot, Paris (1952)
Trowbridge, A.C.: Dictionary of Geological Terms. Garden City, New York: Dolphin Reference Books 1967
Trudgill, S.T.: The marine erosion of limestone on Aldabra Atoll, Indian Ocean. Z. Geomorphol. N.F., Suppl. *26*, 164-199 (1976)
Ule, W.: Physiogeographie des Süßwassers. Enzyklopädie der Erdkunde. Leipzig-Wien: Fr. Deutike 1925
Urbani, F.: Mineralogia de algunas "Calizas" de la parte centrale de la Cordillera de la Costa. Bol. Inf. Assoc. Venez. Geol. Mineral. Petrol. *12*, 417-423 (1969)
Urbani, F.: Polish-Venezuelan Expedition studies Sarisariñama caves. Natl. Speleol. Soc. News, USA *34/1*, 194-195 (1976)
Urbani, F., Szczerban, E.: Venezuela Caves in noncarbonate rocks. Natl. Speleol. Soc. News, USA *32/12*, 233-235 (1974)
Urbani, F., Zawidski, P., Koisar, B.: Observationes geologicas de la Meseta de Sarisariñama, Gdo. Bolivar, Venezuela. Bol. AVGMP *19/2*, 77-86 (1976)
Verdeil, P.: Principes généraux de la karstification. Spelunca, Mém. 1, Actes 3e Congr. Natl. Spéléol. Marseille, 42-56 (1961)
Verdeil, P.: Les phénomènes d'intermittance dans les réseaux karstiques. 2e Congr. Int. Spéléol. Bari 1958, vol. 1, 62-78 (1962)
Villinger, E.: Beziehungen zwischen den Quellen und Trockentälern im seichten und tiefen Karst der Schwäbischen Alb. 5. Int. Kongr. Speläol., Stuttgart, Vol. 5, Hy 19/1-13 (1969)
Villinger, E.: Seichter Karst und tiefer Karst in der Schwäbischen Alb. Geol. Jahrb. C2, 153-188 (1972)
Volger, O.: Die wissenschaftliche Lösung der Wasser-, insbesondere der Quellenfrage. Z. Ver. Dtsch. Ing. *21* (1877)
Völker, R.: The development of gypsum-caves. 6th Int. Congr. Speleol., Olomouc, ČSSR, Abstr. 90 (1973)
Wagner, G.: Der Karst als Musterbeispiel der Verkarstung. Natw. Monatschr. Dtsch. Naturkdever. *62/9-10*, 193-212 (1954)
Wagner, G.: Einführung in die Erd- und Landschaftsgeschichte. 3. ed. Oehringen: Hohenlohe'sche Buchhdl. F. Rau 1960
Warwick, G.T.: The effect of knick-point recession on the watertable. Z. Geomorphol., Suppl. *2*, 92-99 (1960)
Warwick, G.T.: The origin of limestone caves. In: Cullingford: British Caving, 2nd ed., 1962 a, 55-82
Warwick, G.T.: Cave formations and deposits. In: Cullingford: British Caving, 2nd ed., 1962b, 83-119
Warwick, G.T.: Dry valleys of the Southern Pennines. Erdkunde *18/2*, 116-123 (1964)
Watson, P.J.: The prehistory of Salt Cave, Kentucky. Rep. 16, Illinois, State Museum (1969)

Watson, P.J.: Archeology of the Mammoth Cave Area. New York-London: Academic Press 1974

Weidenbach, F.: Über einige Wasserbohrungen im Jura. Jahrb. Mitt. Oberrheinischen Geol. Ver. N.F. *36*, 54-73 (1954)

Weidenbach, F.: Trinkwasserversorgung aus Karstwasser in der östlichen Schwäbischen Alb. Jahrb. Karst-Höhlenkd. 169-192 (1960)

Well, St.G., Desmarais, D.J.: The Flint-Mammoth Connection. Natl. Speleol. Soc. News *31*, 18-22 (1973)

Westermann: Lexikon der Geographie. III/948 (1970)

Weyl, P.K.: The solution kinetics of calcite. J. Geol. *66*, 163-176 (1958)

White, W.B.: Quartzite karst in south-eastern Venezuela. In: Report on the Angel Falls Expedition. The Netherworld New, USA, *8/3*, 29-31 (1960)

White, W.B., Jefferson, G.I., et al.: Quartzite karst in south-eastern Venezuela. Int. J. Speleol. *2*, 304-314 (1966)

White, W.B., Longyear, J.: Some limitations on speleogenetic speculation imposed by the hydraulics of groundwater flow in limestone. Nittany Grotto Newslett. *10*, 155-167 (1962)

Wiebel, K.W.: Die Insel Kephallenia und die Meermühlen von Argostoli. Hamburg 1874

Wigley, T.M.L.: Analyses of scallop patterns by simulation under controlled conditions; Discussion. J. Geol. *80*, 121-122 (1972)

Wigley, T.M.L.: Speleogenesis: A fundamental approach. Proc. 6th Int. Congr. Speleol. III, Olomouc 1973, 317-324 (1975)

Wigley, T.M.L., Brown, M.C.: The physics of caves. In: Ford, T.D., Cullingford, C.H.D.: The Science of Speleology. London: Academic Press 1976, 329-358

Wilhelmy, H.: Klimamorphologie der Massengesteine. Braunschweig: Georg Westermann 1958

Williams, P.W.: An initial estimate of the speed of limestone solution in County Clare. Irish Geogr. 4, 432-441, 1963

Williams, P.W.: Aspects of the limestone physiography of counties Clare and Galway, West-Ireland, Unpubl. Ph.D. Thesis, Cambridge Univ. (1964)

Williams, P.W.: The Geomorphic Effects of Groundwater. In: Water, Earth and Man. R.J. Chorley (ed.). London: Methuen 1969

Wissmann, H.v.: Der Karst der humiden heißen und sommerheißen Gebiete Ostasiens. In: Das Karstphänomen in den verschiedenen Klimazonen. Erdkunde *8/2* (1954)

Wüst, G.: Verdunstung und Niederschlag auf der Erde. Z. Ges. Erdkunde, Berlin (1922)

Zehender, F., Stumm, W., Fischer, H.: Freie Kohlensäure und pH von Wasser im Calciumkarbonat-Löslichkeitsgewicht. Bull. Schweiz. Ver. Gas- u. Wasserfachmännern *11*, 1-7 (1956)

Zötl, J.: Die hydrogeologischen Verhältnisse im Raume des Buchkogelzuges bei Graz. Steir. Beitr. Hydrogeol. *6*, Graz (1953)

Zötl, J.: Neue Ergebnisse der Karsthydrologie. Erdkunde *11/2*, 107-117 (1957)

Zötl, J.: Beitrag zu den Problemen der Karsthydrographie mit besonderer Berücksichtigung der Frage des Erosionsniveaus. Mitt. Geogr. Ges. Wien, *100*, 101-130 (1958)

Zötl, J.: Zur Frage der Niveaugebundenheit von Karstquellen und Höhlen. Z. Geomorphol., Suppl. *2*, 100-102 (1960)

Zötl, J.: Die Hydrographie des nordostalpinen Karstes. Steir. Beitr. Hydrogeol. *13*, 54-183 (1961)

Zötl, J.: Karsthydrogeologie. Wien: Springer 1974

Subject Index

Aachquelle, Germ. 78, 83, 113 f., 125, 141
Aachtopf, s. Aachquelle
Abisso Enrico Revel, Apuanic Alps, Italy 226
–, Emilio-Comici, Italy 236
–, Michele Gortani, Friaul, Italy 236
Ablösung 144
Active zone 99 f. .
Age determination of sinter 191 f.
Agen Allwed, S. Wales, GB 176, 236
Air circulation 224
–, closed 222
Air in the soil 19 f.
Aire Head Spring, Yorkshire, GB 42
Alternate karst opening (estavelle) 69, 124
Alumada Mine Cave, Mexico 198
Anastomoses 158 f.
Anemolites 194
Anglesite 198
Anhydrite 1 ff., 13
Apatite 179, 198
Aquifer 82, 104 f.
Aragonite 29 f., 181 ff., 184, 188, 198
– caves 237
Ardealite 179, 198
Areuse Spring, Ct. NE, Switz. 124
Aromatic substances (tracers) 140
Atmospheric humidity in caves 214, 226
Attapulgite 198
Augenstein 166
Aulopholite 197
Aven Armand, Causses, Fr. 190 f.
–, d'Orgnac, Fr. 190, 194
Avre (river), Fr. 7
Azurite 198

Bacterium prodigiosum 141
Banded clay 169
Bärenfalle, Tennengebirge, Austria 198
Bärenhöhle, Swabian Alb, Germ. 192
Barite 175, 198
Base level, local 67, 74, 79, 80, 102, 107, 112 f., 116 f., 124 f., 174, 203, 205, 212 f.
Bat guano/excrement 101, 165
Bätterich, Lake of Thun, Switz. 129

Beatushöhle, Lake of Thun, Switz. 159
Bedding cave 234
– interstices 200 ff.
– planes 9, 59, 200, 202 f., 211
Bedding-plane passage 144 ff.
Bell-shaped dome 151, 156, 190
Bergschlag/bump 144
Berome Moore Cave, MO, USA 197
Beudantite 198
Bewetterung 214
Big Springs, MO, USA 125
Binkley's Cave System, IN, USA 236
Bioherm 9, 11
Biolith 9
Biostrom 9, 11
Biphospammite 198
Birnessite 198
Blanchard Mines Cave, NM, USA 198
Blautopf, near Ulm, Germ. 88 f., 124 f.
Blind passage, pocket passage 151 f.
Blind valley 67 f.
Blocksturz 144
Bloedite 198
Blue Grotto, Capri, Italy 233
Blue Spring Cave, IN, USA 236
Bone Breccia 177 f.
Bone Soil 178
Bonheur River, Causses, Fr. 123
Boulder debris 165
Boyden Cave, CA, USA 164
Breakdown/incasion 144 ff.
– caves 237
Breathing Cave, VA, USA 205 f.
Brenztopf, Swabian Alb 124 f.
Brochantite 198
Broken Hill Cave, Rhodesia 198
Brunnengrotte, St. Kanzian, Yugosl. 187
Brushite 198
Bump 144, 147
Butler Cave, VA, USA 201, 205
Butler Sinking Creek System, WV, USA 236

$CaCO_3$-dissolution = limestone dissolution 28–49
$CaCO_3$-gel 181

Calcarenites 7, 10 f.
Calcareous mud 180
Calcareous sinter 42, 47, 75, 101, 180, 184, 185–195
– tufa (Calc-tufa) 47, 180, 184 f.
Calcite 29 f., 181, 186 ff., 195, 197, 243
– clusters (roses) 181, 187
–, floating 187
– ledges 181, 187
California Cave, USA 198
Calumet Cave, AZ, USA 199
Camenitsa/solution pans 27, 57, 60
Canyon 67, 82, 102, 151, 153 f., 205, 209 f.
Carbonate apatite 198
– rocks 1, 4–11
Carlsbad Caverns, NM, USA 13, 45, 196, 198, 209
Castleguard Cave, B.C., Canada 224 f.
Caumont Caves, Fr. 7
Cave 74, 111
–, Aragonite 237
– atmosphere 214
–, bedding 234
–, bell-shaped 151, 156, 190
–, breakdown 237
–, chalk 234
– clay 43, 102, 165, 167, 242
–, commercialized 232
–, contact 234
–, corrosion 233
–, definition of 74, 231
– development 200–213
–, dripstone 237
–, dynamic 215, 224
–, endogenous 233
–, exogenous 233
– fertiliser 178 f.
–, giant (cavern) 235
–, glacier 227, 234
–, granite 234
– grooves 159 f., 169
–, gypsum 15, 234, 237
–, horizontal 98, 231, 234
–, ice 165, 166, 218, 224, 227–230, 237
–, joint 234
– karren 42, 153, 158 f.
–, karst 233 f.
– lakes 117
–, large 235
–, lava 74, 232
– levels 76, 116 ff., 204
– – of the piezometric-surface type (evolution level) 117 f.
–– of the river-bed type 118
– marble = cave onyx 186
–, medium 235

– meteorology 214
– minerals 165, 197 ff.
–, one-cycle 207
– onyx 186
–, open-joint 155, 233
–, overcovering 232
– pearls 186, 194 f.
–, phosphate 237
– phosphates 165, 178 f.
– pinnacles 160
–, primary 232 f.
–, quarzite 234
–, reef 232
– rills 159
– river 75, 82, 102, 124, 241
–, river-bank 233
– river theory 82, 99, 102, 116 f.
–, secondary 233 f.
– sediments 165–197
– –, allochthonous, autochthonous 165
– –, coarse clastic 166
– –, fine clastic 166–177
–, shelter 233
– signs 238–243
–, small 235
– spring 120, 124, 242
–, static 215, 225
–, statodynamic 215
–, subcutaneous 115, 166
–, surf (wave-cut) 233
–, swelling 237
– system 231
– –, dentritic 207
– – network 206
–, tectonic 233 f.
– temperature 223–226
–, tufa 185, 232
–, two-cycle 208, 212
– wind 214 f., 218, 221, 223, 233, 243
Caves, classification of 231
Cavitation 97
Cavity, formation of 42, 74, 211 ff.
–, initial (Urhohlraum) 74
–, primary 73
–, syngenetic 73
^{14}C-dating 184, 191 f.
Ceiling channels 159
– dents 159, 161
– dimples 159
– halftube 40, 155
– – passage 151, 155, 213
– karren 158
– pockets 159 ff.
Cenote 50, 62, 64
Cerussite 198
Chalk 7

Chalk Cave 234
Chamber (Hall) 151, 156
Chimney 151, 156, 241
Chlorite 175
Christobalite 198
Chromite 175
Chromobacterium violaceum 141
Cimolite 198
Clay furrows (surge marks) 176
— minerals 174 f., 198 f.
— scales 175 f.
Clints 50, 59
CO_2, biogenic 19, 52, 56, 66
—, combined 33
—, diffusion coefficient 19, 25
—, free 33
—, mechanochemical 45 f.
—, partial pressure 18, 22
Cockpit 50, 65
— karst 65, 71
Coelestine 175, 198
Cold period 174
Collapse shaft 156
Condensation water 75 f., 101, 159, 217, 226
Contact cave 234
Conulite 175, 191
Convection currents 214, 216 f.
Copper Queen Cave, AZ, USA 198
Coral formation 192 f.
Corrosion = limestone dissolution 28–49, 100, 151, 160, 164, 212, 217, 233
— caves 234
—, cooling 38–41, 117, 152
—, forms of 158
—, inverse 153
—, lateral 70
—, mixing 35–41, 44 f., 47, 64, 74, 110, 117, 152 f., 158, 161, 201, 211
—, normal 28–35, 41–44, 158
—, paradox of mixing 44
— plains 50, 66, 70
—, pressure 41 f.
—, retrogressive 55
—, tables of 58
—, thermal mixing 40 f.
—, upward 153
Crandallite 179, 198
Crevice Cave, MO, USA 235
Csoklovina Cave, Romania 179, 198
Cueva de los Cristales, Venezuela 198
— del Indio, Venezuela 198
— de los Ursulas, Venezuela 199
— de los Verdes, Lanzarote 232
Culverson Creek Cave System, WY, USA 236

Cumberland Caverns, TN, USA 235
Curve of discharge 136
Cyanotrichite 198
Cycles, theory of (Davis) 103, 211

Dachstein Caves, Austria 99, 153, 199, 205, 224, 227, 229, 236
Dachstein Ice Caves 224, 227, 229
Danube sinks, water loss/seepage, Swabian Alb, Germ. 78, 113, 138 f., 141
Deckenbruch 144
Deckensturz 144
Deckenzapfen = pendants 158
Deep phreatic 105 f., 110, 212
Demänovské Jaskiňe, ČSSR 116, 118, 194
Detergents 140
Diadochite 179, 198
Diana Cave, Romania 199
Diapir = salt dome 3
Discharge 135
Dissolution, chemical 12, 15 ff.
—, physical 12 ff.
Dobšinské L'adová Jaskyňa, ČSSR 160, 215
Doline 60–63, 65, 74
—, alluvial 61
—, asymmetrical 62 f.
—, bowl-shaped 62
—, collapse 3, 50, 61
—, density of 61
—, funnel-shaped 62, 63
—, karren 62
—, kettle-shaped 62
—, large 63
—, secondary 62 f.
—, shaft 62
—, small 63
—, solution 61
—, subsidence 3 f., 61
—, symmetrical 62
—, trough-shaped 62
Dolomite 1 f., 4–11, 198
Dolomitic sand 2, 6
Dome 151, 156
Dome Home Cave, KY, USA 160
Domica Baradla Cave, ČSSR/Hungary 44, 191, 236
Double passage 151 f.
Douglas Cave, N.S. Wales, Australia 174
Drachenhöhle, Mixnitz, Austria 178, 199
Drainage, diagram of 127
Drapery 187, 189
Drifting material 140 f.
Dripstone 185, 187–195
— cave 237
— column 191

Dripstone, hanging 187
–, standing 190 f.
Dry valley = Karst dry valley 7, 50, 66 ff.
Dürschrennenhöhle, Switz. 198

Easegill Caverns, GB 235
Ebb and Flow Spring, MO, USA 126 f.
Eccentrics (Excentriques) 193 ff., 243
Eisenblüte (Iron blossoms) 184
Eisriesenwelt, Salzburg, Austria 45, 99, 166, 187, 198, 205, 224, 227, 229 f., 235
El Capote Cave, Nueva Leon, Mexico 199
Elaterite 198
Elliptical passages 152, 154, 213
Emergence 123
End siphon 242
Endellite 198
Endless Caverns, VA, USA 159
Endokarst 73–76, 224
Epsomite 197 f.
Equation of continuity 85–89
Erosion 49, 116, 144, 152, 160 f., 164, 212, 233, 234
Estavelle, alternate karst opening 69, 97, 124, 157
Evaporite 1 ff.
Evolution level 116 f., 205, 207
Exokarst 50–72, 74
Exsurgence 123
Exterior stalactite 189

Facets 14
Feeder 99, 101, 169, 205, 208 f., 212
Feengrotten, Saalfeld, GDR 198
Ferghanite 198
Fern Cave System, AL, USA 236
Fictitious table 83
Firn cave 228
Flint 7 f., 166
Flint Mammoth Cave, KY, USA 10, 44, 84, 116, 118, 204 f., 235
Flint Ridge Cave, KY, USA 154, 196, 199
Flowstone 186 ff.
Fluorapatite 179, 198
Fluoresceine 139
Fluorite 198
Fontaine de Fontestorbes, Ariège, Fr. 128
Fontaine Ronde, Jura, Fr. 128
Fontaine de Vaucluse, Fr. 121, 124 f., 157
Foot cave 57
Frost debris 165 f.

Galenite 198
Garland passage 151 f.
Garma Ciaga-Sumidero de Cellagua, Santander, Spain 236

Gelber Brunnen, Lake of Thun, Switz. 142 f.
Geldloch, Ötscher, Austria 228
Geoisotherms 224
Geological organ 58
Geothermal gradient 224
– heat 225
Ghetural de la Scarisvara 227
Giant caves 235
Giobertite 198
Glacial karst 50, 59, 68
Glacier cave 227, 234
Glacière 227 f.
Goethite 175, 198
Gorge-shaped passage 153
Gouffre André Touya, Pyr., Fr. 236
– d'Aphanié, Fr. 236
– Berger, Vercors, Fr. 236
– d'Enversac, S. Fr. 134
– de la Pierre St. Martin, Pyr., Fr. 236
– de Poudak, Htes. Pyr., Fr. 128
– du Cambon de Liard, Pyr., Fr. 236
– Touya de Liet, Pyr., Fr. 236
Gours 187
Gran Caverna de Santo Tomás, Cuba 236
Grand Caverns, VA, USA 191
Granite caves 234
Granulometry 168, 171
Gravitational channel 204, 207
– flow 82
Great Cave, KY, USA 199
Greenbrier Caverns, WV, USA 8, 44, 235
Grikes 58, 68
Grooving 169
Grotta del Monte Cucco, Umbria, Italy 236
– del Vento, Prov. Ancona, Italy 176, 197
– di Castellana, Apulia, Italy 61, 160, 190
– di Padriciano, Triest, Italy 226
– Gigante, Triest, Italy 226, 235
– Smeralda, Amalfi, Italy 233
Grotte de Bèze, Jura, Fr. 135
– de Bramabiau, Causses, Fr. 123 f.
– de la Clamouse, Fr. 188
– des Demoiselles, S. Fr. 191
– La Cigalère, Ariège, Fr. 197
– di Frassassi, s. Grotta del Vento 176, 197
– des Grès, S. Algeria 199
– de Han, Fr. 191
– di Minerve, Fr. 199
– de Remouchamps, Belgium 119
– du Bournillon Vercors, Fr. 124 f.
Groundwater 102, 111–114
– theory 83, 99, 102, 107, 111, 117
Gruberhornhöhle, Salzburg, Austria 236
Guanite 198
Guano 177, 178
Guanovulite 198

Subject Index   265

Gypsum 1 ff., 13 ff., 111, 165, 195 ff., 198, 243
− cave 14, 234, 237
− − flowers (oulopholites) 196
−, deposition of 14
− karren 3
− karst 13
− stalactites 196
− starbursts 196

Halloysite 198
Halotrichite 198
Havasu Canyon Cave, AZ, USA 199
Heat, the earth's 214
Heavy minerals 174 f.
Heimkehle, GDR 14
Helictite 181, 193
Heligmite 193
Hematite 198
Hemimorphite 198
Herman Smith Cave, IL, USA 199 ff.
Hexahydrite 198
Hibbenite 198
High-mountain karst 41
Highwater zone 93, 99 ff., 102, 117, 135 f., 148, 169, 205
Hjulström Curve 168, 171, 173
Hoar-frost 226
Hochleckengroßhöhle, Höllengebirge, Austria 236
Höhlensystem der Sieben Hengste 236
Höllgrotte, Baar, Switz. 185, 232
Hölloch, Central Switz. 10, 37, 39, 42 ff., 45, 53, 80, 84, 93−98, 110, 116 ff., 120, 125, 135, 137, 143, 147 f., 150, 152, 157, 159 ff., 167, 170 f., 174−177, 187, 196, 198, 203, 205, 207, 218 ff., 224, 235 f.
Hopeite 179, 198
Horizontal caves 98, 152, 231, 234
Horseshoe falls effect 55
Hoyos del Pilar, Spain 236
Hudson Bay Cave, B.C. Canada 199
Hum (pl. Humi) 69
Humus-water grooves 56
Hungerbrunnen, Swabian Jura, Germ. 121, 128 f.
Huntite 198
Hydrocalcite 198
Hydrographische Wegsamkeit 74
Hydrological perviousness 74, 77, 201 f.
Hydromagnesite 198
Hydroxylapatite 179, 198
Hydrozincite 198

Ice 198, 227−230
Ice cave 165 f., 218, 244, 227−230, 237

−, dynamic 228 f.
−, static 228
−, statodynamic 230
Ice-water grooves 160
Idzuk Spring, Bihar Mountains, Romania 128
Illite 175
Inactive zone 99 f.
Incasion/breakdown 65, 118, 144−150, 151 f., 156, 158, 205, 212, 233
− debris 166
Ingleborough Cave, GB 191
Initial cavities 74
− network 203 f.
− phase 42, 74, 212
Interface 15, 17, 25 f., 180
Interior valley 71
Interstice 99, 153, 187, 201 ff., 208 f.
−, capillary 200, 202
−, widening of 208−211
− passage, steep 213
Inverse ceiling channels 159
Inverse solution pockets 160, 242
Ion product, limestone 17 f.
Ions, other in solution 33 ff., 182, 184
Island Ford Cave, VA, USA 198

Jarosite 198
Jasper Cave, WY, USA 199
Javořičko Caves, ČSSR 198
Jewel Cave, SD, USA 235
Joint 10, 200, 211
− cave 234
− groundwater 112
− interstice 201, 203, 213
− network 211
− passage 144 f., 151 f., 154 f., 213
− −, fictitious 206, 212
− tectonic 149

Karren 7, 14, 50−60
− cave 158 f.
−, cavernous 58
−, ceiling 158 f.
−, covered 57
−, debris 59
− field 50, 52, 59
−, First (Austrian) 54
−, free 53−56
−, half-exposed 56
−, heel-print (Trittkarren, Germ.) 52, 55
−, meandering 56
−, Rillen- (German) 54
−, root 58
−, round 58
− spires 50

Karren, surf 59
– table = corrosion table 48, 58
–, undercut 50, 57
–, wall 50, 56
–, wave 58
Karst, alpine 48
–, bare 47, 53–56, 58
– barré 114
–, blocked 115
– cave 234
– cavities 116
–, cone 57, 65, 71
–, cuesta-like 50, 68
–, cycle of development of 211
–, deep 67, 81 f., 99, 106, 117
– denudation 46–49, 58 f.
–, dry-valley 67 f.
–, endo- 73–76
–, exo- 50–72, 74
–, fluvial 50, 66 f.
–, glacial 50, 59, 68
–, green 44, 117
–, groundwater 82 f., 106, 110, 112 f.
– gulf 64
–, high-mountain 48
–, hydrologically coupled 97 f.
– levels, underground 116–119
– phenomenon, underground 80
– plains, marginal 66
–, pseudo- 12, 73
– river 66 f.
–, shallow 67, 81 f., 99, 101, 113
–, silvan 42, 48, 58, 63
– spring 7, 67, 78, 81, 120–137
– –, rising 88
– –, submarine 102
–, stepped pavement 50, 68
–, superficial 50–72
–, syngenetic 73
– trough 3, 66
– valley 50, 66 f.
–, volcanic 73
– water 102, 111–114
– – body 67, 83 ff., 98, 101, 103, 107, 110, 112, 120, 126
– – surface, permanent 75, 81, 203, 234
– window 64
Karst-hydrological contrast 120
Karst-hydrologically active 74, 77, 99, 114, 137, 207
Katerloch, Styria 187, 192
Kettle-shaped dolines 62
Keyhole profile 153 f.
Kievskaja ou Kilsi, Pamir, USSR 236
Köhbrunnen, Hallstättersee, Austria 129

Kotliči 62
Krka Falls, Yugosl. 66, 185

Labyrinth 14, 151 f., 203
La Calmouse, S. Fr. 194
La Cigalère, Ariège, Fr. 198
Lamprechtsofenhöhle, Lofer, Austria 107, 236
Large caves 235
Lauiloch, Muotatal, Switz. 53
Lava-Tunnel, Lava Cave 74, 232
Lavabeds National Monument, CA, USA 232
Law of outflow 88 ff.
Leconite 198
Lee Cave, KY, USA 198 f.
Lehman Cave, NY, USA 194
Lehman Caves, NV, USA 198
Le Moulis, S. Fr. 199
Leonhardshöhle, Homburg, Germ. 199
Lettenmayerhöhle, Austria 233
Leucoxene 175
Lilburn Cave, CA, USA 128, 186, 199
Lime balance 47 f.
Limestone 1, 4–11
– deposits 180 ff.
–, freshwater 6
–, porous 10, 101
– removal/denudation 46–49, 58 f.
– table 58
– tufa 184
Limonite 196, 198
Lithophorite 198
Lookout Cave, WA, USA 198
Long Cave, KY, USA 208
Lost River Cave, IN, USA 175
Lublinite 181
Luray Cave, USA 178
Lycopodium spores (tracer) 140 f.

Magnesite 5, 199
Mahorčičeva Jama, Slov., Yugosl. 113
Malachite 199
Malograjska Jama, Slov., Yugosl. 124
Mammoth Cave, KY, USA 158, 171 f., 175, 178, 196 f., 199
Markasite 196
Marl 7
Martinite 179, 199
Mass transfer 15, 24
Massenkalke (unbedded limestone masses) 15
Matlock, England 198
Maturity, phase of 212
Medium cave 235
Medusa dripstone 187
Meerschaum 199
Meigen test 181

Melanterite 198
Micrite 9, 11
Microgours 187
Miller Cave, TX, USA 198
Miller Spring, MO, USA, 127
Minervite 199
Mirabilite 197, 199
Misenite 198
Mixing corrosion 35–41, 44 f., 47, 64, 74, 110, 117, 152 f., 158, 161, 201, 211
–, paradox of 44
–, thermal 40 f.
Moaning Cave, CA, USA 192
Mondloch, Lucerne, Switz. 195
Monetite 179, 199
Monohydrocalcite 199
Montmorillonite 175, 199
Mud stalagmite, mud pyramid 175

Negra Cave, Puerto Rico 199
Nesquehonite 199
Network caverns 206
– of cavities 112
New Cave, Ireland 191
New Cave, NM, USA 178, 191
Newberyite 179, 199
Nicajack Cave, TN, USA 199
Nitrammite 199
Nitrate soils 165
Nitrocalcite 199
Nitromagnesite 199
Noiraigue, Neuenburg, Switz. 135
Norman Bone Cave, WV, USA 192
Numburghöhle, Kyffhäuser, GDR 14

Ogof Agen Allwed, S. Wales, G.B. 176, 236
Ogof Ffynnon Ddu, S. Wales, G.B. 176, 235
Ojo Guareña, Burgos, Spain 235
Old age, phase of 212 f.
Olivenite 199
Ombla Spring, Dubrovnik, Yugosl. 124
One-cycle caverns 207
Opal 2, 7, 199
Open joint 99, 200, 202 f., 211
– – caves 155, 234 f.
Optimističeskaja Peschtschera, Podolia, USSR 14, 235
Oulopholites 196
Overcovering caves 232
Ozernaja Peschtschera, Pod., USSR 14, 235

Pajares Cave, Puerto Rico 198
Palm stem stalagmite 191, 192 f.
Palygorskite 199
Parahopeite 199

Parau, Iran 236
Passage 151–156
–, bedding-plane controlled 152
–, blind (pocket) 151 f.
–, ceiling half-tube 151, 155, 213
– cross-section 44, 117, 151, 206 f., 213
–, development of 213
–, double 151 f.
–, elliptical 152, 154, 213
–, Garland 151 f.
–, joint 151 f., 154, 213
– network 203
–, rectangular 153
–, shaft 151, 153 f., 213
–, steep interstice 213
Pendants 158
Peneplain 118, 204
Percolation zone 136
Permafrost 115
Permeability barrier 104 f.
Perviousness, primary 74
Phosphate cave 237
– soils 177, 179 f.
– weathering 179
Phreatic conditions 102, 113
– zone 82 ff., 99–110, 111, 113 f.
– –, shallow 102
Phytogenic sediments 177 f.
Pickeringite 199
Piedras Cave, Honduras 198
Piezometric surface 83, 100 f., 104, 135, 171
– tubes 83, 100, 102, 110, 124, 129, 171 f.
Pig Hole Cave, VA, USA 199
Pigotite 199
Pinargözü Cave, Turkey 215, 221
Pinnacles 50, 59
Pit 156
Pivka Jama 177
Platteneckeishöhle – Bergerhöhle, Austria 236
Pocket caves, descending 228
– valley 67 f.
Polje 50, 65, 68–72, 97 f., 157
–, basin 71
–, high-surface 71
–, marginal 71
–, open 69
–, over-flow 71
–, para- 72
–, river 67
–, semi- 71
–, storage 71
–, trough 71
–, valley 71
Pollen analysis 184
Ponor 7, 66, 69, 124

Pores 10, 111 f.
Port Miou, S. Fr. 130
Postojnska Jama, Yugosl. 45, 47, 119
Potholes 233
– due to erosion 161, 242
Pressure flow 82 f.
– gradient 211
–, loss of 89–97
– -sheet 147
– /tension dome 146, 150
– wave 135
Primary caves 232
– interstices 43 ff.
Prinzenhöhle, Swabian Alb, Germ. 192
Pseudorinnenkarren 12
Psilomelane 199
Puits du Pot II, Fr. 235
Puits Lépineux, Fr. 235
Pyrite 196 f., 199
Pyrrhotine 199

Quartz 199
– pebbles 166
– sandstone 7
Quartzite 1, 7, 73, 166
– cave 1, 234
– karst 1

Radio isotopes (tracer) 139 f.
Rainbow Spring, FL, USA 125
Rampert Cave, Grand Canyon, AZ, USA 178
Raufarholshellir, Iceland 232
Rectangular passages 153 f., 164, 213
Reef cave 12, 232
– limestone 11, 73
Removal 168
Réseau Ded, Chartreuse, Fr. 236
– de la Dent de Crolles, Isère, Fr. 235
– de la Pierre St. Martin, Pyr., Fr. 236
– des Aiguilles, Dévoluy, Fr. 236
– du Vernau, Fr. 236
– Félix Trombe, Hte. Garonne, Fr. 236
Resurgence 123 f.
Rhumequelle, Harz, Germ. 125
Rimstone 186 f.
– pool deposits 186 f.
– pools 186
River-bank caves 233
Rock salt 1, 3 f., 13 f.
– scales 147
Roof created by dissolution 14
Runoff channels 161
Rushmore Cave, SK, USA 188
Rutile 175

Salmoite 199
Salt 199
– dome 3
– karst 4, 12
– -water swallow-hole 131 ff.
Saw-tooth sinter 189
Scallops 102, 158 ff., 161–164, 242
Schleichender Brunnen, Muotatal, Switz. 80, 84, 93, 97, 120, 124 f.
Schlenkendurchgangshöhle, Salzburg, Austria 76
Scholzite 179, 199
Schwyzer Schacht, Muotatal, Switz. 18
Sea Cave, Cornwall, GB 199
Sea Mills, Argostoli 131 ff., 133
Secondary cave 233
Secondary minerals 198
Sediment profile 173 f.
Sedimentation, conditions of 168–173
Sediments, biogenic 101
–, cave 165–199
–, chemical 165, 180–199
–, clastic 165–177
–, coprogenic 165, 178
–, phytogenic 165, 177
Seeping water 74, 82, 99, 101, 113, 164
–, zone of 112, 136
Segeberg, Germ. 2
Selenite 199
Senility (phase of) 212
Sepiolite 199
Serratia marcescens 141
Shaft 1
–, collapse 156
–, piezometric 157
–, spring 157
–, stepped 156
–, swallow 157
Shallow phreatic zone 102 f., 110, 116, 211
Shelter cave 233
Shields 192 f.
Shor-Su Cave, Usbekistan, USSR 199
Sierra de los Organos, Cuba 71
Silicate rocks 1, 12
Siliceous limestone 7
Silicic acid 1, 7 f.
Silvan karst 42, 48, 58, 63
Silver Spring, FL, USA 47, 110, 125
Sima de la Pena Blanca, Santander, Spain 236
Sinter, calcareous 42, 118, 160, 165, 185 ff., 191, 243
– draperies 189 f.
– flags 189 f.
–, sawtooth 189
– tubes 188

Sinter, white stripes 187
Siphon 124, 160, 170, 208, 222, 242
Skelskoy Caves, Krim, USSR, 198
Skewness 171 f.
Skipton Cave, Australia 199
Škocjanske Jame, Slov., Yugosl. 61, 113, 138
Sloans Valley Cave System, KY, USA 235
Small caves 235
Smithsonite 199
Snieznaja Schachta, Caucasus, USSR 236
Sodaniter 199
Solution flutes 8, 54 f.
– grooves/Rinnenkarren 50, 56 f., 60
– notches 57
– – (in caves) 101
– – (on surface) 57
– pans (Camenitsa) 27, 57, 60
Sotano de los Golondrinas, Mexico 235
– de Tinaja, Mexico 192
Spangolite 199
Sparite 9
Speak-easy Cave, MO USA 8
Speleogenesis 144, 200–213
Speleometeorology 214–226
Speleomorphology 151–164
Speleothems 101, 153, 161, 185 ff., 192
Spencerite 179, 199
Sphalerite 199
Splash-cup stalagmite 190 f.
Spluga della Preta, Alpi Lessini, Italy 236
Sponge reef 10
Spring alluvial karst 121
–, ascending 121
–, bedding 121
–, brackish-water 130 ff.
–, cave 120, 124, 242
–, contact 121
–, descending 121
–, discharge of 134 f.
–, ebb and flow 121, 125–128
–, episodic 121
–, fracture 121
–, freshwater 130
–, groundwater 124
–, high-water relief 124 f.
–, intermittent 121, 124, 127
– on bedding joints 121
–, overflowing 121
–, perennial 121
–, periodic 121, 125 f.
–, rock 121
– shafts 157
–, subaqueous 129–134
–, sublacustrine 124, 129
–, submarine 124, 129–133

–, talus 124
– tufa (calc-tufa) 184
–, vauclusian 123 f.
–, waller (bubbling) 129
Stalactite 130, 187–189, 193
–, exterior 189
–, gypsum 196
–, maccaroni 188
–, silicic 1
Stalagmite 102, 190 f.
–, mud 175
–, negative 191
–, palm-stem 191 f.
–, splash (pile d'assiettes) 190 f.
Storeyed structure of underground karst levels 116
Struvite 170, 199
Submarine karst cave 130
– – spring 102
Sulforhodamine 139
Sulphur 199
Sumidero de Cellagua, Santander, Spain 236
Summer-ice theory 227
Surf caves 233
Surface, piezometric 83, 90, 100, 110, 117, 155
Swallow-hole, sink 69, 75, 78, 97, 106, 107, 121, 124
Swelling cave 237
Syngenite 199
Systema cavernario de los perdidos, Cuba 236

Tafonis 12
Tantalhöhle, Salzburg, Austria 99, 101, 205, 236
Taranakite 179, 199
Tarbuttite 179, 199
Temperature gradient 39
Tenorite 199
Tension dome 148, 150
– joints 148, 233
Teufelsbrunnen near Opatija, Yugosl. 132
The Hole, WV, USA 236
Thenardite 197, 199
Thermal spring 39, 77, 110
Through-cave 215
Timavo, Triest, Italy 125, 138 f.
Tintic Cave, UT, USA 198 f.
Tinticite 179, 197 f.
Titus Canyon Cave, CA, USA 198
Tosbeckenhöhle 233
Tourmaline 175
Tracer 138–141
Travertine 185 f.
Trihydrocalcite 199

Tritium 139
Tube-flow 112
Tufa barriers 185
Turanite 199
Two-cycle cavern 208, 212
Tyuya Muyun Cave, USSR 197 f.
Tyuyamunite 199

Uranine 131, 139, 142
Urhohlräume (initial cavities) 74

Vadose zone 81, 99 f., 101 f., 113
Vanadinite 199
Variscite 179, 199
Varved clay 169
Vaterite 29, 181 f.
Vauclusian spring 25 f., 88, 110, 123 ff.
Velocity of flow, lineal 78 f., 93, 107, 142
Verbruch 144
Vermiculations 176 f.
Versturz 144
Vertical cave 234
– – parts 156, 235
– drainage 99
– shaft 157, 161 f., 207
Vesicles 232
Volcanic karst 74
Voûte mouillante 82
Vrulje 130–133

–, brackish-water 130 f.
–, fresh-water 130
Vrtača = doline 61
Vypustek Cave, Mähren, CSSR 179

Waldschmiede, GDR 3
Wall pockets 159
Water, condensation 75 f., 101, 159
– in the soil 75
–, precipitated 75
–, seeping 74, 82, 99, 101, 112, 164
Waterfall 222 f., 242
Webers Cave, IA, USA 198
Welbeck Colliery, GB 199
Whitlockite 179, 199
Wielka Sniezna, Tatra, Pol. 236
Wilson Cave, NV, USA 199
Wind cave 233
Wind Cave, SD, USA 197 ff., 235
Wind velocity 220
Winter-ice theory 227
Wyandotte Cave, IN, USA 197 f.

Youlgrave Cave, GB 198
Youth, phase of 212 f.

Zbražow Cave, ČSSR 199
Zirkon 175
Zwerglöcher, GDR 3

Plates

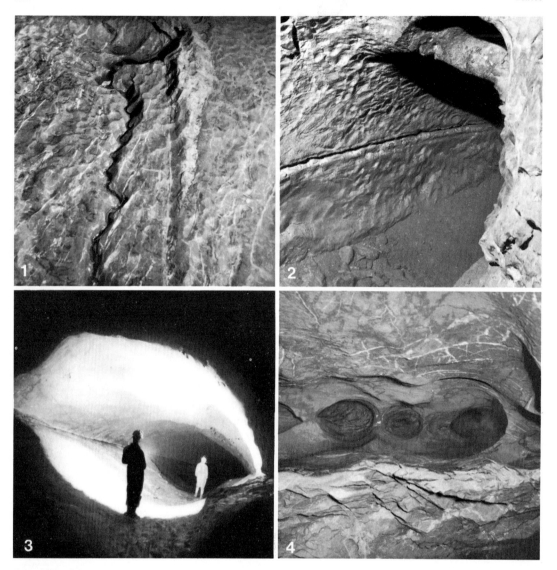

Plates 1.1–1.4

1 Normal corrosion in a drainage channel on the floor of Rabengang. The water has flowed under phreatic conditions for approx. 1 km to the point of emergence without dissolving any more limestone. In contact with the cave's atmosphere (0.035%–0.12% CO_2) it absorbs CO_2 and corrodes the rock. Hölloch (Switzerland), 800 m below the surface

2 Corrosion notch of a former cave lake 400 m below the earth's surface. Here, too, the water arrived after passing through phreatic conditions and therefore without having dissolved limestone (closed system Me-CO_2-H_2O). CO_2 is absorbed from the cave's atmosphere through the water's surface, so that the top layers of water have the highest concentration of CO_2 and therefore the strongest corrosion. Hölloch, 1001 Nacht

3 Bedding-plane passage dropping with a dip of 15% in the direction shown. Corrosion on the ceiling, also, thus without air and hence not normal corrosion. The passage is explicable only by mixing corrosion. Hölloch, SAC-passage 700 m beneath the earth's surface

4 Inverse solution pockets (ceiling pockets) are created by mixing corrosion on joins where joint water rich in lime emerges into passage water with a lower lime content. Hölloch, Titanengang, 600 m below the surface

Plates

Plates 2.1–2.4

1 Rillenkarren or solution flutes with solution levels at the lower end (Muota Valley, Ct. Schwyz, Switzerland), 2000 m above sea level, Schrattenkalk (Urgonian)
2 On the edge where no used water flows over the limestone surface from above: solution flutes; under them 5–10 times wider solution flutes of the second order, gathering together several of the smaller ones. Belonging to the second morphogenic type of corrosive effect. Mallorca near Lluc, Early Tertiary limestone
3 Heelprint karren with solution level. Glattalp in the hindmost catchment area of the Muota (Ct. Schwyz, Switzerland), 1850 m above sea level, Schrattenkalk (Urgonian)
4 Meandering karren, created by run-off water without area-wide sprinkling. Twärenen (Muota Valley, Ct. Schwyz, Switzerland), 1900 m above sea level, Schrattenkalk (Urgonian)

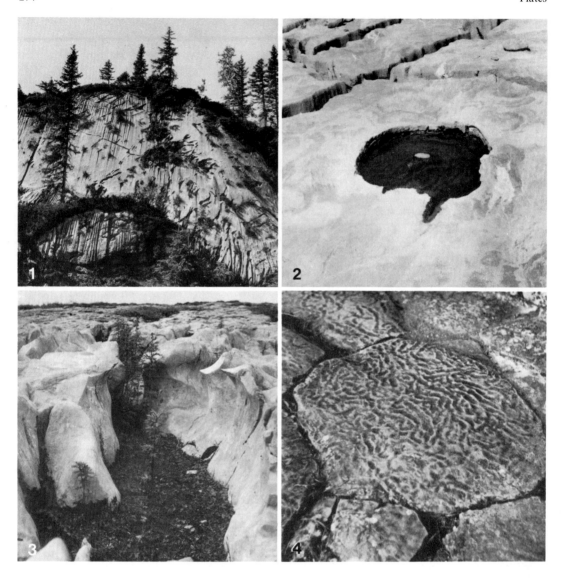

Plates 3.1–3.4

1 Wall karren, caused by run-off water with or without area-wide sprinkling. Bödmerenwald (Muotatal, Ct. Schwyz, Switzerland), 1600 m above sea level, Schrattenkalk (Urgonian)

2 Solution pan (Kamenica) in Quintnerkalk (Upper Malm) of Glattalp (hindmost Muotatal, Ct. Schwyz, Switzerland), 1800 m above sea level

3 Undercut karren characterized by overhanging flanks, caused by humus in the ground with a high content of biogenic CO_2 in the soil's air. Glattalp (hindmost Muota Valley, Ct. Schwyz, Switzerland), 1850 m above sea level, Malm limestone (Quintnerkalk)

4 Root karren, root marks on marble slabs created by corrosion, in the Tempel region of Paestum south of Salerno, (Italy)

Plates 4.1–4.4 (see pp. 276)

1 Round karren form from solution grooves under a covering of soils as well as in other ways – in the picture they are combined with undercut karren, Bödmerwald (Muota Valley, Ct. Schwyz, Switzerland), 1650 m above sea level, Seewerkalk (Upper Cretaceous)

2 Pinnacles created by the cutting away of the side walls by adjacent solution grooves and grikes. They are rare in the limestone Alps, on the other hand frequently found in Pleistocene karst regions which have not known glaciers. Up to 2 and more m high. Kaiserstockkette (Misthaufen, 2230 m above sea level) in the vicinity of the peak. Schrattenkalk (Urgonian)

3 Clints, the base of pinnacles which have been broken off by glaciers. Subsequently various small karren such as solution flutes and heelprint karren among others have been superimposed on them. Schrattenfluh (Ct. Lucerne, Switzerland). 1800 m above sea level

4 Network of grikes in the Quintnerkalk (Upper Malm) of the Märenberge (2200 m above sea level) north of the Klausen Pass (Ct. Uri, Switzerland)

Plates 5.1–5.4 (see pp. 277)

1 Stepped pavement karst with large joint which has been opened by corrosion. In thickly layered Quintnerkalk (Malm) narrow shaft dolines (karren dolines), in thinly layered, easily weathered limestone, funnel-shaped dolines. Märenberge, 2250 m above sea level, north of Klausen Pass (Ct. Uri, Switzerland)

2 Karren tables in the cuesta-like karst of the Märenberge (see Plate 5.1)

3 Funnel-shaped dolines in the alluvials overlying the Quintnerkalk of the Charetalp (Muota Valley, Ct. Schwyz, Switzerland), 1850 m above sea level

4 Stepped pavement karst develops in ± horizontally layered limestone. Quintnerkalk (Upper Malm) of Glatten Mountain to the north of Klausen Pass (Ct. Uri, Switzerland), approx. 2200 m above sea level

Plates 4.1–4.4 (Captions see p. 275)

Plates 5.1–5.4 (Captions see p. 275)

Plates 6.1–6.4

1 Incasion. The layers of the ceiling break under their own weight; the first phase develops as tension cracks. Place of measurement in Berom Moore Cave, on the edge of the Mississippi Valley, carboniferous limestone, Missouri, USA

2 Junction effect; breakdown of the ceiling at the intersection of large passages in horizontal limestones until a dome-like form of equilibrium is reached. In the middle a ceiling crack as preliminary stage of a new process of incasion. Large hall in the Historic Trip of Mammoth Cave, Kentucky, USA

3 Under the pressure of the overlying rock, pressure plates are created, which break off independently of the position of the bedding, or are forced away as bumps. Bump slab in the Schrattenkalk of Hölloch (commercial part). 400 m of overlying rock which corresponds to a pressure of 1000 t/m²

4 The rock layers on the right are remainders of pressure slabs which have fallen down; proof of previous incasion. Hölloch, Rabengang, 800 m below surface

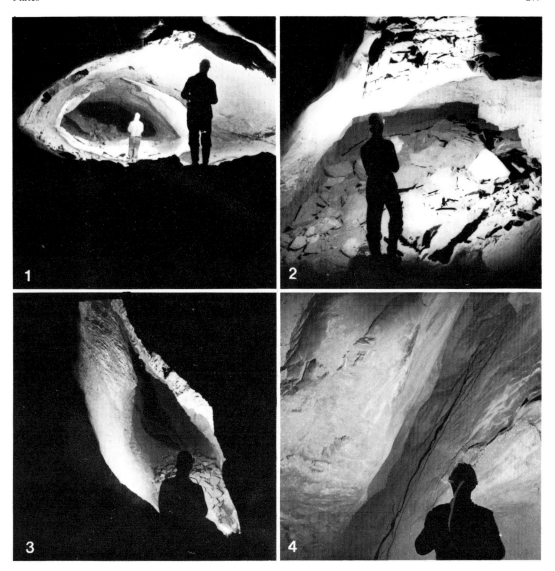

Plates 7.1–7.4

1 The formation of pressure slabs and bumps does not hinder retention of the elliptical form of the passage's cross-section, but the upper half is more arched than the lower. Occasionally there is breakdown debris lying on the floor. Hölloch (Switzerland), Lehmschollengang, approx. 700 m below surface

2 If the ceiling caves in under its own weight (rooffall), the original elliptical cross-section is destroyed. In addition, breakdown debris covers the floor's primary forms. Hölloch, SAC-passage, approx. 600 m below surface

3 Joint passage formed under phreatic conditions: steep elliptical cross-section. Hölloch, close to Otterkamin, lower level

4 Joint, widened under vadose conditions to a joint passage. Hölloch

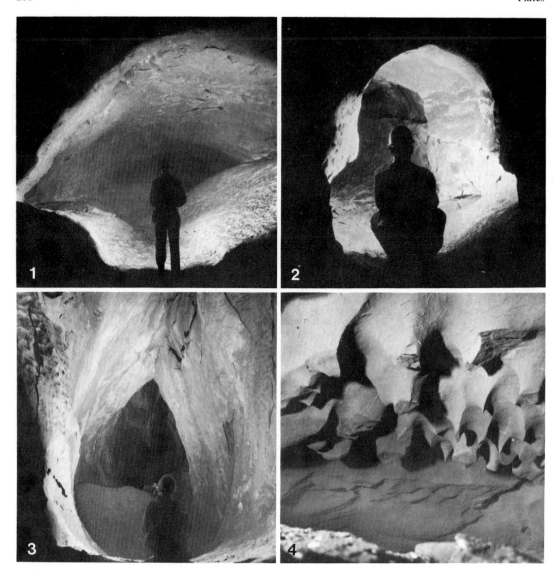

Plates 8.1–8.4

1 Two-cycle passage's cross-section. In a later vadose phase a stream cut into the floor of a bedding passage with a phreatic elliptical cross-section. Such a cut can be many meters deep if the cave stream has been active for a long time (canyon). Hölloch (Switzerland), SAC-Passage

2 Ceiling half-tube passage. Superimposed on the phreatic ellipse is a semi-circular ceiling half-tube, created by corrosion resulting from compression of air bubbles carried in the water. The ceiling half-tube passage in the picture is extremely strongly developed. Hölloch, passage branching off at the Geröllschloß

3 Joint passage with elliptical cross-section of phreatic origin. The water carried sand and gravel and widened the lower parts of the cross-section with them by erosion. Gotischer Gang in Tantal Cave, Hagengebirge, Land Salzburg, Austria

4 Pendants on the roof in the Schrattenkalk near the Sphinx, Hölloch

Plates 9.1–9.4

1 Erosion pothole cut open, with central axis, which can only form under phreatic conditions – no aspiration of air. Hölloch (Switzerland), commercial part

2 The Wasserdom in Hölloch fills up during high water as high as the *dark horizontal line* in the background. The waterfall plunging down onto the water's surface from a height of more than 20 m creates a whirlpool with a horizontal axis. With the aid of sand and gravel the foot of the wall has been hollowed out

3 Scallops. They are formed by erosion as well as by corrosion. The steep side (*shaded*) of the scallop points in the direction the water flows away, thus from left to right; they are therefore an aid in determining the direction in which water formerly flowed. Oasis in Hölloch

4 Shrinkage cracks in cave clay, which loses water through evaporation even when the air's humidity is over 95%. Hölloch, rear SAC-passage

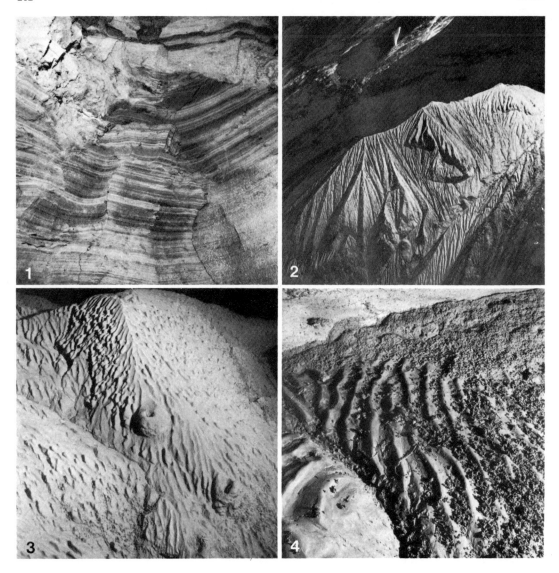

Plates 10.1–10.4
1 Banded clay, partly clayey silt (*dark*), partly silty clay (*light*). Every pair of layers indicates a high-water phase. Hölloch (Switzerland), Papageienkammer
2 Deposits of silty cave clay on a boulder. The flutes were created during sedimentation by the circulation of the water resulting from differences in density (varying amounts of material in suspension). Hölloch, high system, rear east passage
3 Mud stalagmites, created by dripping water from the ceiling of the passage, which splashes outward the mud deposited even in the splash-cup by every high-water phase. Hölloch, rear SAC passage
4 Ripple marks with superimposed worm heaps. The mud deposited during the high water is consumed by lumbricides (*Octolasium transpadanum* ROSA) and the indigestible parts are excreted as worm heaps. Hölloch, Schlundgang section

Plates

Plates 11.1–11.4

1 In motionless cave ponds with water rich in lime clusters of calcite (calcite roses) can form with larger scalenohedral crystals, if CO_2 is given off into the cave's atmosphere. Hölloch (Switzerland)

2 Ideally formed pair of dripstones; above the stalactite of the second type, below the stalagmite of the candle type (constant cross-section from top to bottom). Rübli, the carrot, Hölloch, Himmelsgang

3 *Left* sinter tubes with few eccentrics; *middle* sinter tubes overgrown with eccentrics; *right* a stalactite of the second type which developed around a sinter tube. Hölloch, Einsamkeit

4 Eccentrics. Water is carried to the end by a capillary canal and thus in such a small amount that evaporation can keep pace with it. No drops form so that the forces of the capillary and of crystallization outweigh the weight of the water. When drops form, sinter tubes are created which grow downward vertically. Hölloch, Rollgang

Plates 12.1–12.4

1 Stalagmite of the first type (no central sinter tube) with draperies and sinter band. Ostgang in the high system of Hölloch (Switzerland)
2 The combination of sinter deposits under stalactitic and stalagmitic conditions results in, among others, especially frequent forms of the meduse-sinter type. Meduse dripstones occur especially frequently in warmer climates. France, Causses, Grotte de Dargilan
3 Combination of Meduse dripstones with flowstone on a wall. Grotte de Dargilan, Causses, France
4 Dripstone room in the Grotta de Castellana south of Bari, Apulia

a related journal

Environmental Geology

ISSN 0099-0094   Title No. 254

Editorial Board: E. E. Angino, Lawrence, KS, USA; W. von Engelhardt, Tübingen, W. Germany; W. L. Fisher, Austin, TX, USA; J. C. Frye, Boulder, CO, USA; M. G. Gross, Baltimore, MD, USA; G. R. Harvey, Woods Hole, MA, USA; M. K. Hubbert, Reston, VA, USA; R. G. Kazmann, Baton Rouge, LA, USA; P. E. LaMoreaux, Tuscaloosa, AL, USA; F. B. Leighton, Irving, CA, USA; W. F. Libby, Los Angeles, CA, USA; G. Müller, Heidelberg, W. Germany; W. A. Pryor, Cincinnati, OH, USA; S. F. Singer, Charlottesville, VA, USA; M. A. Sozen, Urbana, IL, USA; H. A. Tourtelot, Denver, CO, USA; L. J. Turk (Editor in Chief), Austin, TX, USA; C. R. Twidale, Adelaide, S. Australia; K. Young, Austin, TX, USA

Environmental Geolgy is an international journal concerned with the interaction between man and the earth. Its coverage of topics in earth science is necessarily broad and multidisciplinary. The journal deals with geologic hazards and geologic processes that affect man; management of geologic resources, broadly interpreted as land, water, air, and minerals including fuels; natural and man-made pollutants in the geologic environment; and environmental impact studies. Environmental geology is a field that has grown out of an urgent social need to broaden and increase the applications of the earth sciences to the many, varied, and increasingly complex problems arising out of an industrial society's use of the earth. It comes out of a simple truism – society can do a better job of managing the earths resources if it knows something about the earth – combined with the hard fact that as society puts more and more pressure on the earth to supply its needs for food, energy, materials, and recreation, and as the complexity and vulnerability of its engineering and life-support systems increase, the margin for error decreases. In more drastic terms, the probability of disaster – a substantial loss of human life and property – increases.

Subscription information and/or sample copies upon request.

Please send your order or request to your bookseller or directly to:
Springer-Verlag, Journal Promotion Department, P. O. Box 105 280, D-6900 Heidelberg, W.-Germany

North America: Springer-Verlag New York Inc., Journal Promotion Department, 175 Fifth Avenue, New York, NY 10010, USA

Springer-Verlag
New York
Heidelberg
Berlin

U. Förstner, G. T. W. Wittmann

Metal Pollution in the Aquatic Environment

With Contributions by F. Prosi, J. H. van Lierde
Foreword by E. D. Goldberg

1979. 102 figures, 94 tables. XIV, 486 pages
ISBN 3-540-09307-9

Distribution rights for India, Sri Lanka, Nepal, Bangladesh, Pakistan: Narosa Publishing House, NewDelhi, India

Contents: Introduction. – Toxic Metals. – Metal Concentrations in River, Lake, and Ocean Waters. – Metal Pollution Assessment from Sediment Analysis. – Metal Transfer Between Solid and Aqueous Phases. – Heavy Metals in Aquatic Organisms. – Trace Metals in Water Purification Processes. – Concluding Remarks. – Appendix. – References. – Subject Index.

During the past decade public attention has been drawn to heavy metal pollution, especially of inland water resources. Scientists from fields such as botany, chemistry, geochemistry, geology, microgeology, sedimentology and toxicology have employed various media and analytical techniques to determine the extent and effects of heavy metal pollution. The authors propagate the determination of heavy metal enrichment by sedimetological investigations, developing earlier research. The book emphasizes the importance of determining interactions between sediment and water interfaces and possible interrelationships between different heavy metal species.
Physicochemical influences in the mobilization of heavy metal species from sediments, and on the uptake and toxicity of heavy metals are treated in detail. Various processes for removing heavy metals from sewage effluents and sludges are considered. The authors conclude from their own research and the evaluation of more than 1,500 recent papers in this field, that such waters must be drastically reduced to conserve the environment.
Ecologists and other scientists engaged in combatting heavy metal pollution will find this book to be a valuable contribution to the literature in this field.

Springer-Verlag
Berlin
Heidelberg
New York